POWER SYSTEM OPTIMIZATION

POWER SYSTEM OPTIMIZATION

LARGE-SCALE COMPLEX SYSTEMS APPROACHES

Haoyong Chen
South China University of Technology, P. R. China

Honwing Ngan
Asia-Pacific Research Institute of Smart Grid and Renewable Energy, Hong Kong

Yongjun Zhang
South China University of Technology, P. R. China

with contributions from

Xifan Wang and Xiuli Wang
Xi'an Jiaotong University, P. R. China

Tony C.Y. Chung
University of Saskatchewan, Canada

Fuqiang Zhang
Washington University, USA

Zechun Hu
Qinghua University, P. R. China

Min Xie
South China University of Technology, P. R. China

WILEY

This edition first published 2016

Registered office

John Wiley & Sons Singapore Pte. Ltd., 1 Fusionopolis Walk, #07-01 Solaris South Tower, Singapore 138628.

For details of our global editorial offices, for customer services and for information about how to apply for permission to reuse the copyright material in this book please see our website at www.wiley.com.

Library of Congress Cataloging-in-Publication Data applied for

ISBN: 9781118724743

Set in 11/13pt, TimesLTStd by SPi Global, Chennai, India.

Printed and bound in Singapore by Markono Print Media Pte Ltd

1 2016

To our parents

Contributors

Haoyong Chen Department of Electrical Engineering, South China University of Technology, Guangzhou, Guangdong, P. R. China

Tony C.Y. Chung University of Saskatchewan, Saskatoon, Canada

Zechun Hu Qinghua University, Beijing, P. R. China

Honwing Ngan Asia-Pacific Research Institute of Smart Grid and Renewable Energy, Kowloon, Hong Kong

Xifan Wang Xi'an Jiaotong University, Xi'an, Shaanxi, P. R. China

Xiuli Wang Department of Electric Power Engineering, Xi'an Jiaotong University, Xi'an, Shaanxi, P. R. China

Min Xie South China University of Technology, Guangzhou, P. R. China

Fuqiang Zhang Washington University in Saint Louis, Saint Louis, USA

Yongjun Zhang Department of Electrical Engineering, South China University of Technology, Guangzhou, Guangdong, P. R. China

Contents in Brief

Contents

Foreword

This book, in short, is a valuable assessment that presents profound knowledge and ideas about electric power systems and their functions. Having served our society for a long time, these infrastructures have now gained increased complexity. Further developments, such as smart grids and renewable energy technologies, have made new contributions to how power is gathered. Whether for knowledgeable students or accomplished researchers already in the field, the contents of this book are sure to contribute to a heightened viewpoint of the changing system.

During 2008, when I served as the President of the IEEE Power and Energy Society, I guided the IEEE Smart Grid New Initiatives Project involving many IEEE Societies through a successful year. In that time, I also made a decision to change the Society's name from its former title of IEEE Power Engineering Society in order to reflect the industry's continuing evolution.

This text is no exception in addressing the growth of the energy sector. *Power System Optimization* provides thought-provoking revelations into understanding and approaching these large-scale complex systems. The theories and interpretations in this book are presented in a detailed way and with clear meaning that make the realization of these concepts much easier to grasp.

Readers will find the explanations in this book useful in merging current applications to meet increasing advances in the field of energy and power. As the integration of unique systems becomes apparent, a broader understanding of theories and practices will serve useful in achieving optimal success.

Seeking to do just that, this book gathers approaches from different disciplines, such as systems engineering, operations research, and microeconomics. Presented in a unified manner are the vast topics of power system optimization, including: power system planning, operation, reactive power optimization, and electricity markets. The economic implication of the duality principle in mathematical programming is discussed first. In later chapters, the applications of their theories and methods to the components of the power system are explained in great detail. Theories of large-scale systems optimization are surveyed, and several theories used in microeconomics – such as general equilibrium theory, game theory, and mechanism design – are linked to provide contextual approaches. Decomposition–coordination approaches are also introduced, with an emphasis on the Lagrangian relaxation method and coevolutionary approach. The source code of the coevolutionary algorithm is given in the appendix and readers can further develop their own applications based on it.

In becoming familiar with this book, readers will gain an improved insight into the uses of the changing power designs such as electric power systems. They are characterized by large-scale engineering systems, coupled with market systems (electricity markets) and communication and control systems, along with other various systems. Developing advanced optimization approaches is crucial in ensuring that these structures remain intact.

This book is designed to be especially suitable for researchers and students in electric power engineering and related studies, as it can provide new understandings for electric power engineers. Through examples and models shown in the text, an observer can better their perspective on how to integrate existing knowledge with emerging ideas in this field. For those who seek to gain deeper insights into energy development, I am confident that this book proves to be a vital perspective on power system optimization.

Wanda Reder
IEEE Fellow

Preface

The approaches of large-scale system optimization have long been applied to power system planning and operation, and there is extensive literature on such optimization. On the other hand, optimization is also the basic tool for electricity markets, and is often used with microeconomic models. However, people seldom look at physical power systems and economic market systems in microeconomics from a unified system point of view. In fact, both are large-scale distributed systems, and there are intrinsic connections between optimization approaches of power systems and microeconomics (Figure 0.1). In general, a power system (an engineering system composed of generators, loads, and transmission lines) and a microeconomic system (a social system composed of producers, consumers, and markets) have many common characteristics, such as the following:

1. they both consist of subsystems interconnected together,
2. more than one controller or decision-maker is present, resulting in decentralized computations,
3. coordination between the operation of the different controllers is required, resulting in hierarchical structures, and
4. correlated but different information is available to the controllers.

Many optimization approaches have been developed for power system planning and operation, such as linear programming, nonlinear programming, integer programming, and mixed integer programming. Decomposition and coordination techniques such as Dantzig–Wolfe decomposition, Benders' decomposition, and Lagrangian relaxation are often used. On the other hand, mathematical optimization is essential to modern microeconomics, which is the theory about optimal resource allocation, defined as "the study of economics at the level of individual consumers, groups of consumers, or firms … The general concern of microeconomics is the efficient allocation of scarce resources between alternative uses but more specifically it involves the determination of price through the optimizing behavior of economic agents, with consumers maximizing utility and firms maximizing profit" (from the *Economist's Dictionary of Economics*). Because the market system can also be regarded as a large-scale system containing many subcomponents (buyers and

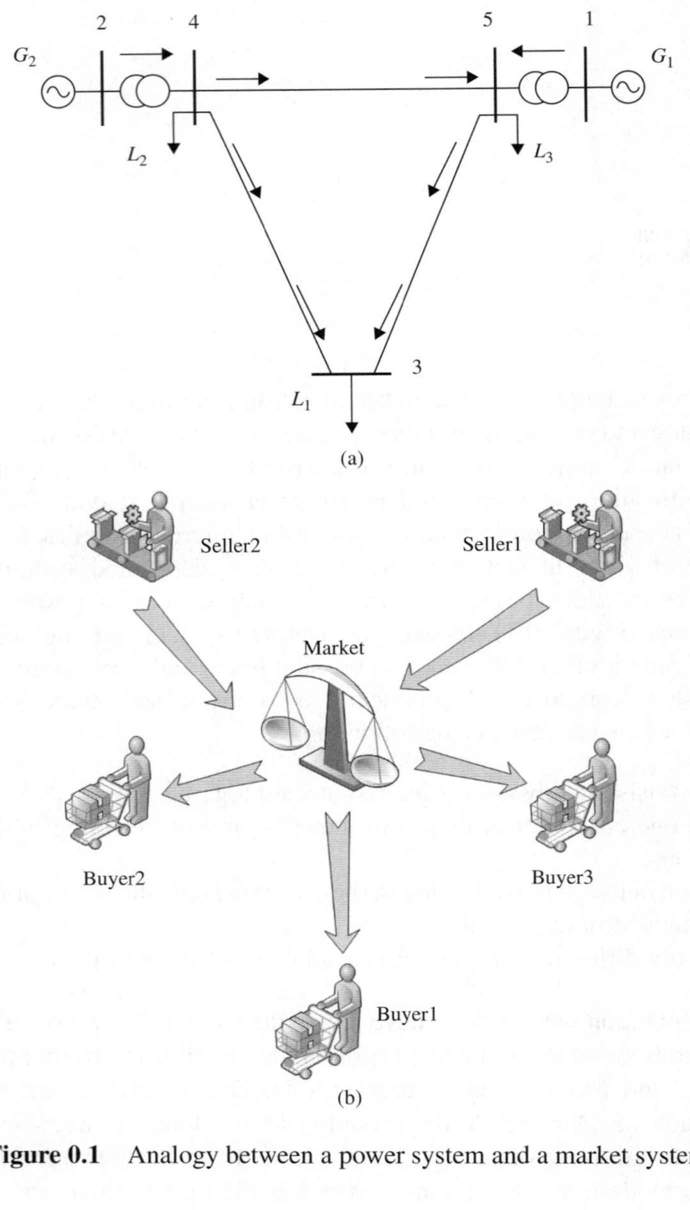

Figure 0.1 Analogy between a power system and a market system.

sellers), the decomposition and coordination principle are also adopted. *Then a unified view of optimization for power systems/electricity markets can be established from the large-scale complex systems perspective*. This is the starting point of this book.

Here, as an example, we take the unit commitment (UC) problem, which is a classic optimization problem in power system operation. Consider a thermal power system with I units. It is required to determine the start-up, shut-down, and generation levels of all units over a specified time horizon T. The objective is to minimize the total cost subject to system demand and spinning reserve requirements, and other individual unit constraints. The notation to be used in the mathematical model is defined as follows:

T	time horizon studied, in hours (h);
I	number of thermal units;
$p_i(t)$	power generated by unit i at time t, in megawatts (MW);
$x_i(t)$	state of unit i at time t, denoting the number of hours that the unit has been ON (positive) or OFF (negative);
$u_i(t)$	decision variable of unit i at time t, 1 for up, 0 for down;
$C_i(p_i(t))$	fuel cost of unit i for generating power $p_i(t)$ at time t;
$S_i(x_i(t-1), u_i(t))$	start-up cost of unit i at time t;
$p_d(t)$	system demand at time t, in megawatts (MW).

The objective function of UC is to minimize the total generation and start-up cost:

$$\min F = \sum_{t=1}^{T}\sum_{i=1}^{I}[C_i(p_i(t)) + S_i(x_i(t-1), u_i(t))] \tag{1}$$

The system power balance constraint is

$$\sum_{i=1}^{I} p_i(t) = p_d(t), t = 1, 2, \ldots, T \tag{2}$$

The individual unit constraints include: unit generation limit, minimum up/down-time, ramp rate, unit spinning reserve limit, etc.

Here we only give a simplified model description, and the detailed formulation of UC will be given in the later chapters.

Different solution methods, such as priority list, dynamic programming, mixed integer programming, and Lagrangian relaxation, have been proposed by researchers. We take the Lagrangian relaxation method as an example. The basic idea of Lagrangian relaxation is to relax the systemwide constraints, such as the power balance constraint, by using Lagrange multipliers, and then to decompose the problem into individual unit commitment subproblems, which are much easier to solve. Lagrangian relaxation can overcome the dimensional obstacle and get quite good suboptimal solutions. By using the duality theory, the systemwide constraint (here referring to the power balance constraint) of the primal problem is relaxed by the Lagrangian function (3). Then the two-level maximum–minimum optimization framework shown in Figure 0.2 is formed. The low-level problems (4) solve the optimal commitment of each individual unit. The high-level problem (5) optimizes the vector λ of Lagrange multipliers, and a subgradient optimization method is often

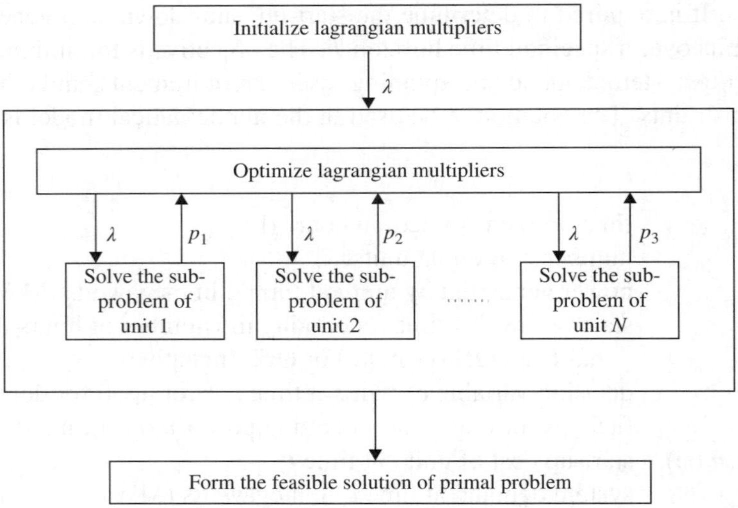

Figure 0.2 Illustration of Lagrangian relaxation.

adopted. When λ is passed to the subproblems, each individual unit will optimize its own production p, namely to minimize its cost or maximize its profit. In this procedure, λ serves as the function of market prices to coordinate the production of all units to reach the requirement of system demand. The optimization of Lagrange multipliers is in fact the price adjustment process in the market.

The Lagrangian function is

$$L = \sum_{t=1}^{T} \left\{ \sum_{i=1}^{I} [C_i(p_i(t)) + S_i(x_i(t-1), u_i(t))] + \lambda(t) \left[p_d(t) - \sum_{i=1}^{I} p_i(t) \right] \right\} \tag{3}$$

where $\lambda(t)$ is the Lagrange multiplier associated with demand at time t.

The individual unit subproblems are

$$\min L_i = \sum_{t=1}^{T} \{[C_i(p_i(t)) + S_i(x_i(t-1), u_i(t))] - \lambda(t)p_i(t)\} \tag{4}$$

subject to all individual unit constraints.

The high-level dual problem is

$$(P-D)\max_{\lambda} L(\lambda)=\sum_{t=1}^{T}\left\{\sum_{i=1}^{I}[C_i(p_i(t))+S_i(x_i(t-1),u_i(t))]\right\} + \sum_{t=1}^{T}\left\{\lambda(t)\left[p_d(t)-\sum_{i=1}^{I}p_i(t)\right]\right\} \tag{5}$$

We can compare this optimization procedure with a market economy. Consider an economy with I agents $i\in\Phi=\{1,\dots,I\}$ and L commodities $l\in\Gamma=\{1,\dots,L\}$. A bundle of commodities is a vector $x\in\mathbb{R}_+^L$. Each agent i has an endowment $e^i\in\mathbb{R}_+^L$ and a utility function $u^i:\mathbb{R}_+^L\to\mathbb{R}$. These endowments and utilities are the primitives of the exchange economy, so we write $\epsilon=((u^t,e^t)_{i\in\Phi})$. Agents are assumed to take as given the market prices for the goods. The vector of market prices is $p\in\mathbb{R}_+^L$; all prices are nonnegative.

Each agent chooses consumption to maximize his/her utility given his/her budget constraint. Therefore, agent i solves

$$\max_{x\in\mathbb{R}_+^L} u^i(x) \quad \text{s.t.} \quad p\cdot x\le p\cdot e^i \tag{6}$$

The consumer's "wealth" is $p\cdot e^i$, the amount he/she could get if he/she sold his/her entire endowment. We can write the budget set as

$$B^i(p)=\{x : p\cdot x\le p\cdot e^i\} \tag{7}$$

A key concept of a market system is equilibrium. Market equilibrium refers to a condition where a market price is established through competition such that the amount of goods or services sought by buyers is equal to the amount of goods or services produced by sellers. There are two kinds of equilibrium considered in microeconomics, namely, competitive equilibrium and Nash equilibrium.

A competitive (or Walrasian) equilibrium for the economy ϵ is a vector $(p,(x^i)_{i\in\Phi})$ such that the following hold.

1. Agents maximize their utilities: for all $i\in\Phi$,

$$x^i\in\arg\max_{x\in B^i(p)} u^i(x) \tag{8}$$

2. Markets clear: for all $l\in\Gamma$,

$$\sum_{i\in\Phi}x_l^i=\sum_{i\in\Phi}e_l^i \tag{9}$$

The above model (6) and (9) is apparently a decentralized large-scale optimization model, which is similar in form to power system optimization problems such as the above-mentioned unit commitment. Clearly, we can see that the utility maximization problem (6) of each agent corresponds to the individual unit subproblem (4) except for the opposite sign. At the solution of the high-level dual problem (5) or the primal problem (1), the items with Lagrange multiplier

$$\lambda(t)\left[p_d(t) - \sum_{i=1}^{I} p_i(t)\right], t = 1, 2, \ldots, T$$

will tend to zero, and this is just the market clearing condition (9).

The Nash equilibrium is widely used in economics as the main alternative to competitive equilibrium. It is used whenever there is a strategic element to the behavior of agents and the "price taking" assumption of competitive equilibrium is inappropriate. Nash equilibrium is a core concept of game theory, which is the study of mathematical models of conflict and cooperation between intelligent rational decision-makers. The mathematical approaches of game theory belong to another kind of decentralized optimization, which also has analogs in power system optimization.

In fact, from the perspective of large-scale system optimization, we shall find in later chapters that the solution method of competitive equilibrium is related to the interaction balance method (or nonfeasible method) and the solution method of Nash equilibrium is related to the interaction prediction approach (or feasible method) of large-scale systems.

The authors' work over a decade has focused on the application of large-scale optimization to power system planning and operation, and also on the application of microeconomics and game theory to electricity markets. The authors have made significant achievements in these research areas. Based on previous research, this book will make a more systematic investigation on large-scale complex systems approaches to power system optimization. The authors believe that this book not only brings a new point of view to power system optimization for the reader, but also provides many approaches suitable for practical application.

Acknowledgments

From Professor Fuqiang Zhang, from Olin Business School, Washington University, St. Louis, USA, I got many valuable suggestions. The book was supported in part by the China National Funds for Excellent Young Scientists (51322702) and in part by the National Natural Science Foundation of China (51177049).

H.C.

List of Figures

List of Tables

Acronyms

AC	alternating current
ACCPM	analytic center cutting plane method
ACE	agent-based computational economics
AGA	annealing genetic algorithm
AGA	genetic algorithm with annealing selection
AGC	automatic generation control
AVC	automatic voltage control
BES	battery energy storage
CACD	cost of adjusting control device
CCA	coevolutionary computation approach
CCA	cooperative coevolutionary algorithm/approach
CCEM	coevolutionary computation applied to the electricity market
CCGA	cooperative coevolutionary genetic algorithm
CCHP	combined cooling, heat, and power
CEA	coevolutionary algorithm
CGA	cataclysmic genetic algorithm
CGA	coevolutionary genetic algorithm
CHP	combined heat and power
CRF	capital recovery factor
CSF	conjectured supply function
DAI	distributed artificial intelligence
DC	direct current
DG	distributed generation
DisCo	distribution company
DMS	distribution management system
DP	dynamic programming
DSR	demand-side response
EA	evolutionary algorithm
EEF	equivalent energy function
EENS	expected energy not served
EP	evolutionary programming

EPEC	equilibrium problem with equilibrium constraints
ES	energy storage
ES	evolution strategy
ESS	entirety selection scheme
GA	genetic algorithm
GenCo	generation company
GGDF	generalized generation distribution factor
IAEA	International Atomic Energy Agency
ICM	intelligent communication manager
IEEE	Institute of Electrical and Electronics Engineers
IMC	iterative Monte Carlo
IPCPM	interior point cutting plane method
IPD	iterated prisoners' dilemma
ISO	independent system operator
IWM	individual welfare maximization (algorithm)
JASP	Jiaotong Automatic System Planning Package
KDM	knowledge data manager
KKT	Karush–Kuhn–Tucker
KT	Kuhn–Tucker
LCS	layer control structure
LI	Lerner Index
LMP	locational marginal price
LOLC	loss of load cost
LOLP	loss of load probability
LR	Lagrangian relaxation
LSF	linear supply function
LSFE	linear supply function equilibrium
MAS	multi-agent system
MCP	mixed complementarity problem
MIOPF	mixed integer optimal power flow
MIP	mixed integer programming
MPEC	mathematical program with equilibrium constraints
MW	megawatts
NCP	nonlinear complementarity problem
NCS	network control structure
NGO	nongovernmental organization
NPV	net present value
OLTC	on-load tap changer
OO	ordinal optimization
OPC	ordered performance curve
OPF	optimal power flow
ORPC	optimal reactive power control
ORPD	optimal reactive power dispatch

ORPP	optimal reactive power planning
PD	prisoners' dilemma
PES	Power and Energy Society of IEEE
PHEV	plug-in hybrid electric vehicle
PHS	pumped hydro storage
PoolCo	(independent power) pooling company
PSO	particle swarm optimization
PV	(solar) photovoltaics
PX	Power Exchange
RES	renewable energy source
RMS	reason maintenance system
RO	robust optimization
RTO	regional transmission organization
RTOSE	real-time operating system extensions
SA	simulated annealing (algorithm)
SCADA	supervisory control and data acquisition
SCED	security-constrained economic dispatch
SCCP	stochastic chance constrained programming
SCPM	simplex cutting plane method
SCUC	security constrained unit commitment
SFE	supply function equilibrium
SGA	simple genetic algorithm
SGA	standard genetic algorithm
SP	stochastic programming
SVM	support vector machine
TEAM	transmission economic assessment methodology
TMM	time map manager
UAC	unit adjustment cost
UC	unit commitment
VAR	volt-ampere reactive
VCM	volt/VAR control mismatching
WASP	Wien Automatic System Planning Package
WECS	wind energy conversion system

Symbols

a_{gj}	generalized generation distribution factor of unit g in branch j
a_g, b_g, υ_g	coefficients of the quadratic production cost function of unit g
x_j	impedance of branch j
L_j	flow capacity of branch j
J	branch set
g	index of unit
G_T	set of thermal units
Nt	number of hours within the planning period
$F(\cdot)$	production cost function
$S_u(\cdot)$	start-up cost function
$K(g, m_g)$	cost of the interval m_g of the piecewise start-up cost function of unit g
$\overline{p_g}$	upper output limit of unit g
$\underline{p_g}$	lower output limit of unit g
$r(g, t)$	spinning reserve provided by unit g in period t
$R(t)$	spinning reserve requirement of the system in time period t
R_u	ramp-up rate limit of unit
R_d	ramp-down rate limit of unit
$T_u(g)$	minimum ON time of unit g
$T_d(g)$	minimum OFF time of unit g
GW	set of wind farms
$D(t)$	system demand in period t
$I(g, t)$	commitment state of unit g in period t
$p(g, t)$	generation of unit g in period t in predicted scenario
s	index for wind power generation scenario
St	set of scenarios in period t
$p(s, g, t)$	generation of unit g in period t in scenario s
t	index for time

1

Introduction

Optimization theories and approaches have been extensively applied to power system planning and operation problems. This is a rather traditional and ongoing research area [1]. With the complication of power systems, the deregulation of the power industry, and the development of smart grids, many new problems have emerged and new methods have been developed. Many optimization theories and approaches have acquired industrial application and introduced technical and economic benefits. The mathematical optimization methods applied in power systems include linear programming, nonlinear programming, mixed integer programming, dynamic programming, artificial intelligence, stochastic programming, etc. This book focuses on the advanced theories and approaches from the perspective of large-scale complex systems, rather than the traditional ones. However, to begin with the fundamentals, we will first review the basic optimization applications in power system planning and operation.

The aims of this chapter are as follows:

1. To present a broad review of mathematical optimization applications to power system planning and operation, which is the foundation for the theories and approaches presented in the subsequent chapters.
2. To explain the basic concepts to those interested in the optimization field, but unfamiliar with power system problems and terminology. It is hoped that this chapter may motivate some people to become involved in the challenging power field.
3. To summarize the results of traditional power system research, to allow the reader to understand the differences among them and the more advanced approaches presented in books, and to encourage new development and further research.

To give the reader a unified mathematical description of different power system optimization problems, the generalized notation used in this book, such as x and u for variables, and f, h, and g for functions, and their power system meanings are

Power System Optimization – Large-scale Complex Systems Approaches, First Edition.
Haoyong Chen, Honwing Ngan and Yongjun Zhang.

explained. Vectors (lower-case) sometimes and matrices (upper-case) usually are in bold face; and matrix transposition is indicated by a superscript T, such as $\mathbf{A}^{\mathrm{T}}$.

The problems discussed include generation, transmission, and distribution expansion planning, optimal operation problems such as hydrothermal unit commitment and dispatch, optimal load flow and volt-ampere reactive (VAR) optimization, and optimization models of electricity markets based on theories of microeconomics.

Numerous important works have appeared on these topics in books and journals all over the world. It is an impossible task to discuss all of them. Since the objective of this chapter is to introduce the basic concepts and methods of power system optimization, we will lay the emphasis of our discussion on research reported by IEEE papers in *IEEE Transactions on Power Systems* and Technical Meetings.

1.1 Power System Optimal Planning

Power system expansion planning is traditionally decomposed into load forecasting, generation planning, and transmission planning. Load forecasting is the basis for power system planning, which provides the basic data for calculation of electric power and energy balance. Although generation planning and transmission planning are essentially indivisible, these two issues have to be decomposed and solved separately and further coordinated due to their different focuses and the difficulty in solving them as a whole.

Traditional power system planning is based on scheme comparison, which selects the recommended scheme from a few of the viable options with some technical and economic criteria. However, because this approach is empirical, the final result is not necessarily optimal. With the fast development of power technologies, the rapidly growing demand for electricity, and the increasingly diversified energy resources used in power generation, the generation mix becomes increasingly complicated. On the other hand, large-scale interconnected systems across different areas have been formed gradually. All these factors have brought difficulties to the economic and technical assessment of power system planning schemes, and traditional planning approaches are difficult to adapt to these challenges. Fortunately, the development of computer science, systems engineering, operational research, and other research areas has provided new means for the optimization of power system planning. Theory and practice in power system optimal planning have made considerable progress in recent years. A number of commercial planning software packages have emerged and their benefits have been affirmed in the power industry.

The objective of power system planning is to determine what schemes are the most beneficial from the overall and long-term perspective. This requires us to choose the best planning scheme from all possible choices. The application of power system optimal planning theories and methods not only can have more accurate and comprehensive technical and economic evaluation, but also can evaluate the impacts of various uncertainties by sensitivity analysis, so that the planning results are produced with a higher referential value.

1.1.1 Generation Expansion Planning

The objective of generation expansion planning is to choose the least expensive expansion scheme (type, number, capacity, and location of generating units), in terms of investment and operation costs, that satisfies certain constraints. The key constraints are electric power and energy balance, which means that the total power and energy produced by all the generating units can meet the requirement of demand. Other technical constraints, such as limitation of resources, also need to be met. Generally, generation expansion is carried out over a planning horizon of many years, which turns into a dynamic optimization problem.

Several key issues should be analyzed quantitatively in generation expansion planning, such as: annual investment flow and operating cost, quantity of primary energy resources used in generation, reliability of electric power supply, etc. The investment cost of building a particular plant in a given year is independent of the other decisions in a given scheme. However, the operating cost is much more complicated, and is related to the generation mix, system load, generating unit outages, transmission network losses, availability and cost of energy from neighboring systems, fuel costs, etc. Some influencing factors are intrinsically random, such as generating unit outages. The fact that units must be added in discrete sizes presents a further complication. Considering all these conditions, the mathematical model of generation expansion planning is large-scale, nonlinear, discrete, and stochastic, which is a very difficult problem to solve.

Generation expansion planning has long been of interest to researchers, and many sophisticated and effective techniques have been developed. The approaches differ in the questions they are intended to answer, the model details, and the optimization methods.

The early work often used linear programming models [2, 3]. The objective functions takes the following form:

$$J = c^t x + d^t u$$

where x denotes the capacities of different types of generating units installed in each year and u specifies the energy produced by each power plant (or plant type). A number of different load levels are considered here. The investment cost c and the operating cost d should be calculated with the method of technological economics. The load levels related to d and u are obtained by dividing estimates of the load duration curves into a number of discrete segments (Figure 1.1). The variables u and x are related through linear constraints so that a plant cannot produce power exceeding its installed capacity. Other constraints limit the capacity of certain types of units and require total capacity to exceed expected load. This formulation is a high-dimensional optimization

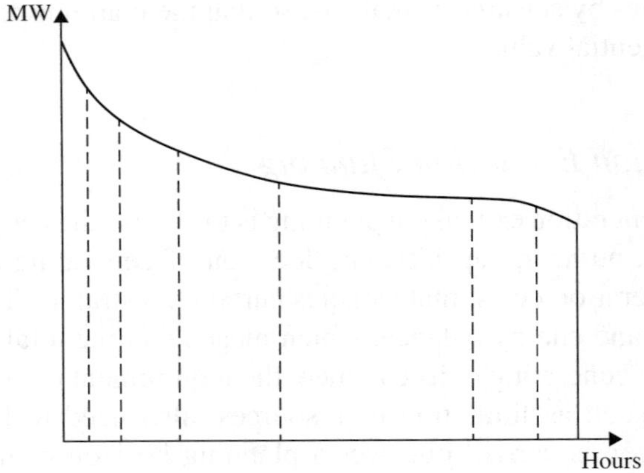

Figure 1.1 Typical load duration curve.

problem. Decomposition techniques such as Dantzig–Wolfe decomposition may be needed to solve it.

A dynamic programming based model of generation planning has been presented by Booth [4, 5]. The method can handle integer variables and nonlinear constraints. The random variables are treated with a probabilistic approach. As a significant innovation, the expected outage rates for various units are considered by modification of the load duration curve. The problem is formulated as: choose $v(t)$ (capacity additions in year t) to minimize

$$J = E\left\{\sum_{t=1}^{T} f(t, x(t), v(t), \dots)\right\} \tag{1.1}$$

where

$$x(t+1) = x(t) + v(t)$$

The function $f(\cdot)$ is related to probabilistic load models, fuel models, etc. A variety of technical and economic constraints are considered.

The problem is decomposed into a series of forward dynamic programming problems. A pretreatment is employed to dynamically reduce the dimensionality of the problem. However, the computational burden is still heavy.

A more advanced generation planning model JASP has been proposed by Chen [6], which decomposes the generation planning problem into a high-level power plant investment decision problem and a low-level operation planning problem and solves them by a decomposition–coordination method. Lagrangian relaxation is used to solve the power plant investment decision problem, and a probabilistic production simulation based on the equivalent energy function method is used to solve the operation planning problem. Simulation results show that JASP can not only overcome the "curse of dimensionality" but also find an economical and technically sound generation planning scheme.

1.1.2 Transmission Expansion Planning

Transmission expansion planning is an important part of power system planning, whose task is to determine the optimal power grid structure according to the load growth and generation planning schemes during the planning horizon to meet the requirements of economic and reliable power delivery. In general, transmission planning should answer the following points:

- where to build a new transmission line,
- when to build a new transmission line, and
- what type of transmission lines to build.

Transmission expansion planning is closely related to generation planning. It is based on generation planning, and in turn has some impact on the latter. In generation planning, the influence of geographical distribution of power plants and transmission costs are generally not considered or just considered cursorily, and it is possible that the original generation planning scheme should be modified during transmission planning. Therefore, generation planning and transmission planning should be decomposed and further coordinated so that the whole power system planning can be optimal.

The basic principle of transmission planning is to minimize the power grid investment and operating costs under the premise of ensuring safe and reliable electric power delivery to the load center. Compared with generation planning, transmission planning is more complex. First, the transmission planning should consider the specific network topologies, and each line in the rights of way must be treated as an independent decision variable. Thus the dimension of transmission planning decision variables is higher than that of generation planning. Second, transmission planning should satisfy very complex constraints. Some constraints are related to nonlinear equations, and even related to differential equations. Third, many factors that are either random or difficult to predict are extremely important, such as future load growth in various areas. Therefore, it is difficult to establish a perfect transmission model and even more difficult to solve it.

Transmission planning has two kinds of formulations: static and dynamic. Static transmission planning is concerned only with the planning scheme in a future target year, and it is not necessary to consider the planning scheme transition, which is also known as a level-year plan. Because static transmission planning does not answer the question *when* to build new transmission lines, it is not necessary to consider the time value of capital. With a longer planning period, the planning horizon needs to be divided into several level years and the scheme transition between the level years needs to considered. In this case, we must determine when and where to build new transmission lines. This kind of planning is called long-term or dynamic transmission planning.

An early static formulation was presented in Puntel [7], which attempts to design the optimal network structure for a specified future time. The present network, available rights of way, costs, and future loads and generation levels are assumed known.

The problem is to choose x, the susceptance installed in the rights of way, to minimize

$$J = c^t x + \sum_{k=1}^{M} d_k (P_k / P_{c_k})^{2n} \tag{1.2}$$

The first term $c^t x$ is the investment cost of transmission capacity; the second term is the overload penalty on each of the M transmission lines or transformers. The load P_k on element k is a nonlinear function of x, the given load and generation conditions, even when the network model is linear. There are also inequality constraints on each element of x.

Optimization is performed by first computing a gradient vector $\partial J/\partial x$ through use of an adjoint network. A band around this gradient in the portion of the x space bounded by the limits of additions in each right of way is searched for integer solutions. A cost J is computed for each such x found. The x that minimizes J is the starting point for the next iteration and for the computation of the next $\partial J/\partial x$. The search ends when no improving solution is found.

Once the network has been designed, a sensitivity analysis is performed to determine which contingencies would be the most critical. Each serious outage is tested individually and a composite $\partial J/\partial x$ is computed by summing the gradient of all tests. As in the normal situation, this gradient guides the addition of new lines. The process terminates when as many as possible overloads caused by the contingencies have been eliminated by appropriate network additions.

Garver [8] considered the static optimization problem of designing a network to meet a specific load. The problem is formulated as a power-flow problem. Linear programming is used to find the most direct route from generation to loads: all rights of way can transmit power, but those without transmission lines are penalized to encourage flow through the existing network. A line is added to the right of way with the biggest overload and then a new linear flow is computed. The process terminates when all overloads have been eliminated. The estimated flows on the final network are fairly close to those computed by a standard load flow.

The linear program minimizes an approximation to cost of new facilities, as the penalty term of an overload can be related to the cost of constructing a line in that right of way.

An improved genetic algorithm (GA) approach to optimal multi-stage (dynamic) transmission network planning is presented in a more recent work [9]. The multi-stage planning of a transmission system has to consider not only how to form the network schemes of every stage but also how to coordinate the network schemes of every stage.

Because of the combinatorial nature of the transmission network planning, it has not been well solved by conventional optimization techniques. In the proposed improved GA approach, a fitness function that includes investment and an overload constraint is constructed. The overload is checked by DC load flow. A concise codification model called "redundant binary-coded technique" is proposed. Using this technique, the crossover operation can be executed inside the gene so that the re-combinatorial and search functions of the crossover operator are well utilized. The simulated annealing selector is used to adjust the fitness function in the evolution process. Some improvements are employed to speed up the convergence of the algorithm, such as keeping excellent seeds, mutation in pairs, etc. Based on the proposed model, a computational program has been developed. Three case studies are applied to demonstrate the usefulness and effectiveness of the suggested multi-stage transmission network planning model.

All of these approaches are suboptimal and involve important approximations and simplifying assumptions. The static approaches can handle larger networks, but at the cost of ignoring growth dynamics. The dynamic approaches often suffer from the "curse of dimensionality".

1.1.3 Distribution System Planning

The current development of power systems puts increasing emphasis on distribution systems, which are the lower-voltage networks supplying power from the high-voltage transmission systems to the loads. Distribution system planning has different considerations along with transmission planning, and several special concepts and factors need to be addressed.

Distribution systems tend to be less geographically extensive than transmission systems. However, the network structures are often more complicated and consist of more branches and nodes. On the other hand, the operation of distribution systems may be easier than transmission systems and stability constraints are often not included. Reliability and power quality are usually the main concerns of users and should be taken into consideration in distribution system planning.

The objective of distribution system planning is to determine the optimal construction scheme under the premise of meeting load growth and securing a reliable supply of electricity based on the results of load forecasting during the planning horizon and existing network status, so that the investment and operating costs of the distribution system are minimal. Distribution system planning also has static (one-stage) and dynamic (multi-stage) formulations. In dynamic distribution system planning, the correlations of decision variables among different stages in the planning horizon should be taken into account. The mathematical models are often discrete, nonconvex, nonlinear, and large-scale, and various optimization techniques have been employed.

The early work [10] treats the maximum flow-through capability of the network as the sum of the component capacities and does not explicitly consider whether this capability is actually exploited. This simplification makes the constraints much easier to handle. The planning scheme obtained with this simplification may be optimal for a problem with complex constraints, or it may be used as a bound or goal in optimization subjected to realistic network constraints. Adams and Laughton [10] emphasized the security aspect of the problem by determining which of the possible expansion states at each time period of a planning horizon are feasible, in the sense that they do not produce overloads of circuits or high-voltage transformers. The overall costs (capital and losses) are minimized by determining the proper sequence of states of planning time periods, and dynamic programming is used in optimization.

In a more recent work [11], a GA approach is applied to the optimal multi-stage planning of distribution networks. The objectives for distribution system planning are related to providing the designs and associated implementation plans necessary for an orderly expansion of facilities, minimizing new facility installation costs and operation costs, as well as achieving an acceptable level of reliability. Complex operational constraints such as voltage drop and line thermal limits are considered. GA allows the representation of nonlinearities that are hard to include in mathematical programming methods, and produces multiple solutions that enhance the opportunity for multi-criteria decision-making.

Currently, deregulation in the power system industry and the invention of new-generation technologies have led to innovations in distribution system planning [12]. Distributed generation (DG), with many attractive economical and technical features, in medium- and low-voltage parts of the grid, is one of the alternatives to reinforce distribution systems. In Naderi *et al.* [12], a dynamic distribution system planning model was proposed that considers DG integration into a network as an option to meet the load growth in the planning horizon. An optimal power flow (OPF) is proposed to minimize capital costs for network upgrading, operation and maintenance costs, and the cost of losses for handling the load growth in the planning horizon. A year-dependent decision variable is attached to each investment alternative and results in a dynamic planning scheme, which reduces total planning costs by determining the best timing schedule for investment in network upgrading. A modified genetic algorithm is used to find the optimal topology solution.

1.2 Power System Optimal Operation

1.2.1 Unit Commitment and Hydrothermal Scheduling

Power system optimal operation is a very complex system optimization problem, which is hard to solve as a whole and so is often decomposed into a series of subproblems dealt with separately [13]. For short-term generation scheduling, it is often decomposed into unit commitment, hydrothermal scheduling, power exchange planning, and fuel planning, etc. The purpose of unit commitment and hydrothermal scheduling is to determine which units should be on-line and their generation levels in a power system during a given scheduling horizon (usually a day or a week).

The objective is to minimize total operational costs during the scheduling horizon while satisfying the constraints of power system security and power quality. Unit commitment is a very basic optimal operation problem, whose benefit is generally greater than that of economic dispatch.

Wood [14] describes the basic mathematical models and solution methods of unit commitment and economic dispatch. Unit commitment is a high-dimensional, non-convex, discrete, and nonlinear optimization problem, and it is difficult to find the theoretically optimal solution. However, because it can bring significant economic benefits, various methods have been extensively studied to solve this problem, such as the heuristic, priority list, dynamic programming, integer programming and mixed integer programming, branch-and-bound method, and Lagrangian relaxation method. Artificial intelligence has also be introduced, such as expert systems, artificial neural networks, simulated annealing, genetic algorithms, etc. Sheble and Fahd [15] survey the solution methods and related references of unit commitment.

The primary energy sources used for electric power generation can be broadly classified as renewable and nonrenewable resources. Fossil fuels such as coal, natural gas, oil, and nuclear fuel are nonrenewable resources, which are used for electricity generation in thermal power plants. The most widely used renewable resource for electricity generation is hydro power. Nowadays there are fast developments with other renewable sources such as wind power, solar energy, marine energy, and biomass.

In a thermal plant, electric power is generated as a result of mechanical rotational energy produced by either steam turbines or combustion turbines. From an economic operational point of view, our concern is the relation of fuel cost to the active power generation of the unit, which is an efficiency type model. The fuel cost is equal to the fuel quantity consumed multiplied by the fuel price. The heat rates are often used to express generation efficiency, which include the average heat rate (dividing the fuel quantity by the generation quantity) and incremental heat rate (differentiating the fuel quantity with respect to the generation quantity). The fuel cost curve is often modeled as a quadratic function, and the values of coefficients can be obtained by statistical estimation from experimental data of heat rates. The treatment so far has been static: for the unit commitment problem, dynamics must also be considered. For example, thermal units are shut down and started up at various instants. There are costs related to these operations and time constraints that do not allow plants to be switched in and out frequently (minutes or a few hours).

In a hydro plant, turbines convert the water potential energy into kinetic energy, which in turn is converted into electricity by generators. Hydroelectric installations are classified into two types: conventional and pumped storage. The conversion type is further classified into two classes: storage and run-of-river. The water system modeling problem basically involves a water balance equation at each reservoir, which relates the inflow of water to the reservoir, the water volume in the reservoir, and the rate of usage and spillage. The inflow could be from natural sources (rain, snow, unregulated rivers), or from a reservoir system in which upstream reservoirs release water. In this last case, the dynamics of the reservoir system are important. On the rate of usage and spillage, the question of whether only one plant or a series of cascading plants will use the water must be considered. There are also pumped-storage plants where water is pumped into a higher-level reservoir at time periods when energy from efficient plants can be used for this purpose, and water is discharged through a hydro plant to generate energy at peak periods of consumption where the more efficient plants are at their maximum output.

The most important constraint of unit commitment and hydrothermal scheduling is power balance condition, which is that the sum of the electric power generated by all units must equal the system load plus network loss at every scheduling period. Other constraints include reserve requirements for the system as a whole and for certain specified areas. There are minimum up-time and down-time constraints for each unit. Constraints on hydro energy and power should also be respected.

The solution methods can be classified into heuristic methods, mathematical optimization, artificial intelligence, etc. The earliest ones are heuristic methods such as local optimization and priority list, which are empirical without a strict theoretical foundation, but are often useful in practice. Mathematical optimization methods are those with a strict mathematical basis, among which the most successful ones are dynamic programming and mixed integer programming.

Dynamic programming (DP) is a mathematical method to solve the multi-stage decision optimization. In the enumeration of various possible combinations of states, this method cleverly abandoned solutions that need not be considered. In the UC problem, the entire scheduling horizon is divided into several periods, and each period is a stage of dynamic programming. The states of each stage are all the possible combinations of the unit ON/OFF states. From the initial stage, the cumulative cost (including the fuel cost of generation and start-up cost) to reach each stage is calculated forward, and then, from the state with the smallest cumulative cost in the final stage, a backtracking procedure is carried out, by sequentially recording the state with the total cumulative cost in each stage, and then the optimal UC scheme can be obtained. The problem of dynamic programming is that the computational burden will increase dramatically with the number of units and scheduling periods, which results in the "curse of dimensionality". To overcome this difficulty, different skills are used to limit the number of states. Dynamic programming has been widely used in unit commitment and hydrothermal scheduling, and has been incorporated into some practical scheduling/dispatch software packages.

Mixed integer programming (MIP) includes both integer and noninteger variables, and it can be further divided into linear and nonlinear mixed integer programming according to the type of other variables except for the integer variables. MIP is a very difficult problem, and the commonly used methods include branch-and-bound, Benders' decomposition, generalized Benders' decomposition methods, and so on. In the branch-and-bound method, a branch-and-bound tree is formed and the root node is a relaxation of the original problem. For example, the integer variables are replaced with continuous variables in a certain interval. Its child nodes are also a series of relaxation of the original problem, which can be obtained by fixing the values of some integer variables at the root node. The relaxed problems corresponding to the child nodes have disjoint solution spaces, and the union of these solution spaces is just the solution space of the root node. Each child node is further decomposed into a series of subproblems, and this process is repeated until the leaf nodes of the tree. Because the solution space of a node of the tree is the relaxation of the solution spaces of all its descendant nodes, in a minimization problem, the lower bound of its solution must therefore be the lower bound of the solutions of all its descendant nodes. The key to the branch-and-bound method is that, in the process of calculation, if the solution (or its lower bound) of a node is bigger than a known feasible solution of the original optimization problem, its descendant nodes need no longer be considered. This allows unnecessary calculations to be eliminated.

Benders' decomposition method solves problems in the following form:

$$\max_{x,y} \{c^{\mathrm{T}}x + f(y) \mid \mathbf{A}x + F(y) \leq b,\ x \in R^p,\ y \in S \subseteq R^q\} \tag{1.3}$$

where R^p and R^q are respectively p- and q-dimensional Euclidean spaces; S is any subset of R^q; $\mathbf{A}$ is an $m \times p$ matrix; $f(y)$ is a scalar function and $F(y)$ is an m-component vector function, which are defined on S; and b and c are constant vectors defined on R^m and R^p. Benders' decomposition method decomposes the problem into two subproblems: one is a programming problem defined on S (which may be linear, nonlinear or discrete, etc.), and the other is a linear programming problem defined on R^p. The two subproblems are solved through a multi-step iterative procedure.

MIP solves the mathematical models of unit commitment/hydrothermal scheduling directly, without adding too many restrictions or assumptions. MIP is the mainstream solution method of scheduling problems in the electric power industry.

Power systems are typical large-scale systems, and the optimization and control theories of large-scale systems can be applied. The decomposition and coordination approach of large-scale systems started from Dantzig and Wolfe's decomposition for linear programming [16], and the Lagrangian relaxation (LR) method has been employed in unit commitment. LR is a class of optimization algorithms for solving complex integer and combinatorial optimization problems, which is based on the following ideas. Many difficult integer programming problems are composed of a series of subproblems interrelated by some constraints, which are relatively easy to resolve. Based on this characteristic, the Lagrangian problem is formed by adding a penalty term to the objective function, which equals the sum of products of constraint violation amounts and their respective dual variables. The Lagrangian problem is relatively easy to solve, and for the minimization problem, its optimal value is the lower bound of the optimal value of the primary objective function. The Lagrangian relaxation method will be discussed in detail in Chapter 2.

In the Lagrangian relaxation method for unit commitment, all constraints are divided into two categories: one category includes the systemwide constraints, such as the load balance constraints and spinning reserve constraints; the other category includes the individual unit constraints, such as the generator output power limit constraints, the minimum up-time and minimum down-time constraints, ramp rate constraints, spinning reserve constraints, etc. The systemwide constraints can be written as the penalty term of the objective function, and then the Lagrangian function can be formed. The Lagrangian function can be decomposed into a series of individual unit subproblems, which are generally solved using the dynamic programming method, and the dual problem is often solved by the subgradient method. Application of Lagrangian relaxation in unit commitment began in the 1970s and there is a great deal of literature.

Unit commitment and hydrothermal scheduling based on Lagrangian relaxation have been studied systematically [17–19]. In Guan *et al.* [17], dynamic programming is used to optimize the subproblems and optimization of multipliers is carried out with an adaptive step-size subgradient algorithm. The advantages of the proposed approach include that the generator power outputs do not need to be discretized, and there is a systematic approach to handle ramp rate constraints, as well as an effective initialization procedure. The method is further extended to the hydrothermal scheduling problem in Yan *et al.* [18], where the thermal and hydro power plants are coordinated through Lagrange multipliers, and the hydro power plant subproblems are solved with the priority list method. In Guan *et al.* [19], where the dynamic characteristics of a pumped-storage power station have been considered, the scheduling problem in the whole horizon is solved by dynamic programming on the basis of optimization of single-period operation.

Lagrangian relaxation is a kind of integer and combinatorial optimization algorithm with a mature theoretical foundation and is particularly suitable for solving large-scale system optimization problems. It has the following advantages in unit commitment and hydrothermal scheduling: with the increase of unit number, the computational burden increases almost linearly, which means the dimensional obstacle can be overcome; and the method is very flexible, which means not only that it can successfully solve the unit commitment problem, but also that it can be extended to hydrothermal scheduling problems. However, the algorithm also has some disadvantages: owing to the nonconvexity of the objective function, a duality gap exists and some special measures should be adopted to construct the feasible solution of the primal problem on the basis of the optimal dual solution; and the iterative process of the algorithm may oscillate.

With the development of computer science and artificial intelligence, many new methods have appeared, the most successful of which are the genetic algorithm, simulated annealing, tabu search, expert systems, and so on.

1.2.2 *Economic Dispatch*

The objective of economic dispatch is to minimize the fuel cost of thermal power plants, assuming that hydro generation has been given. The unit commitment scheme of thermal units is also supposed known. Many power systems today are operated under economic dispatch with calculations made every few minutes. Under normal circumstances, control signals are sent to generating units to adjust their power output in accordance with optimization results. In interconnected systems, optimization results are further adjusted by a load frequency control (automatic generation control, AGC) process, which aims at keeping deviation of frequency and total power interchange with neighboring utilities within preset values. The objective function is the sum of thermal fuel costs as a function of generation power output P. The most basic constraint is the power balance equation. Inequality constraints place limits on P. The mathematical formulation is thus as follows:

$$\min f(x)$$

subject to

$$\begin{aligned} h(x) &= 0 \\ \tilde{x} \leq x &\leq \bar{x} \end{aligned} \tag{1.4}$$

where

$$x = (x_1, \ldots, x_n)^{\mathrm{T}}$$

The solution of this problem is found by observing the Kuhn–Tucker optimality conditions:

$$\begin{gathered} \frac{\partial f(x)}{\partial x} + \lambda \frac{\partial h(x)}{\partial x} = 0 \\ h(x) = 0 \\ x \text{ within limits} \end{gathered} \tag{1.5}$$

More recent work deals with more complicated formulations. A particular interest of researchers is to introduce power flow equations directly as equality constraints or as implicit functions. In [20], nodal voltages are used as the state variables of the system. As all other variables can be expressed in terms of nodal voltages, the constraints can be set up based on them for real and reactive generation, voltage magnitudes, flows in transmission lines, etc. Then the economic dispatch problem becomes

$$\min f(x) \tag{1.6}$$

subject to

$$\begin{aligned} h(x) &= 0 \quad (k \text{ equations, } k < n) \\ g(x) &\geq 0 \quad (m \text{ inequalities}) \end{aligned} \tag{1.7}$$

where

$$x = (x_1, \ldots, x_n)^{\mathrm{T}}$$

In this formulation, it is necessary that all network equations appear directly as equality constraints or as a part of $g(x)$ in order for the formulation to be complete.

This is the standard nonlinear programming problem. In [20], it is solved by various penalty function methods of constrained optimization.

With the continuing search for alternatives to conventional energy sources, it is necessary to include wind energy conversion system (WECS) generators in the economic dispatch problem. The uncertain nature of the wind speed is represented by the Weibull distribution in Hetzer *et al.* [21]. In addition to the classic economic dispatch factors, factors to account for both overestimation and underestimation of available wind power are also included. The optimization problem can be numerically solved by Lagrangian relaxation, evolution algorithms, simulated annealing (SA), etc. The solution of the economic dispatch problem via the model presented is dependent on the values of many coefficients, such as the scale factor in the Weibull distribution function, the reserve cost for overestimating the wind energy, and the penalty cost for underestimating the wind energy. The level of wind power scheduled from a particular WECS is strongly dependent on the values of the reserve and penalty cost factors associated with the WECS. If the reserve cost coefficient is increased, the scheduled amount of wind power will be reduced, because it becomes more costly to overestimate the amount of wind power available. Conversely, if the penalty cost coefficient is increased, it becomes more costly to underestimate the amount of wind power available, and the system operator has an incentive to increase the scheduled amount of wind power. Economic dispatch considering uncertainty in renewable generation is still an open topic to be further investigated.

1.2.3 *Optimal Load Flow*

The most rigorous steady-state electric power system network model is provided by the load flow equations, which refers to the problem of solving the network equations. Generally, two groups of data are given for each node: P, Q for load nodes, and P, V for generator nodes, according to the physical quantities that can be easily controlled at those loads. There is a generation node that is treated differently and given by V, σ. The value of σ is set to zero as a reference for the load flow equations and P is left to balance the power of the whole network. This node is called a slack node. Mathematically, the load flow problem can be described as the following n equations:

$$h(x) = 0, \quad x = (x_1, \ldots, x_n)^{\mathrm{T}}. \tag{1.8}$$

The x or voltage variables are called the state variables of the system. Once they are known, all other electric quantities, such as reactive generation, slack-bus generation, and power flows in transmission lines, can be directly calculated.

Given a set of loads, different load flow solutions can be obtained by varying other input data. An optimal load flow problem is one that incorporates this exact model in the formulation, which refers to an operating state or load flow solution where some power system quantity is optimized, subject to constraints on the problem variables and on some functions of these variables. Optimal load flow has received more and more attention from power system planning, operation and control areas since the 1960s. At present, the application of advanced control equipment has brought new motivation to optimal load flow research. Furthermore, as the power industry moves into a more competitive and commercial environment, optimal load flow becomes even more important.

Optimal load flow is similar to economic dispatch in form. However, economic dispatch uses formulations of far lower dimension and sophistication. Some relevant variables such as generator voltage magnitudes are not included in optimization. As a result, constraints imposed by considerations of system security are not easily handled by procedures using power balance or other traditional models. The advantages of optimal load flow lie not so much in higher accuracy – more important is its ability to include security constraints in the formulation. A number of problems can be defined by different choices of decision variables, objective functions, and constraints. Some of these problems will be described below together with the techniques used to solve them.

In the approach of Dommel and Tinney [22], two sets of variables are defined: the state variables of the load flow problem, the x variables; and the load flow data control variables, the u variables. The network equations relate the x and u variables. Other load flow quantities that remain fixed, such as power demand at load buses, can be expressed as functions of x and included equality constraints. All inequality constraints not directly on control variables u are called "functional constraints" and are used to penalize the cost function. The problem can be formulated as follows:

$$\min f(x, u) \quad (x = (x_1, \ldots, x_n)^{\mathrm{T}}) \tag{1.9}$$

subject to

$$\begin{aligned} h_1(x) &= 0 \quad (k \text{ equalities},\ k < n) \\ g(x, u) &\geq 0 \quad (m \text{ functional inequalities}) \\ \tilde{u} \leq u &\leq \overline{u} \quad (n - k \text{ control variables}) \\ h_2(x, u) &= 0 \quad (n - k \text{ network equations}) \end{aligned} \tag{1.10}$$

An $F(x, u)$ function is defined by penalizing $f(x, u)$ with violated functional inequality constraints; n equations $b(x, u)$ can be defined by joining the k equality constraints $h_1(x) = 0$ and the $n - k$ network equations $h_2(x, u) = 0$. The Kuhn–Tucker optimality conditions are as follows:

$$\begin{aligned} &h(x, u) = 0 \\ &\frac{\partial F}{\partial x} + \left(\lambda^t \frac{\partial h}{\partial x}\right)^t = 0 \quad \text{or} \quad \left(\frac{\partial h}{\partial x}\right)^t \lambda = -\frac{\partial F}{\partial x} \\ &\frac{\partial F}{\partial u} + \left(\lambda^t \frac{\partial h}{\partial u}\right)^t = 0 \quad \text{or} \quad \nabla_u F = 0 \ \text{(reduced gradient)} \end{aligned} \tag{1.11}$$

Choosing the feasible values for u, the first set of equations is solved for x. This is the load flow problem. If Newton's method is used with triangular factorization techniques to solve the load flow, the calculation of λ involves a repeat solution with the triangularized Jacobian $\partial h/\partial x$, using the second set of equations. The third group defines the reduced gradient of F with respect to only u variables. This gradient can be used to calculate corrections for u variables

$$\Delta u = \alpha \nabla_u F \tag{1.12}$$

According to the Kuhn–Tucker optimality conditions, violations of inequality constraints on u variables due to the above correction Δu should be prevented by setting u to its limit. At this point, a new iteration is performed, and this process is continued until $\nabla_u F = 0$ for all off-limit u variables.

In a power system, there are strong couplings between real power flows and voltage angles or reactive power flows and voltage magnitudes. However, the P, δ and Q, V variable sets are weakly coupled. Billinton and Sachdeva [23] discussed these coupling effects and suggest a decomposition technique for solving the economic dispatch. The proposed approach decomposes the complete problem into two halves, i.e. the optimum voltage evaluation, and the real power optimization with optimally determined voltages. Both problems are solved by the Fletcher–Powell technique. This decomposition approach is still used as reference for optimal load flow of large-scale power systems.

1.3 Power System Reactive Power Optimization

Reactive power optimization is a kind of large-scale nonlinear optimization problem with multi-variable, multi-restriction, and multi-extreme mathematical characteristics. It has been a hot issue in global electrical fields for years. A great deal of deep research has been carried out and there are thousands of publications on reactive power optimization models and algorithms for power grids with different characteristics.

In the power system planning stage, reactive power optimization means optimal reactive power planning (ORPP). That is an off-line problem concentrating on the optimal allocation of reactive power compensation devices in order to enhance the controllability of power systems. In the power system operation stage, reactive power optimization means optimal reactive power dispatch (ORPD) or optimal reactive power control (ORPC). That is an on-line problem concentrating on the optimal regulation of reactive power and voltage control devices in order to improve the economy and safety of power system operation. On-line problems have very high requirements on the effect of optimization and the calculation speed of solution.

The ORPD problem can be described as an objective function and a set of constraint conditions. The characteristics of the ORPD problem are listed as follows:

1. Multi-objective
2. Number of constraint conditions of various types
3. Nonlinear objective function and constraint conditions

4. Uncertainty of load and operation mode
5. Discrete control variables that cannot be regulated frequently
6. Nonconvexity and multi-extreme
7. Objective function is hard to describe using formula including control variables
8. Solution space lacks connectivity

ORPD encounters two major problems in engineering applications:

1. The discrete control devices cannot be regulated frequently under the change of operation state of power systems at different times, which leads to the time coupling characteristic of ORPD.
2. The control of different regional power networks influences each other when adopting a hierarchical and zoning control strategy, which leads to the space coupling characteristic of ORPD.

The scale of power systems is expanding constantly, which results in higher requirements on a reactive power optimization algorithm. Some characteristics of an algorithm, such as whether it can converge to the optimization solution reliably and rapidly, and whether it can detect and solve infeasible problems, become more and more important. Because the feasible and infeasible subspaces of the ORPD solution space of complicated power networks interlace with each other, solving the ORPD problem becomes very difficult.

The interior point method is one of the most applied mathematical methods for the reactive power optimization problem. It is fast and has good convergence, and its calculation time is insensitive to the scale of the problem. However, how to probe and deal with the infeasible solution in the solution process is one of the obstacles of the interior point method. In addition, solving a high-order correction equation is still a bottleneck in the calculation for the interior point method.

Artificial intelligence algorithms, such as genetic algorithm, particle swarm optimization algorithm, and so on, make searching the global optimal solution in large-scale space possible because of their characteristics of organization, adaptation, self-learning, and parallelism. However, acquiring the theoretical optimal solution usually consumes an amount of computing memory and needs an unacceptably long time to calculate. It is hard for the artificial intelligence algorithms to meet the engineering requirements of real-time voltage and reactive power control of large-scale power systems.

Since optimal reactive power dispatch of large-scale power systems involves very complicated characteristics, such as having multiple objectives, multiple variables, multiple constraints, discreteness, and real-time responsiveness, the present algorithms cannot balance the opposing demands of optimization and computational efficiency. Therefore, in engineering applications, it comes down to the inevitable choice of voltage and reactive power control to reduce the dimensionality of the high-dimensional ORPD problem. There are two methods to reduce the dimensionality, i.e. (1) hierarchical and zoning space decoupling for ORPD of complicated power systems, and (2) time decoupling for reactive power optimization in each time interval. However, the ORPD problem couples in space and time in nature. On the one hand, the results of volt/VAR control of each regional power network interact because of the electricity connection of through tie lines or transformers. On the other hand, the regulating times of volt/VAR control devices are easily too frequent when the aim of reactive power optimization is to seek the minimum network power loss at each profile independently. Therefore, the modeling research of space decoupling and time decoupling for reactive power optimization should be enhanced.

A problem that has received some attention is the optimal control of reactive sources by minimizing reactive generation while observing voltage and network constraints. Nonlinear programming has been used to solve this problem [24, 25].

1.4 Optimization in Electricity Markets

A market is a basic mechanism to realize the optimal allocation of resources. Thus optimization techniques are naturally adopted in various electricity market research topics. Here we focus only on microeconomics-related topics. A general market model has been introduced by Weber and Overbye [26, 27]. The market primarily includes an independent system operator (ISO), a generation company (GenCo), and a consumer, all of which have their own optimization models. The basic models are introduced as follows.

1.4.1 Strategic Participants' Bids

A strategic participant may be a GenCo or a consumer – in other words, the strategic participant controls the strategic bid of a GenCo or consumer to maximize its own profit. The bids generally take one of the two formats shown in Figure 1.2. Most of the literature on market equilibrium analysis uses the continuous bidding format because of the restriction of solution algorithms. However, the block bidding format is often employed in the real electricity market due to the technical features of the power industry.

Suppose there are I GenCos in the market and each GenCo has a strictly convex quadratic generation cost function as

$$C_i(P_{Gi}) = \frac{1}{2}a_i P_{Gi}^2 + b_i P_{Gi}$$
$$(\underline{P}_{Gi} \leq P_{Gi} \leq \overline{P}_{Gi};\ a_i > 0;\ i = 1, 2, \ldots, I) \tag{1.13}$$

where P_{Gi} is the generation output; a_i and b_i are coefficients; $\underline{P}_{Gi}$ and $\overline{P}_{Gi}$ are respectively the lower and upper limits of P_{Gi}; and I is the number of GenCos. Hence, its

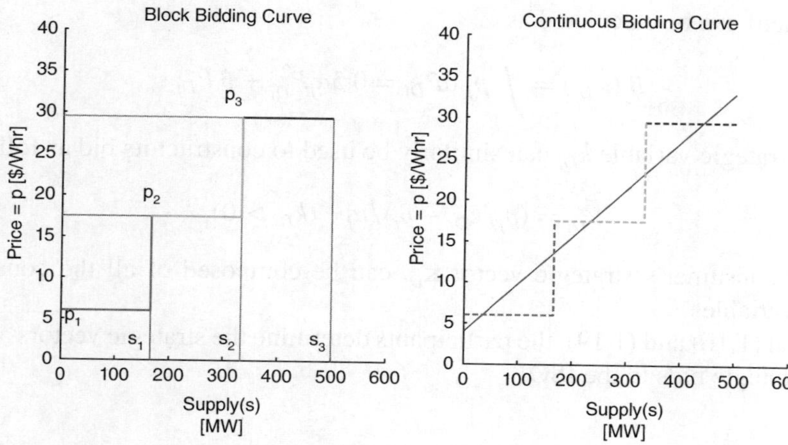

Figure 1.2 Block bidding and continuous bidding curves.

marginal cost can be calculated as

$$\frac{\mathrm{d}C_i(P_{Gi})}{\mathrm{d}P_{Gi}} = a_i P_{Gi} + b_i \tag{1.14}$$

The bids are supposed to be linear functions in the case of the continuous bidding format, so the GenCos are assumed to construct their bids in the form of a linear supply function (LSF) as

$$\widetilde{P}_{Gi} = (p_i - \tilde{b}_i)/\tilde{a}_i \quad (a_i > 0) \tag{1.15}$$

where $\widetilde{P}_{Gi}$ is the submitted quantity, p_i is the bid price for GenCo i, and $\tilde{a}_i$ and $\tilde{b}_i$ are the coefficients (strategic variables) of the LSF. If we suppose that GenCo i chooses $\tilde{a}_i$ and $\tilde{b}_i$, subject to the condition that the two variables have a fixed linear relationship, a strategic parameter k_{Gi} can be used to vary the bid from the true marginal cost function as

$$\widetilde{P}_{Gi} = (p_i/k_{Gi} - b_i)/a_i \quad (k_{Gi} > 0) \tag{1.16}$$

Hence, the strategic parameters of all the GenCos form the GenCos' strategic vector $\mathbf{k}_{Gi}$, in which k_{Gi} is one element.

On the other hand, suppose there are J consumers in the market and the true marginal benefit for each consumer bid is also defined as a linear function as

$$\begin{gathered} P_{Dj} = (p_j - \beta_j)/\alpha_j \\ (\underline{P}_{Dj} \le P_{Dj} \le \overline{P}_{Dj};\ \alpha_j > 0;\ j = 1, 2, \ldots, J) \end{gathered} \tag{1.17}$$

where P_{Dj} is the quantity demanded; p_j is the locational marginal price (LMP); α_j and β_j are coefficients; $\underline{P}_{Dj}$ and $\overline{P}_{Dj}$ are respectively the lower and upper limits of P_{Dj}; and J is the number of customers.

The benefit can be calculated as

$$B_j(P_{Dj}) = \int p_j \, dP_{Dj} = 0.5\alpha_j P_{Dj}^2 + \beta_j P_{Dj} \tag{1.18}$$

Also the strategic variable k_{Dj} can similarly be used to construct its bid as follows:

$$\widetilde{P}_{Dj} = (p_j/k_{Dj} - \beta_j)/\alpha_j \quad (k_{Dj} > 0) \tag{1.19}$$

Thus, the consumer's strategic vector $\mathbf{k}_{Di}$ can be composed of all the consumers' strategic variables.

Based on (1.16) and (1.19), the participants determine the strategic vectors and submit the strategic bids to the ISO.

1.4.2 Market Clearing Model

After receiving the bids from all participants, the ISO uses a security-constrained economic dispatch to determine the supplies, the demands, and the LMPs, based on the DC-based optimal power flow. In the market clearing process, the bids are treated as the true costs and benefits for participants, so the optimization problem to maximize a quasi-social welfare, subject to the transmission and generation constraints, can be formulated as follows:

$$\max \ \Gamma(\mathbf{K}_G, \mathbf{K}_D) = \sum_{j=1}^{J} \mathbf{K}_{Dj}\mathbf{B}_j - \sum_{i=1}^{I} \mathbf{K}_{Gi}\mathbf{C}_i \tag{1.20}$$

$$\text{s.t.} \qquad \mathbf{H}\theta = \mathbf{P}_G - \mathbf{P}_D \tag{1.21}$$

$$\underline{F}_l \le F_l \le \overline{F}_l \ (l = 1, 2, \ldots, L) \tag{1.22}$$

$$\underline{P}_{Gi} \le P_{Gi} \le \overline{P}_{Gi} \tag{1.23}$$

$$(i = 1, 2, \ldots, I)$$

where $\mathbf{K}_G$ is the GenCo strategic vector, with $\mathbf{k}_G$ as an element; $\mathbf{K}_D$ is the consumer strategic vector, with $\mathbf{k}_D$ as an element; $\mathbf{B}_j$ is the benefit vector for consumer j, with B_j as an element; $\mathbf{C}_i$ is the cost vector for GenCo i, with C_i as an element; $\mathbf{H}$ is the susceptance matrix; $\boldsymbol{\theta}$ is the vector of bus voltage angles; $\mathbf{P}_G$ is the vector of bus generation outputs; $\mathbf{P}_D$ is the vector of bus demands; F_l is the power flow on line l; $\underline{F}_l$ and $\overline{F}_l$ are the lower and upper flow limits on line l; and L is the number of lines in the system. In a perfectly competitive market, there are no strategic behaviors of GenCos or consumers, and all $\mathbf{k}_G$ and $\mathbf{k}_D$ are equal to 1.

In (1.20), the first equality constraint is the DC power flow equation, the second inequality constraint is the transmission constraint, and the third one is the generation constraint for each GenCo.

For the continuous bidding case, gradient-based approaches such as the interior point method are applicable, and the Lagrange multiplier of the relative power flow constraints in (1.20) are LMPs.

1.4.3 Market Equilibrium Problem

Each rational participant maximizes its profits by choosing and adjusting its strategies based on the market state variation. Therefore, the optimization problem for each participant t (t represents GenCo i or consumer j) could be formulated as the following two-level mathematical program:

$$\max\ R_t(\mathbf{k}_{Gi}, \mathbf{k}_{Dj}) = \sum_{i=1}^{I}(\lambda_{Gi}P'_{Gi} - C_i(P'_{Gi})) + \sum_{j=1}^{J}(B_j(P'_{Dj}) - \lambda_{Dj}P'_{Dj}) \tag{1.24}$$

$$\text{s.t.}\quad \underline{k}_{Gi} \leq k_{Gi} \leq \overline{k}_{Gi} \tag{1.25}$$

$$\underline{k}_{Dj} \leq k_{Dj} \leq \overline{k}_{Dj} \tag{1.26}$$

where λ_{Gi} and λ_{Dj} are LMPs of the GenCo and consumer, respectively, which should be solved through (1.20); P'_{Gi} is the awarded generation for GenCo; P'_{Dj} is the quantity demanded by the consumer; $\underline{k}_{Gi}$ and $\overline{k}_{Gi}$ are the lower and upper limits of the generation strategic variable; and $\underline{k}_{Dj}$ and $\overline{k}_{Dj}$ are the lower and upper limits of the consumer strategic variable.

In a perfectly competitive market, the market equilibrium is called Walrasian equilibrium [28]. However, this is only an ideal case. In reality, the strategic behaviors of market participants should be considered. The participant's profit is dependent not only on its own strategy, but also on the strategies of its opponents. Then this situation should be explained with game theory. The market equilibrium is the point from which each player does not have any incentive to deviate unilaterally (Nash equilibrium), i.e.

$$R_t(\mathbf{k}^*_{Gi}, \mathbf{k}^*_{Dj} \mid \mathbf{k}^*_{-Gi}, \mathbf{k}^*_{-Dj}) \geq R_t(\mathbf{k}'_{Gi}, \mathbf{k}'_{Dj} \mid \mathbf{k}^*_{-Gi}, \mathbf{k}^*_{-Dj}) \tag{1.27}$$

where $\mathbf{k}^*_{Gi}$ and $\mathbf{k}^*_{Dj}$ are strategic vectors of participants in the equilibria; $\mathbf{k}^*_{-Gi}$ and $\mathbf{k}^*_{-Dj}$ are strategic vectors of the opponents of participants in the equilibria; and $\mathbf{k}'_{Gi}$ and $\mathbf{k}'_{Dj}$ are arbitrary strategic vectors of participants.

The left-hand side of (1.27) is the profit of participant t in the equilibria and the right-hand side means the profit in the case that participant t chooses an arbitrary strategic vector with its opponents holding the equilibrium strategies.

2

Theories and Approaches of Large-Scale Complex Systems Optimization

Power systems are typical large-scale complex systems. With the development of smart grids and renewable energy, there has been a continuous evolution along this direction. In this chapter, we will give an outline of theories and approaches of large-scale complex systems optimization. The large-scale complex systems have multi-dimensional, highly interacting, and frequently stochastic characteristics [29]. Though the concept of large-scale systems emerges repeatedly for understanding and application, there is no precise definition for largeness nor generally acceptable quantitative measures of scale. However, there is still a clear distinction between large-scale systems and "small-scale systems".

The following essential differences between large-scale systems and small-scale systems are clarified in Ho and Mitter [30].

1. Presence of more than one controller or decision-maker, resulting in decentralized computations.
2. Correlated but different information available to the controllers and possibly at different times.
3. Requirements of coordination between the operation of the different controllers, resulting in hierarchical structures.
4. Availability of only aggregated and highly imprecise models.

Power System Optimization – Large-scale Complex Systems Approaches, First Edition.
Haoyong Chen, Honwing Ngan and Yongjun Zhang.

5. Systems that may operate as a "team" or in an essentially conflicting manner; thus, there may be a single-objective function, multi-objective functions, or conflicting objective functions.
6. It may be more reasonable to adopt a "satisfying" strategy rather than an optimizing strategy.

There are more recent discussions on complex systems and complexity. A special issue of *Science* magazine (vol. 284, no. 5411, 2 April 1999) expounded more contemporary understandings of complex systems from biology, chemistry, economics, and meteorology. In the editorial entitled *Beyond reductionism*, a complex system is defined as one whose properties are not fully explained by an understanding of its component parts, thus the predominantly reductionist approach cannot be applied to it [31].

Power systems, as the world's largest man-made machines, have attracted special interest from researchers in the field of large-scale systems. They are combinations of various kinds of large-scale systems. Physically, power systems are complex industrial systems composed of generators, loads, transformers, and transmission lines. On the other hand, power systems are microeconomic systems composed of producers, consumers, and markets because the operation of electricity markets relies on them. With the development of smart grids, much information and communication equipment will be installed in the power systems, which introduces even more complex interactions between energy and information.

2.1 Basic Theories of Large-scale Complex Systems

Large-scale physical systems such as power systems are often characterized by geographical separation, so that issues such as the economic cost and reliability of control have to be taken into account, thus providing impetus for a decentralized scheme. The technological advances and reduced cost of computer and communication networks provide another technological force for distributed computation. Thus, for economic and possibly reliability reasons, there is a trend toward decentralized decision-making, distributed computation, and hierarchical control.

A formalization approach has been introduced [32] to establish a general mathematical theory of multi-level systems. To explain the key ideas behind this approach, it is intuitive to consider the structure of organizational type systems. In a traditional type organization, whether of economic or biological or managerial nature, the decisions are made in a pyramid-like form and the goals vary in complexity. For example, problems at the base level of the organization are fairly simple, although the number of problems may be huge. Each of these problems can be solved subject to some decision rules (or relative to some intervention parameters), which should be reached (or manipulated) by other higher-level problems. This model of parameterized subproblems repeats itself over many levels within the organization. A group of decision problems that performs similar kinds of activities constitutes a level in the structure. At the top of the whole organization there is one decision problem upon which the overall objective of the system depends. In this sense, a multi-level system is a hierarchy of subsystems or decision problems. The multi-level systems theory was developed to solve the control problems of large-scale systems based on the organizational approach. The basic idea of this theory is to organize the control of small parts of the system (subsystems) so that this decentralized control is as good as its centralized equivalent in an overall sense.

2.1.1 Hierarchical Structures of Large-scale Complex Systems

In dealing with large-scale complex systems, three basic types of hierarchical multi-level structures can be identified [29]. It should be noted that, for a real-world large-scale system, its hierarchical structure may be a combination of the above three. Complexity is generally recognized in the interaction among model parameters and decision variables, the description of various behavioral aspects, operation in an uncertain environment, and the existence of many conflicting objectives.

2.1.1.1 Multi-stratum Hierarchical Structure

The first type of hierarchical structure is concerned with modeling the various aspects and phenomena involved in complex systems. Thus the levels of this type of structure, which are often called strata, have their own descriptive variables and laws, respectively. For example, a power system can be described in three strata:

first stratum a physical power system composed of electrical equipments,
second stratum a controlled system from the perspective of information and control, and
third stratum an electricity market system consisting of producers, consumers, and marketplace.

2.1.1.2 Multi-layer Hierarchical Structure

This hierarchy deals with the complexity of decision-making processes in large-scale systems. The levels of this structure are termed layers. Operation of multi-layered systems depends on the two-way communication between level subsystems and includes determination of the strategies to be used, reduction of the uncertainties by appropriate learning methods, and selection of the influencing function, which are in turn applied to the subprocesses. Taking frequency and active power control of power systems as an example, generally three layers of control are used:

first layer a local control layer, namely the primary frequency control, whose goal is to adjust the governor quickly when loads change and reduce the deviation of the frequency,

second layer a systemwide control layer, namely the second frequency control, whose goal is to adjust the setting values of prime movers to achieve a nondifferential frequency regulation, to keep the inter-area power exchanges tie line close to the setting values, and

third layer the tertiary frequency control, namely economic dispatch, whose goal is to allocate power to generating units optimally while keeping power balance between generation and load.

2.1.1.3 Multi-level Hierarchical Structure

A multi-level structure is the most general model of hierarchy since it considers the existence of conflict goals and objectives among the decision subproblems. In this form of hierarchical structure, the system consists of several distinctly separable interrelated subsystems. All decision-making units are arranged in a certain hierarchical relation. Each unit is subjected to the intervention of its higher-level units, while it has influence on its lower-level decision-making units. Although each unit has a certain degree of freedom, there may be conflicting objectives of decisions among the same level of decision-making units, and then a higher-level coordination is needed.

In other words, higher levels have the responsibility of relaxing the interactions among lower levels by properly accounting for the couplings and constraints, handling the uncertainties, and resolving the conflict goals. Electric power systems are typical multi-level systems, and Figure 2.1 illustrates a multi-level hierarchy of power systems.

Although the above three hierarchical structures give a description of large-scale complex systems from different perspectives, they have some common points.

1. The higher the level at which the decision units are located, the wider will be their scope of influence on system performance.
2. The higher the level at which the decision units are located, the longer will be their decision period, and they are responsible for solving the slow dynamics problems.
3. The higher the level at which the decision units are located, the greater the changes that they can bring about to the system.
4. The higher the level at which the decision units are located, the more uncertainty they will face, and sometimes it is more difficult to formulate a quantitative model.

The three hierarchical structures may be imbedded within one another when dealing with real-world complex systems. There may be multiple strata in a multi-layer hierarchical structure; or multi-level optimization may be needed in an optimization process of a multi-layer hierarchical structure.

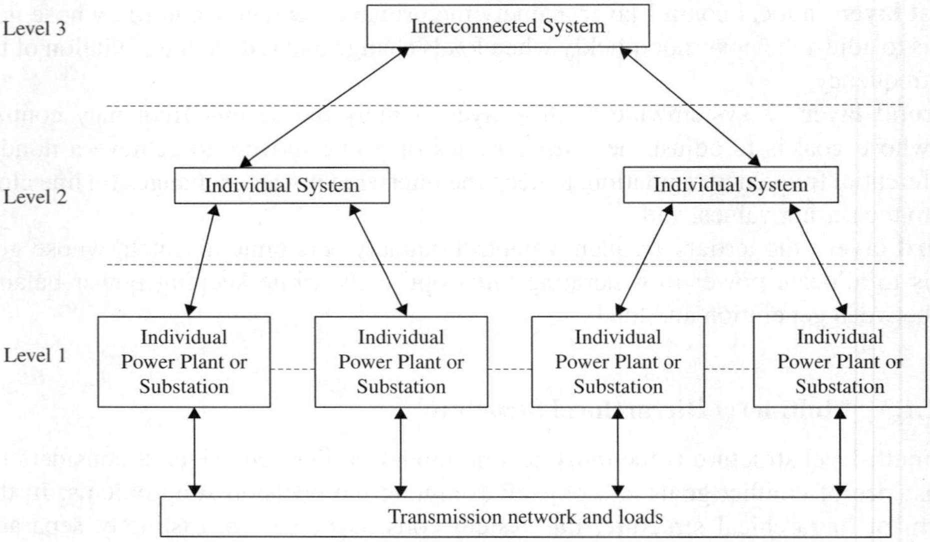

Figure 2.1 Multi-level hierarchical power systems.

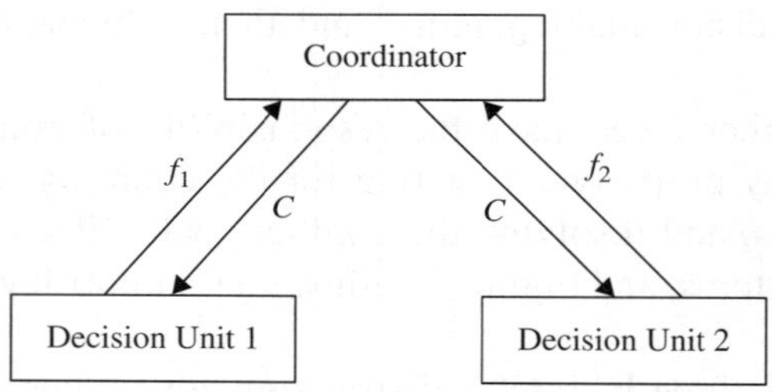

Figure 2.2 Process of coordination.

The successful operation of multi-level systems depend on two basic processes – decomposition and coordination. The decomposition is obvious and natural if the integrated system possesses an inherent hierarchical order, structure, or behavior. Modern industrial processes and large social organizations are typical examples. In fact, sometimes decomposition is almost mandatory. On the other hand, to reduce the system dimension, simplify the system model, or partition the phenomena involved, decomposition is carried out in an arbitrary manner and may take alternative forms. This is generally the basis for decentralization in systems engineering fields. Having defined the subsystems, the important question that immediately arises is how to attain the overall problem solution from the local solutions. This is our next topic to discuss.

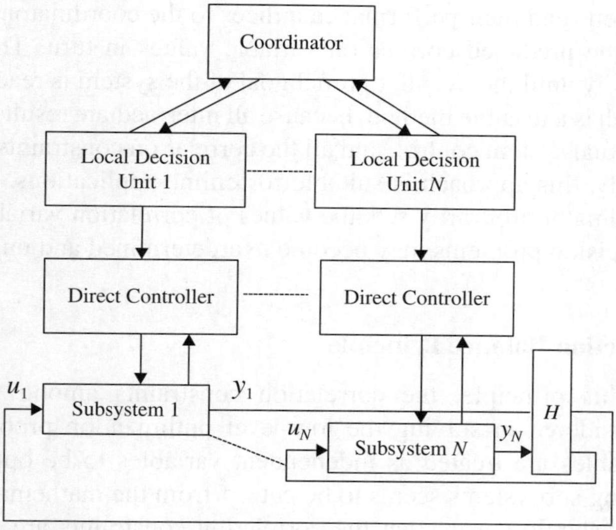

Figure 2.3 Static hierarchy of large-scale systems.

2.1.2 Basic Principles of Coordination

Coordination is a basic concept commonly adopted in large-scale systems control. As in the two-level structure shown in Figure 2.3, the high-level coordinator controls the two low-level decision units, which have their own control models and goals of the respective subsystems. The task of the coordinator is to ensure that the decisions made by the low-level units can meet the requirements of the overall system control target through necessary intervention. Thus the coordinator has to constantly exchange information with the low-level decision-making units. On the one hand, it will send the intervention signal C to the low-level units; on the other hand, it will receive signal f, which includes the decisions and related performance indicators of the low-level units. The coordinating task is accomplished through the intervention signal. There are many different coordinating approaches, but most of them are based on two basic principles:

- interaction prediction principle (in feasible decomposition), and
- interaction balance principle (in nonfeasible decomposition).

2.1.2.1 Interaction Prediction Principle

According to this principle, the input and output variables for correlation of each subsystem should be predicted by the coordinator. The low-level decision-making units solve their decision problem according to predicted values of the correlation variables, and then send their performance indices to the coordinator. Then the coordinator revises the predicted correlation variable values in turn. This process will continue iteratively until the overall optimal goal of the system is reached. This coordinating approach is a feasible method, because all intermediate results can be directly applied to the actual system control, and all the correlation constraints are always satisfied. Apparently, this approach is suitable for online applications. The problem is that, if the coordinator arbitrarily sets the values of correlation variables, sometimes the low-level decision problems may become overdetermined and unsolvable.

2.1.2.2 Interaction Balance Principle

According to this principle, the correlation constraints among the subsystems will not be considered in solving the low-level optimization problems. Then the correlation variables are treated as independent variables to be optimized and the correlation among subsystems seems to be cut off from the mathematical viewpoint. The coordinator should ensure that the correlation constraints are satisfied finally, by correcting the optimization objectives of the decision-making units through intervention signals (coordination variables). The correction terms of the objective functions tend to zero at last and the optimal value of the original objective function will be obtained. Obviously, this is a nonfeasible approach and the intermediate results of all iterations cannot be applied to the actual system; thus only the final result can be used in actual control.

2.1.3 Decomposition and Coordination of Large-scale Systems

A large-scale industrial production process is complex, involves many interrelated subsystems, and contains a lot of input variables, output variables, and internal correlation variables. These variables are usually limited by some constraints. There are difficulties in computation and communication to determine the optimal operating point of the whole system by using a single controller and the conventional steady-state mathematical programming model. A new approach was proposed to solve such a complex system optimization problem based on multi-level and hierarchical control [32]. The whole complex system is decomposed into several interrelated subsystems, and each has its own control objective and controllers, which constitute the lower-level control layer and directly control the production process. At the same time, they are coordinated by a higher-level coordinator, which ensure the overall optimization in steady state by coordinating the control performance indices of subsystems or correlation variables among them. Thus a hierarchical control structure is formed as shown in Figure 2.3. The low-level local decision units can work independently and in parallel. Because the number of variables is significantly reduced for each subsystem after decomposition, the optimization problem becomes relatively simple and can be solved with conventional mathematical programming.

As a result of the decomposition process, whether natural or conceptual, a number of uncoupled subsystems are generated along with their coupling relations. Apparently, the independent subsolutions would not result in the overall solution due to the interaction of the subsystems being unaccounted for. The multi-level systems theory has contributed to the comprehensive treatment of subsystems interaction. In this respect, coordination can be defined as the task of the high-level controllers, which attempt to bring about a harmonious operation of the low-level systems by handling their interactions, resolving their conflicts, and adjusting the goal and model interventions. The intervention parameters through which the high-level controller influences the local decision-making units are termed "coordination parameters" and are classified into two types. The first type modifies the subsystems control model and is called "model coordination", and the second type changes the goal of subsystems and is called "goal coordination". The actual coordinating method depends on how the coordination parameters are selected, adjusted, and utilized. Coordination principles and their application have been systematically investigated by Mesarovic and his colleagues [32–40]. It should be noted that the coordination principles (whether prediction or balance) depend on the successful choice of suitable coordination parameters. In this aspect, Mesarovic *et al.* introduced the concept of "coordinability". A system is "coordinable" by a coordination principle if the principle is applicable and there exists a coordination parameter such that the corresponding coordination condition can be satisfied.

In general, to use the basic aspects of multi-level systems theory, a two-step procedure is always considered. First, the overall control problem should be decomposed into a number of control subproblems. In this respect, there are three different ways to carry out such a decomposition [29].

1. *Decomposition based on the system structure*. This means the partitioning of the system into subsystems according to their physical behavior or operational phases. Control system problems are typical examples.
2. *Decomposition based on the amount of influence*. This implies the arrangement of subsystems in several levels, with upper levels holding higher priorities than lower levels. Management systems and econometric models could be treated this way.
3. *Decomposition based on the nature of control*. This refers to the application of different methodologies to control of multifaceted characteristics of the system. The interconnected power systems are perfect models for such a decomposition.

The second step is to synthesize the overall objective by coordinating the actions of the subsystems. Although there are many well-established techniques for coordination, they are essentially modifications or combinations of the goal and model coordination methods. Fundamental ideas from related fields, such as mathematical programming, economics, and decision analysis, are useful in designing coordination scheme, which are inherently iterative.

2.2 Hierarchical Optimization Approaches

The general theory of large-scale systems has been applied extensively to the development of optimization techniques for static as well as dynamic systems. These techniques have different characteristics, vary in alternative aspects, and exploit multi-level ideas with different depth and breadth. This book will focus on static optimization techniques of large-scale complex systems and their application to power systems.

Multi-level static optimization techniques for interconnected systems have been studied for half a century, with interesting applications in power systems, complex industrial plants, chemical engineering, and process control. A generalized description of the large-scale system optimization problem can be given as follows.

Suppose a large-scale system consists of N subsystems coupled together, each of which has an independent objective function, and the overall objective function of the large-scale system is supposed to be the sum of the subsystem objective functions. A diagram of a general subsystem is shown in Figure 2.4; and when the subsystems are interconnected, the overall large-scale system in Figure 2.5 is formed. The formulation of a subsystem is expressed as (2.1)–(2.4) below. Without loss of generality, the constraints of each subsystem are expressed in inequality form:

$$Z_i = T_i(X_i, M_i) \tag{2.1a}$$

$$Y_i = S_i(X_i, M_i) \tag{2.1b}$$

$$X_i = \sum_{i=1}^{N} C_{ij} Z_j \tag{2.2}$$

$$G_i(M_i, X_i, Y_i) \geq 0 \tag{2.3}$$

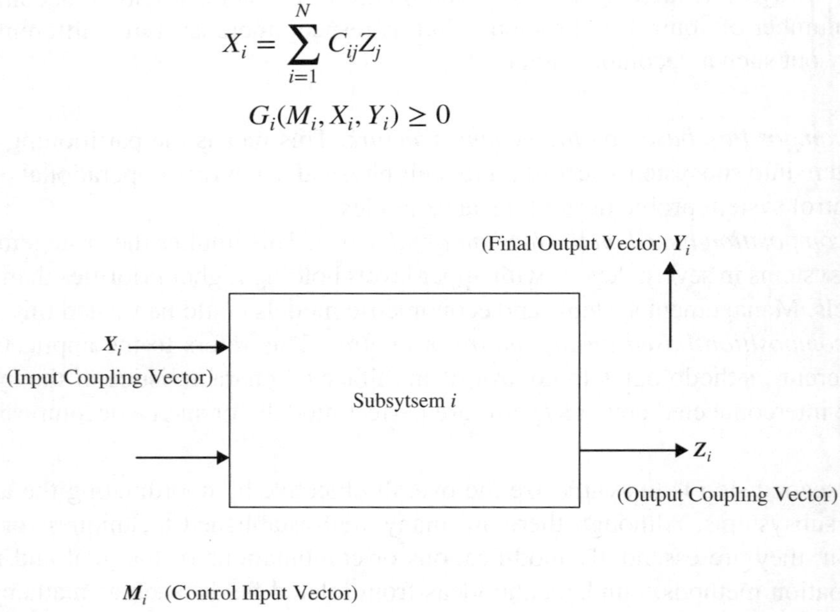

Figure 2.4 A general subsystem representation of the large-scale systems.

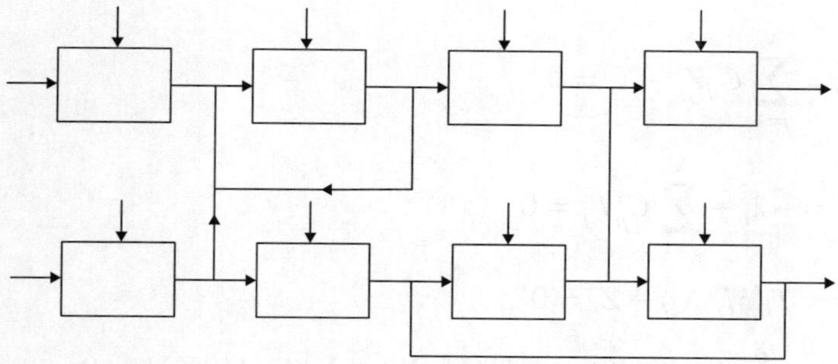

Figure 2.5 The representation of the overall large-scale system.

$$F = \sum_{i=1}^{N} f_i(M_i, X_i), \quad i = 1, \dots, N \tag{2.4}$$

Here, for the ith subsystem, the vector Y_i represents the final output, the vector Z_i stands for the output coupled with other subsystems, the vector M_i is the external decision or control variable, and the vector X_i accounts for the input from other subsystems. The subsystem is illustrated in Figure 2.4. Our objective is to determine the optimal controls $M_1, \dots, M_N$ that minimize (or maximize) (2.4) subject to (2.1)–(2.2) while satisfying the constraints (2.3). Note that the coupling between subsystems is expressed by (2.2).

To solve the optimization problem, we first define an appropriate Lagrangian functional in the form:

$$\varphi \overset{\Delta}{=} \sum_{i=1}^{N} [f_i(M_i, X_i) + \lambda_i^{\mathrm{T}}(T_i(M_i, X_i)) - Z_i] + U_i^{\mathrm{T}} G_i(M_i, X_i, S_i) + P_i^{\mathrm{T}} \left\{ \sum_{i=1}^{N} C_{ij} Z_j - X_i \right\} \tag{2.5}$$

where P_i and λ_i are Lagrange multipliers for the equality constraints (2.1a) and (2.1b) and the U_i are nonnegative Kuhn–Tucker multipliers for the inequality constraints (2.3). (2.5) could be rearranged as

$$\varphi = \sum_{i=1}^{N} \varphi_i(M_i, X_i, Z_i, \lambda_i, U_i, P_i, \dots, P_N) \tag{2.6}$$

where φ_i is the part of the Lagrangian associated with the ith subsystem. Necessary conditions for an extremum are that the Lagrangian be stationary with respect to all its arguments. These conditions yield the following vector equations, which must be satisfied:

$$\varphi_{P,i} = \sum_{j=1}^{N} C_{ij}Z_j - X_i = 0 \tag{2.7a}$$

$$\varphi_{Z,i} = -\lambda_i + \sum_{j=1}^{N} C_{ji}^{\mathrm{T}} P_j = 0 \tag{2.7b}$$

$$\varphi_{\lambda,i} = T_i(M_i, X_i) - Z_i = 0 \tag{2.7c}$$

$$\varphi_{M,i} = \frac{\delta}{\delta M_i}\{f_i(M_i, X_i) + \lambda_i^{\mathrm{T}} T_i(M_i, X_i) + U_i^{\mathrm{T}} G_i(M_i, X_i)\} = 0 \tag{2.7d}$$

$$\varphi_{X,i} = \frac{\delta}{\delta X_i}\{f_i(M_i, X_i) + \lambda_i^{\mathrm{T}}(M_i, X_i) - P_i^{\mathrm{T}} X_i + U_i^{\mathrm{T}} G_i(M_i, X_i)\} = 0 \tag{2.7e}$$

$$\varphi_{U,i} = G_i\{X_i, M_i, S_i(X_i, M_i)\} \geq 0 U_i \geq 0, \tag{2.7f}$$

$$U_i^{\mathrm{T}} G_i = 0, \quad i = 1, \ldots, N$$

We should note that (2.7a) and (2.7b) represent the subsystems' interactions and Lagrange multipliers, through which a subsystem can affect the overall system behavior.

If the conditions $\varphi_{P,i} = 0$ $(i = 1, \ldots, N)$ are relaxed, which means that the Lagrange multipliers should be adjusted iteratively and forwarded to the subproblems in the calculation, then a two-level decomposition–coordination approach is formed. This decomposition is nonfeasible, because the overall system equations are not satisfied except at the solution of the overall optimization problem. The two-level optimization approach based on this method is shown in Figure 2.6. Since (2.7b) is satisfied for the first-level subproblems, then by substituting it in (2.5), the overall Lagrangian gives

$$\varphi = \sum_{i=1}^{N} \{\overline{f}_i(M_i, X_i) + U_i^{\mathrm{T}} G_i(M_i, X_i)\} \tag{2.8}$$

from which we can restate the ith subproblem as follows. Find the optimal input M_i which minimizes the modified objective function $\overline{f}_i$ subject to $G_i \leq 0$. The modified objective $\overline{f}_i$ is defined as

$$\overline{f}_i(M_i, X_i) = f_i(M_i, X_i) + \lambda_i^T Z_i - P_i^{\mathrm{T}} X_i \tag{2.9}$$

From (2.9), it can be recognized that this nonfeasible decomposition approach belongs to the class of goal coordination.

On the other hand, if the conditions (2.7b) are relaxed, which implies that the interaction variables are fixed, this approach of decomposition–coordination is a feasible one, which means that the system equations are always satisfied. The corresponding two-level structure is given in Figure 2.7.

Substituting (2.7a) and (2.7c) in (2.5), we can conclude that the overall Lagrangian retains the original objective function. However, in this case the system model is

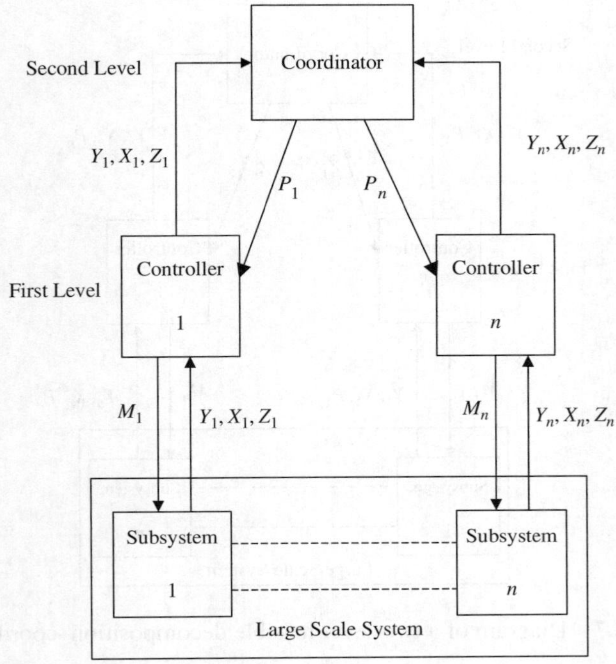

Figure 2.6 Diagram of a two-level nonfeasible decomposition–coordination.

changed due to the iterative values of Z. Thus feasible decomposition belongs to the model coordination group.

Lasdon *et al.* [41–48] developed a nonfeasible decomposition–coordination algorithm based on the excess demand approach by analogy to competitive markets. For feasible decomposition–coordination, a large step gradient technique is developed, and for both cases the convergence behavior has been rigorously analyzed. The dual properties of both methods have been studied [49–51], which led to a mutual understanding of these methods. In general, by similarity to and extension of the work of Dantzig and Wolfe [16], it is shown [52] that the nonfeasible and feasible solutions form a dual pair. This indicates that, at the optimal solution, the dual variables and multipliers are exactly identical to the primal multipliers and variables, respectively.

For the general interconnected dynamical systems, Pearson and Reich [41, 49, 51, 53, 54] investigated thoroughly the decomposition–coordination procedure based on model and goal coordination methods, and they also developed pseudo-model coordination techniques. For special linear quadratic problems, they emphasized the dual features and clarified the advantages and limitations of both model and goal coordination methods. In addition, certain Kuhn–Tucker as well as saddle-point conditions and their relations to multi-level control have been demonstrated. Constrained optimization problems are considered with the assumption that the constraints and the

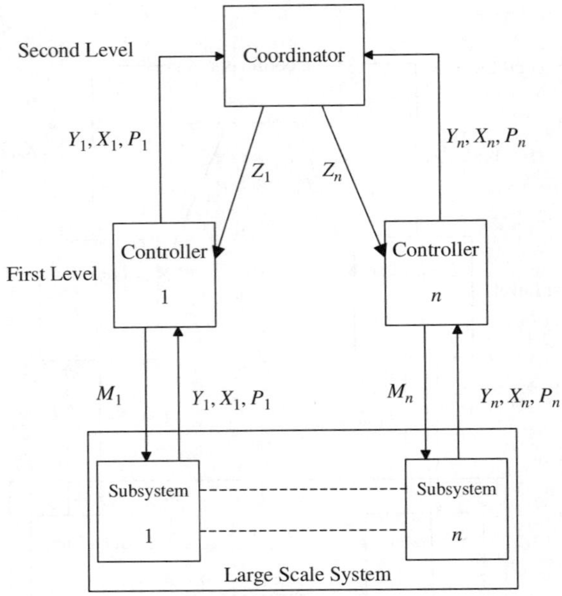

Figure 2.7 Diagram of a two-level feasible decomposition–coordination.

cost function are separable. It should also be noted that Cohen [55] has tackled the problem without these assumptions, but this problem still needs further work.

Here we have discussed two basic two-level optimization approaches, namely feasible decomposition and nonfeasible (dual) decomposition. Generally, they have distinct features with regard to practical applications. While the feasible method guarantees that the interactions are satisfied at each iteration, the optimality relations, originally relaxed for decomposition, are not always achieved. On the other hand, the dual method appears to be most promising from the point of view of computational requirements although only the converged final iteration of the second level yields a physically realizable solution.

An early survey of possible methods that can be used for the practical control of large interconnected dynamical systems is given in Singh *et al.* [56]. The optimal and suboptimal methods are discussed. The general problem formulation is described as follows.

System dynamics: for $i = 1, \ldots, N$, $K = 0, \ldots, P-1$,

$$\begin{aligned} x_i(K+1) &= f_i[x_i(K), u_i(K), K] \\ x_i(0) &= x_{i0} \end{aligned} \tag{2.10}$$

Interconnections:

$$\sum_{i=1}^{N} h_i[x_i(K), u_i(K), K] \leq 0 \quad (K = 0, \ldots, P-1) \tag{2.11}$$

Constraints: for $i = 1, \ldots, N, K = 0, \ldots, P-1$,

$$R_i[x_i(K), u_i(K), K] \leq 0 \tag{2.12}$$

$$Q_I[x_i(P)] \leq 0 \tag{2.13}$$

Performance measure:

$$J = \sum_{i=1}^{N} \left\{ S_i[x_i(P)] + \sum_{K=0}^{K-1} F_i[x_i(K), u_i(K), K] \right\} \tag{2.14}$$

Equations (2.10)–(2.13) represent N coupled subsystems with discrete-time dynamics where x_i and u_i are respectively the state and control vectors.

The objective is to determine the optimal sequences $[\{x_{i*}(K), u_{i*}(K)\}, i = 1, \ldots, N]$ that minimize the performance measure (2.14) subject to the constraints (2.11)–(2.13). This is a general class of constrained optimization problem. Here we will not investigate the different solution techniques, but rather discuss the application of duality theory to the multi-level control problem based on the work of Lasdon [47] and Pearson [52]. Lasdon and Pearson have shown that the introduction of the dual problem results in a natural two-level structure of the optimization problem, in which the first level solves the primal problem and the second level solves the dual problem. Thus the approach involves the formulation and solution of the Lagrangian dual of the above original problem. The dual problem is

$$\underset{\text{w.r.t. } \lambda}{\Phi(\lambda)}, \quad \lambda \geq 0 \tag{2.15}$$

where

$$\Phi(\lambda) \overset{\triangle}{=} \min_{x,u} \varphi(x, u, \lambda), \quad \text{subject to (2.3)–(2.5)} \tag{2.16}$$

where the Lagrangian function is

$$\varphi(x, u, \lambda) = \sum_{i=1}^{N} \left\{ S_i[x_i(P)] + \sum_{K=0}^{P-1} F_i[x_i(K), u_i(K), K] + \lambda_i^{\mathrm{T}h_i}[x_i(K), u_i(K), K] \right\} \tag{2.17}$$

Owing to the fact that the interconnection constraints (2.11) provide the only coupling between the subsystems, (2.17) can be separated into additive subproblems, and as a result N independent Lagrangian minimization problems are defined. This is the first level. On the other hand, (2.15) and (2.16) constitute the second level.

From (2.15)–(2.17), we can see that the gradient of the dual problem is exactly the value of the coupling constraint. It implies that the overall optimum is reached only when the second-level solution converges. On the other hand, since the dual function is concave without any convexity requirements on the original system, the global optimum is assured.

The computational efforts at the first level may still be burdensome, although the dimensions are reduced. However, they can be solved in parallel since the subsystem Lagrangians are independent. Furthermore, it is generally impossible for the first level to get the analytical solution (an explicit function of the multipliers) of the Lagrangian minimization problem, and therefore it is always practical to solve the problem numerically and iteratively.

Dirickx *et al.* [57] examine some relationships between multi-level hierarchical systems theory and four different types of optimization problems: a decomposable mathematical program, a sequential optimization problem, a continuous-time optimal control problem, and a welfare maximization problem involving profit distribution to consumers. In fact, it appears that a wide range of optimization problems can be coordinated using coordination principles from hierarchical systems theory. The applicability of the coordination principles is also discussed.

Electric power systems have long been a perfect area for applications of multi-level systems theory. Dy-Liacco [58] presented a unified control system for generation–transmission reliability via the multi-level approach. The control system is characterized by a combination of man and machine functions, a combination of local logic and central computer processing, and communication via an information system between the system–environment complex and the control center. The structure of the so-called adaptive reliability control system design provides a unifying medium for coordinating various operating control functions so as to achieve the overall objective of service reliability. Many of these functions have been known to the industry in some form or other but their applications have been for isolated objectives and have not necessarily been coordinated in terms of a total system design.

Schweppe and Mitter [59] discussed various concepts of hierarchical multi-level system theory and illustrated these concepts by considering the real-time control of bulk power systems for economic and reliable operation. It was shown how decomposition according to level, time, and mode is a natural way to decompose the complex task of controlling an interconnected power system. Then the concept of information and information structure was discussed. The concept of coordination was introduced and a scale of measurement for coordination was proposed. The relationship that necessarily exists between information structure and coordination (a concept different from that of information) was investigated. Finally, the data flow within the decomposed system and various trade-off considerations were elaborated.

2.3 Lagrangian Relaxation Method

The Lagrangian relaxation method [60, 61] is an optimization algorithm for solving complex integer and combinatorial optimization problems, which can be regarded as an extension of the nonlinear programming Lagrange multiplier method to integers and combinatorial optimization problems. The Lagrangian relaxation method is based on the following idea. Many complex integer programming problems can be considered as a series of complex additional constraints added to a relatively simple problem, and we can define a Lagrangian problem making use of this feature. The complex constraints are replaced by a penalty term of the objective function, which is the product of broken constraints and the dual variables. The Lagrange problem is easier to solve if the primal problem is a minimization problem; a lower bound of the optimal solution of the primal problem can be obtained by solving the Lagrange problem. At the beginning, the Lagrangian relaxation method is used in the branch-and-bound algorithm of integer programming, which is employed to replace the linear programming relaxation to obtain the lower bound of the primal problem. It is found in later study that after some heuristic adjustment of the Lagrange problem solution (i.e. dual solution), a good approximation of the optimal solution of the primal problem can be obtained.

A cost minimization problem in an economic system is taken as an example to illustrate the Lagrangian relaxation method. Suppose that there are m different products on the market, which are manufactured by n different firms, and the market demand is given. Then the problem is to determine the amount of output for each firm, so that the overall production cost can be minimized. Supposing that the amount of products is discrete, the number of products j manufactured by firm i is x_{ij}, and the production cost per unit is c_{ij}, then the total production cost of the economic system is

$$z = \sum_{i=1}^{n} \sum_{j=1}^{m} c_{ij} x_{ij} \tag{2.18}$$

If d_j is the market demand for product j, then we have

$$\sum_{i=1}^{n} x_{ij} \geq d_j, \quad j = 1, \ldots, m \tag{2.19}$$

Additionally, the manufacturer has a limitation on their ability to manufacture each product, i.e.

$$\sum_{j=1}^{m} a_{ij} x_{ij} \leq b_i, \quad i = 1, \ldots, n \tag{2.20}$$

Then the formulation of this problem is

$$\min \quad z = \sum_{i=1}^{n} \sum_{j=1}^{m} c_{ij} x_{ij} \tag{2.21}$$

$$\begin{aligned} \text{s.t.} \quad & \sum_{i=1}^{n} x_{ij} \geq d_j, \quad j = 1, \ldots, m \\ & \sum_{j=1}^{m} a_{ij} x_{ij} \leq b_i, \quad i = 1, \ldots, n \\ & i, j, \quad x_{ij} \geq 0 \end{aligned} \tag{2.22}$$

We note that the objective function is divisible with regard to n firms. The first type of constraints, namely the market demand constraints, are coupled constraints related with n firms; whereas the second type of constraints, namely the production capacity constraints of firms, are related only to an individual firm. So it is easy to think that, if the first type of constraint can be relaxed, this optimization problem can be decomposed into a series of individual firm production optimization problems, which can be solved one by one. Based on this idea, we relax the first type of constraints to form the Lagrangian problem. To this end, m nonnegative Lagrange multipliers $\lambda_1, \dots, \lambda_m$ need to be introduced, and the penalty term should be added to the objective function. The Lagrange problem is as follows:

$$\begin{aligned} \min \quad z_d(\lambda) &= \sum_{i=1}^{n}\sum_{j=1}^{m} c_{ij}x_{ij} + \sum_{j=1}^{m} \lambda_j \left(d_j - \sum_{i=1}^{n} x_{ij} \right) \\ &= \sum_{i=1}^{n}\sum_{j=1}^{m} (c_{ij} - \lambda_j)x_{ij} + \sum_{j=1}^{m} \lambda_j d_j \\ \text{s.t.} \quad & \sum_{j=1}^{m} a_{ij}x_{ij} \le b_i, \quad i = 1, \dots, n \\ & i,j, \quad x_{ij} \ge 0 \end{aligned} \tag{2.23}$$

where λ denotes the vector $(\lambda_1, \dots, \lambda_m)^T$. Because a nonnegative item is added to the objective function, the objective value of the Lagrangian problem will be a lower bound of the primal problem. The Lagrangian problem $\min z_d(\lambda)$ is a function of λ. Define

$$Lr(\lambda) = \min z_d(\lambda) \tag{2.24}$$

After the market demand constraint is relaxed in the Lagrangian problem, the objective function is divisible for the individual firms, and the constraints for each firm are also independent, so that, when λ is given, the Lagrangian problem can be decomposed into a series of subproblems for individual firms and dealt with separately.

The individual subproblem of firm i is $(i = 1, \dots, n)$

$$\begin{aligned} \min \quad z_{di}(\lambda) &= \sum_{j=1}^{m} (c_{ij} - \lambda_j)x_{ij} \\ &= -\left(\sum_{j=1}^{m} \lambda_j x_{ij} - \sum_{j=1}^{m} c_{ij}x_{ij} \right) \\ \text{s.t.} \quad & \sum_{j=1}^{m} a_{ij}x_{ij} \le b_i \\ & i,j, \quad x_{ij} \ge 0 \end{aligned} \tag{2.25}$$

Supposing the optimal solution is $x_i^* = (x_{i1}^*, \dots, x_{im}^*)^{\mathrm{T}}$, and defining

$$Lr_i(\lambda) = \min\ z_{di}(\lambda) \tag{2.26}$$

the Lagrangian problem can be written as

$$Lr(\lambda) = \sum_{i=1}^{n} Lr_i(\lambda) + \sum_{j=1}^{m} \lambda_j d_j \tag{2.27}$$

Now the problem is how to choose λ to make $\min\ z_d(\lambda)$ (approximately) equal to $\min\ z$. Apparently, the optimal value of λ is the solution to the following dual problem:

$$\begin{aligned} &\max_{\lambda}\ \min z_d(\lambda) \\ &\text{s.t.}\quad \lambda \geq 0 \end{aligned} \tag{2.28}$$

Compared with the saddle-point problem in the last section, it can be seen that (2.28) has the same form with the duality of nonlinear programming, but 3.23 does not hold in general. There is a difference between the optimal solutions of the dual and primal problems, which is called the "duality gap", namely

$$\min\ z - \max_{\lambda}\ \min z_d(\lambda) \tag{2.29}$$

It can be proved that it is continuous and concave [62] and then can be solved by the hill-climbing algorithm. However, it should be noted that this function is nondifferentiable where $\min z_d(\overline{\lambda})$ has multiple solutions $\overline{\lambda}$. The subgradient method can be used to solve such problems.

The m-dimensional vector y is called the subgradient of function $Lr(\lambda)$ at $\overline{\lambda}$ if the following relation holds:

$$Lr(\lambda) \leq Lr(\overline{\lambda}) + y(\lambda - \overline{\lambda}),\ \ \lambda \tag{2.30}$$

Apparently $Lr(\lambda)$ is everywhere subdifferentiable. If

$$x^* = (x_{11}^*, \dots, x_{1m}^*, \dots, x_{n1}^*, \dots, x_{nm}^*)^{\mathrm{T}}$$

is the optimal solution at any λ, then the vector

$$\left(d_1 - \sum_{i=1}^{n} x_{i1}^*, \dots, d_m - \sum_{i=1}^{n} x_{im}^* \right)^{\mathrm{T}} \tag{2.31}$$

is the subgradient at λ. Here we define

$$g_\lambda = \left(d_1 - \sum_{i=1}^{n} x_{i1}^*, \dots, d_m - \sum_{i=1}^{n} x_{im}^* \right)^{\mathrm{T}} \tag{2.32}$$

It should be noted that, because there may be multiple solutions at λ, multiple subgradients in the above form may exist, and other subgradients are the convex combination of these basic subgradients. We have the following theorem regarding the subgradient.

Theorem 2.1 *The necessary and sufficient condition for λ^* to be the optimal solution of $Lr(\lambda)$ is that 0 is the subgradient of $Lr(\lambda)$ at λ^*.*

Nondifferentiable optimization methods based on the subgradient have been widely studied and applied. Various methods have been used to solve the dual problem (primal–dual), and the most widely used one is the subgradient method [63].

The subgradient method can be regarded as an extension of the nonlinear programming gradient method to nondifferentiable optimization, where the subgradient is used instead of the gradient as the ascending gradient direction. Given the initial value of λ^0, the sequence $\{\lambda^k\}$ is calculated according to the following relationship:

$$\lambda^{k+1} = \lambda^k + t_k g_\lambda \tag{2.33}$$

where t_k is the positive scalar step size.

Held *et al.* [63] discussed the theoretical convergence and computing performance of the subgradient method. In theory, under the conditions of $t_k \to 0$ and $\sum_{i=0}^{k} t_i \to \infty$, $Lr(\lambda)$ will converge to the optimal solution. The most widely used step size is

$$t_k = \frac{\gamma_k(\overline{Lr} - Lr(\lambda^k))}{||g_\lambda||^2} \tag{2.34}$$

where γ_k is a scalar and satisfies $0 < \gamma_k \leq 2$, $\overline{Lr}$ is an upper bound of $Lr(\lambda)$, which can be obtained by solving the primal problem with a heuristic method. The sequence $\{\gamma_k\}$ is often calculated by the following procedure.

Let $\gamma_0 = 2$, and let γ_k be halved when $Lr(\lambda)$ does not rise within a fixed number of iterations. Although this rule does not meet the sufficient condition of convergence, it is very successful in practical applications.

Unless we get a λ^k such that $Lr(\lambda^k)$ is equal to the objective function value of a known feasible solution value of the primal problem, we cannot prove the optimality of the subgradient method, and so often use a certain number of iterations as the algorithm termination condition. Typically, the initial value of $\lambda^0 = 0$ is the most natural choice, but in some cases we may have a better choice. The framework of the Lagrangian relaxation algorithm is shown later in Figure 3.4.

2.4 Cooperative Coevolutionary Approach for Large-scale Complex System Optimization

Evolutionary algorithms (EAs) have been successfully applied to a wide variety of power system optimization problems, including planning, scheduling, control, and optimal power flow. "Evolutionary algorithm" is an umbrella term used to describe computer-based problem solving systems which use computational models of some of the known mechanisms of evolution as key elements in their design and implementation. The major evolutionary algorithms include genetic algorithms (GAs), evolutionary programming (EP), and evolution strategies (ESs) [64]. They all share a common conceptual basis of simulating the evolution of individual structures via processes of selection, mutation, and reproduction. EAs are general-purpose techniques and especially suited for the cases when there is no other known problem solving strategy, and the problem domain is NP-hard, as often happens in power systems. However, there are two main difficulties that hinder the application of EAs to large-scale problems. First, the algorithms have a strong tendency to converge prematurely when the problem solution spaces become large. Second, the computation time becomes too large to apply the methods practically.

This section introduces the cooperative coevolutionary approach (CCA) for solving complex system problems. The hypothesis underlying the ideas presented here is that, in order to evolve increasingly complex structures, explicit notions of modularity need to be introduced in order to provide reasonable opportunities for complex solutions to evolve in the form of interacting coadapted subcomponents. CCA simulates the evolutionary mechanism in nature and adopts the notion of an ecosystem. Multiple species in the ecosystem coevolve and interact with each other and result in the continuous evolution of the ecosystem. The application domain problem is mapped into the evolving ecosystem and solved when the coevolutionary process ends.

Early work concerning coevolutionary computation includes the application of a model of hosts and parasites to the evolution of sorting networks using a GA [65]. The Evolutionary Computation Laboratory at George Mason University, Fairfax, VA, USA, directed by De Jong has made significant progress in research on cooperative coevolutionary computation [66, 67]. Chen and Wang [68] applied the cooperative coevolutionary model to power system unit commitment and obtained quite encouraging results.

2.4.1 Framework of Cooperative Coevolution

Here we describe a generalized architecture of cooperative coevolution which models an ecosystem consisting of two or more species [66, 67]. As in nature, the species are genetically isolated – individuals only mate with other members of their species. The species interact with one another within a shared domain model and have a cooperative relationship. The basic coevolutionary model is shown in Figure 2.8.

Each species is evolved in its own population and adapts to the environment through the repeated application of an EA. Figure 2.8 shows the fitness evaluation phase of the EA from the perspective of species 1. To evaluate individuals from one species, collaborations are formed with representatives from each of the other species. There

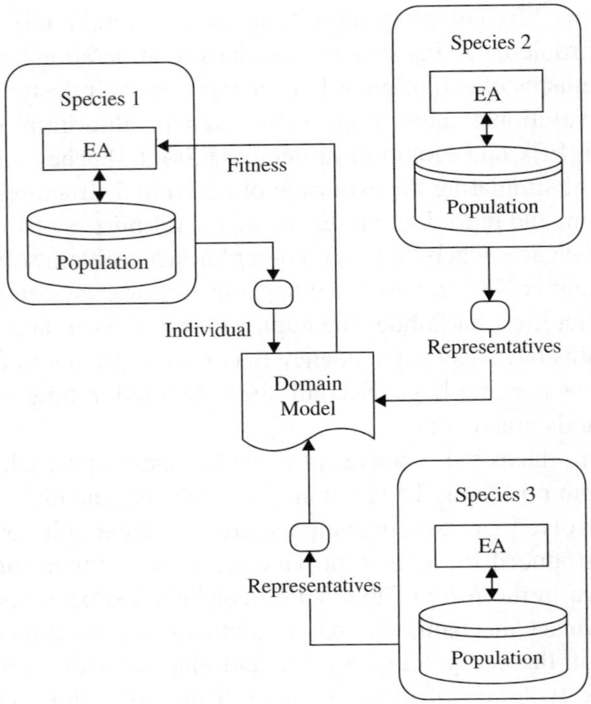

Figure 2.8 Framework of cooperative coevolutionary model.

are many possible methods for choosing representatives with which to collaborate. A natural one is simply to let the current best individual from each species be the representative.

A different framework of CCA is presented in Chen and Wang [68], which combines the basic ideas of the decomposition–coordination optimization of a large-scale system with evolutionary optimization. The basic idea is to relax the systemwide constraints of the primal problem and form the dual problem by the Lagrange multiplier method, then to decompose the dual problem into a number of subproblems. GA is used to solve the subproblems, which are nonconvex, discrete, and nonlinear. A subgradient-based stochastic optimization method is used to solve the dual problem, viz. to coordinate the subproblems. This model uses the Lagrange multipliers as the coordinators of the evolving species. Since Lagrange multipliers can be looked upon as the "prices" in the market, this CCA virtually employs the price coordination mechanism.

Although the current implementations of CCA have utilized a sequential pattern of evaluation, where the complete population of each species is evaluated in turn, the species could also be evaluated in parallel.

2.4.2 *Cooperative Coevolutionary Genetic Algorithms and the Numerical Experiments*

This section uses cooperative coevolutionary genetic algorithms (CCGAs) for function optimization as the example to illustrate CCA [69].

2.4.2.1 Cooperative Coevolutionary Genetic Algorithms

If we think of solving a function optimization problem of *N* variables by CCA, we can map the problem to an ecosystem consisting of a number of species. A natural way is to maintain *N* species each of which optimizes values for a particular variable. One can then assign fitness to a particular individual of a particular species by assembling it along with representatives of the other species to form an *N*-dimensional vector, and using the conventional fitness of the *N*-vector to assign fitness to the individual being evaluated. That is, the fitness of a particular member of a particular species is computed by estimating how well it cooperates with other species to produce good domain solutions.

In CCGAs the evolution of each species is handled by a standard GA. The pseudo-code of a standard GA can be written as follows:

$k = 0$
initialize the population Pop_0
evaluate fitness of each individual in Pop_0
while termination condition = false do
  begin
    select, recombine, mutate from Pop_k *to get* Pop_{k+1}
    evaluate fitness of each individual in Pop_{k+1}
    $k = k + 1$
  end

The pseudo-code of CCGA can be written as:

$k = 0$
for each species s do
  begin
    initialize the species population Pop_0^s
    evaluate fitness of each individual in Pop_0^s
  end
while termination condition = false do
  begin
    for each species s do
      begin
        select, recombine, mutate from Pop_k^s *to get* Pop_{k+1}^s

evaluate fitness of each individual in Pop^s_{k+1}
end
$k = k + 1$
end

2.4.2.2 Results of Numerical Experiments

CCGA is evaluated by comparing its performance with the performance of a standard GA on several function optimization problems. CCGA and standard GA differ only as to whether they utilize multiple species, and all other aspects of the algorithms are the same and are held constant over the experiments.

The experiments are performed on four highly multi-modal functions that have been used for benchmarking evolutionary algorithms, viz. Ackley, Schwefel, Griewangk, and Rosenbrock functions [69]. All the functions have been defined such that their global minima are zero. The primary performance metric used in evaluating the algorithms is the minimum function value found after a fixed number of function evaluations. Each of the results reported is an average calculated over 50 runs. The experimental results are shown in Figure 2.9.

Figure 2.9 shows the minimum value found (best individual) as a function of the number of function evaluations averaged over 50 runs using the four test functions.

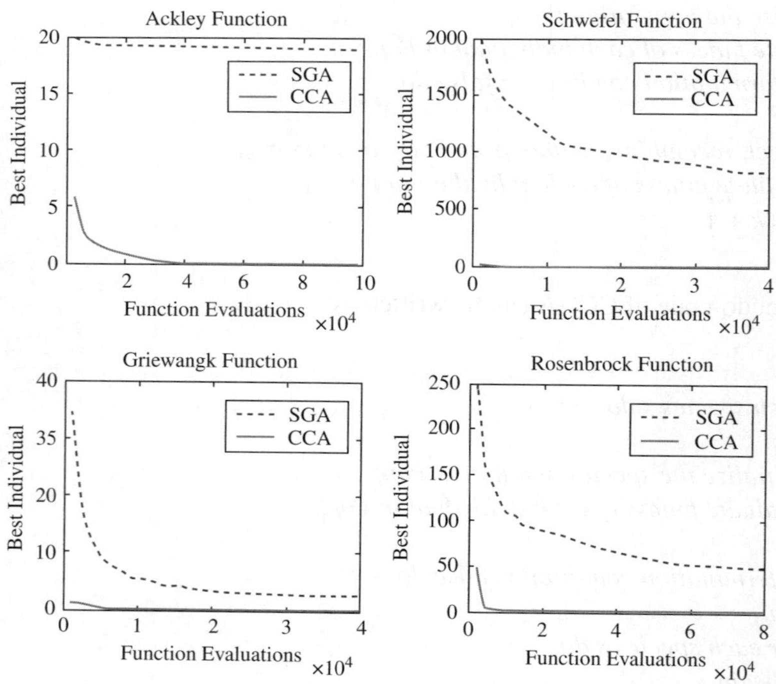

Figure 2.9 Aggregated supply curve with capacity constraint.

The dashed lines represent the convergence curve of standard GA, and the solid lines represent the convergence curve of CCGA. In all cases CCGA significantly outperformed the standard GA both in the minimum value found and in the speed of convergence to zero.

2.4.3 Basic Theories of CCA

CCA simulates the coevolution mechanism in nature, which is defined as "reciprocally induced evolutionary change between two or more species or populations" and breaks through limits of conventional EAs. The theoretical foundation is relatively weak since CCA is a rather new research subject. This section outlines the basic theories of CCA in two aspects.

2.4.3.1 Optimization Theory of CCA

Optimization of a complex system is a basic application domain of CCA. Conventional EAs model a complex optimization problem by a single population, and since the number of problem variables is large, the chromosome of an individual is long. This may cause two primary difficulties. First, the algorithms have a strong tendency to converge prematurely. Second, the computation time becomes too long. But if we decompose the complex problem into a number of subcomponents and evolve such subcomponents by a collection of cooperating species, we can develop an efficient CCA. CCA can overcome the above difficulties since the decision space for each species is smaller. Particularly, the collaboration method among the species is the key to a successful implementation of CCA.

Potter and De Jong [69, 70] investigated the performance of CCGA for function optimization by numerical experiments. They analyzed the performance of CCGA by benchmarking on the test functions referred to in the previous section. The results show that CCGA significantly outperforms GA on these functions. Further improvements of CCGA are presented. Wiegand *et al.* [71] analyzed various types of collaboration mechanisms among species and presented some basic advice about how to choose a mechanism which is appropriate for a particular problem. Their results show that the aforementioned method of assigning an individual a fitness score equal to the value of its best collaboration is generally the best choice.

2.4.3.2 Game Theory of CCA

Since CCA explicitly models the reciprocity among coevolving species, it embodies a dynamic process of strategy choice and interaction, which coincides with the framework of game theory.

The task of understanding the coevolutionary approach is quite difficult. Recently, game theory, especially evolutionary game theory, has begun to be used to analyze the dynamic behavior of the coevolutionary approach [72]. At the Dynamical & Evolutionary Machine Organization laboratory at Brandeis University, Waltham, MA, USA, Ficici and Pollack have done some early work in this aspect [73, 74]. They extended the mixing-matrix and Markov-chain models of EAs to address coevolutionary algorithm dynamics by using game theory and then employ concepts from evolutionary game theory to examine design aspects of conventional coevolutionary algorithms [73]. In another paper Ficici and Pollack explored the relevance of game theory to the analysis of coevolutionary algorithm dynamics [74]. Wiegand, Liles, and De Jong [75] used the multi-population symmetric games of evolutionary game theory to analytically model cooperative coevolutionary algorithms. Using the analysis framework introduced therein [75], we can achieve a better understanding of the degree to which cooperative coevolutionary algorithms can be used for optimization.

2.4.4 CCA's Potential Applications in Power Systems

CCA can be applied to all the domains to which conventional EAs are applied. This section focuses on the discussion of some primary topics concerning power systems and the power market.

2.4.4.1 Basic Methodologies of Applying CCA

CCA has two primary classes of applications in power systems, viz. optimal planning, operation, and control of power systems, and game theoretic analysis of the power market.

Conventional EAs have been successfully applied to a wide variety of power system optimization problems, but the main obstacles hindering their practical applications remain the overlong computation time and their susceptibility to premature convergence. CCA can overcome these difficulties.

One way to apply CCA is based on the framework in Figure 2.8. First, we can decompose the complex power system optimization problem into a number of subsystem optimization problems, and then evolve each subsystem by a single population species, which represents a partial solution. When evaluating the fitness of each individual in a species, we select the best collaborators from other species. The evolution and cooperation process continues repetitively until a satisfactory solution is found. Another way to use CCA is the method employed in Chen and Wang [68]. The basic idea is to relax the systemwide constraints of the primal problem by the Lagrange multiplier method and form the dual problem, then to decompose the dual problem into a number of subproblems. Thus a two-level optimization model is formed in this way. We can design CCA by using EA to solve the low-level subproblems and choosing an appropriate coordinator for the high-level dual problem. We can see that Potter and De Jong's approach [69, 70] belongs to feasible decomposition and Chen's approach [68] belongs to nonfeasible decomposition.

CCA can be used to simulate and analyze strategic gaming behavior in the power market. The key is to represent a player by an evolving species that can optimize the game strategy, and coordinate the species by the game rules. When the individual fitness in a species is evaluated, the strategy represented by this individual is assembled with the strategies represented by its collaborators from other species to form a strategy combination, which determines the payoff of the strategy and thereby determines its fitness score. In the case of a generation market auction design, the market rules are virtually a set of game rules. Using different species to optimize the bidding strategies of different GenCos, and evaluating the fitness of an individual by the profit as the GenCo chooses the bidding strategy represented by it, we can perform the market simulation by CCA. When the coevolutionary process converges, we can get the market equilibrium, which shows the market efficiency.

2.4.4.2 CCA for Power System Optimal Planning and Operation

EAs, especially GAs, have long been applied to power system optimal planning, which can be further divided into generation planning and transmission network planning [9, 76]. Under a competitive market environment, the object of each GenCo for investment decision-making is to maximize its profit, while the objective of the market regulatory entity is to achieve market stabilization through coordination among GenCos [76]. This suggests a CCA that can deal with these conflicting objectives with each species optimizing the investment decision for a GenCo and a coordinator to regulate all the GenCos. GAs can successfully solve the single-area and single-stage transmission system planning problem, but may encounter some difficulties in multi-area and multi-stage transmission system planning problems [9], and that may be where CCA comes into play.

The application of GAs to power system optimal operation is consistently a heated subject [77, 78]. Although GAs are flexible to use and can handle very complicated constraints, they often suffer from overlong computation time when the generator number in the system is large. CCA is designed to overcome this difficulty and has been successfully applied to the unit commitment problem [68]. The results show that CCA has a good convergent property and a significant speedup over traditional GAs and can obtain high-quality solutions. The computational burden is almost linear with the problem scale. Further investigation needs to extend CCA to broader areas of optimal operation problems.

2.4.4.3 CCA for Power Market Auction Design and Bidding Strategy Optimization

Coevolutionary mechanism design is a brand new area of research [79, 80]. Traditionally, researchers design the auction mechanism by studying the analytic properties of different auction models. An alternative is to view a mechanism as the outcome of some evolutionary process involving buyers, sellers, and the auctioneer. Phelps *et al.* applied genetic programming to the development of an auction pricing rule for double auctions in a wholesale electricity marketplace, with buyer and seller strategies coevolving toward the market equilibrium that is used to assess the market efficiency [79, 80].

GA is employed to evolve bidding strategies of GenCos and DisCos (distribution companies) for a multi-round auction market [81]. Trading agents use a GA coupled with various price forecasting techniques to select appropriate bidding strategies for the current market conditions. The bidding strategies evolve as other traders change their trading behavior. Price [82] has used coevolutionary GA to model several standard industrial organization games and a simple model of an electricity pool. Since the single population GA is insufficient for modeling strategic interactions among the players, coevolutionary GAs with multiple populations are used in the competitive game simulations. In each population, the fitness of each individual is evaluated relative to the other populations. An agent's strategies that tend to be good in one generation will affect the mix of strategies in the other agents' populations in the next generations. Simulation results show the technique performs very well.

A coevolutionary approach has been developed by Cau and Anderson [83] to study the dynamic behavior of participants over many trading intervals. Each market participant is represented by a trading agent. The bidding strategy of each agent is modeled as a set of bidding actions. Trading agents coevolve their own populations of bidding strategies using a genetic algorithm. Simulation results show that, through the coevolutionary process with many repeated interactions, participants can learn to play adaptive strategies to improve their trading profit.

The cooperative coevolutionary approach is an extension of conventional single-population EAs. It simulates the mechanism of biological coevolution, and emphasizes the interaction and coadaptation between species in the ecosystem. Numerical experiments show that CCA significantly outperforms traditional GAs in function optimization. Nevertheless CCA is more an adaptive system consisting of a number of coadapted subcomponents than merely an optimizer. It accords with the framework of evolutionary game theory. CCA has achieved successful applications in some areas of power systems, and more promising applications are expected in power system optimal planning and operation, together with power market auction design and bidding strategy optimization.

3

Optimization Approaches in Microeconomics and Game Theory

Microeconomics is a branch of economics that studies the behavior of individuals and small impacting organizations in making decisions on the allocation of limited resources [84]. Typically, it applies to markets where goods or services are bought and sold. Microeconomics examines how these decisions and behaviors affect the supply and demand for goods and services, which determines prices, and how prices, in turn, determine the quantity supplied and quantity demanded of goods and services. Mathematical optimization is a basic tool for microeconomics, and the most important application of microeconomics in power systems is electricity markets.

Here, we will give an example of economic analysis for explanation of the fundamental concepts. In this chapter we will examine a model of a particular market for home energy consumption. Typical energy consumption in the home includes space heating, air conditioning, water heating, lighting, and electric appliances, and they are supplied through electricity or natural gas. In some countries such as the United States, electricity and natural gas are often provided by different companies; and some states such as Texas have established a retail electricity market in which the electricity providers offer different electricity plans for customers to choose. To some extent, electricity and natural gas are so-called substitute goods, which means that one can be used to replace another, so that a consumer having more electricity may require less gas, and vice versa. We will introduce several concepts and tools of microeconomics along the way here.

The Microeconomic System

Economics proceeds by developing models of social phenomena. By a "model" we mean a simplified representation of reality. A model's power stems from the

Power System Optimization – Large-scale Complex Systems Approaches, First Edition.
Haoyong Chen, Honwing Ngan and Yongjun Zhang.

elimination of irrelevant detail, which allows the economist to focus on the essential features of the economic reality that he/she is attempting to understand.

Here we are interested in what determines the price of electricity, so we want to have a simplified description of the electricity market. There is a certain approach to choosing the right simplifications in establishing a model. In general, we want to adopt the simplest model that is capable of describing the economic situation we are examining. Then the complications can be added gradually, allowing the model to become more complex and realistic.

There are two products in the energy market, namely electricity and natural gas. Some energy requirements in the home can be satisfied by both, such as water heating or cooking. However, electricity is generally considered to be more desirable by customers, since it is clean, safe, and easy to use. On the other hand, burning natural gas may release carbon dioxide and other harmful substances. Also, there may be a danger of explosion. So most customers would prefer electricity if they can afford it.

Although there are two markets – electricity and natural gas – we will focus exclusively on the electricity market. The natural gas market should be interpreted as where people who do not use electricity can go for some purpose, and its price is fixed at some known level. We shall be concerned solely with the determination of the price of electricity and how much is to be used.

An economist would describe the distinction between the prices of the two products in this model by saying that the price of natural gas is an *exogenous variable*, while the price of electricity is an *endogenous variable*. This means that the price of natural gas is taken as determined by factors not discussed in this particular model, while the price of electricity is determined by forces described in the model.

But what determines this price? What determines the quantity of electric energy to be consumed? What can be said about the desirability of different economic mechanisms for electricity markets? What concepts can we use to judge the merit of different market transaction results? These are all questions that we want the economic model to address.

Optimization and Equilibrium

Whenever we try to explain the economic behavior of humans, we need to have a framework on which our analysis can be based. In microeconomics we use a framework built on the following two simple principles [85].

Optimization principle People try to choose the best patterns of consumption that they can afford.

Equilibrium principle Prices adjust until the amount that people demand of something is equal to the amount that is supplied.

Let us think over these two principles. If people are free to choose their actions, it is reasonable to assume that they try to choose things they want rather than things they do not want. Of course, there are exceptions to this general principle, but they typically lie outside the domain of economic behavior. Thus the *economic human* is a basic hypothesis in microeconomics as a rational and narrowly self-interested agent who has the ability to make judgments toward his/her own ends. Using these rational assessments, an economic human attempts to maximize utility as a consumer and economic profit as a producer. This principle links the microeconomic theory with mathematical optimization.

The second principle is more difficult to understand. Consumers and producers in a market react differently to price changes. Higher prices tend to reduce demand while encouraging supply, and lower prices increase demand while discouraging supply. Microeconomic theory suggests that in a free market there will be a single price that brings demand and supply into balance, called the "equilibrium price". Both parties require the scarce resource that the other has and hence there is a considerable incentive to engage in an exchange. The equilibrium price is also called the "market clearing price" because at this price the exact quantity that producers take to market will be bought by consumers, and there will be nothing "left over". This is efficient because there is neither an excess of supply and wasted output, nor a shortage – the market clears efficiently. If we view the market as a large-scale system, we can see that the equilibrium principle describes how the system is coordinated with a number of components or subsystems (the individual consumers and producers), and the market clearing prices are the key coordinating variables.

It is worth observing that the definition used for equilibrium may be different in different models. In the case of the simple electricity market we exemplified here, the demand and supply equilibrium idea may be adequate. But in more complex models we will need more general definitions of equilibrium. Typically, equilibrium will require that the economic agents' actions must be consistent with each other. Two extensively investigated market equilibria, namely the general equilibrium in competitive markets, and the strategic equilibrium (Nash equilibrium) in imperfectly competitive markets, will be analyzed in the following sections.

3.1 General Equilibrium Theory

There are two basic multi-agent models used in microeconomics: the general equilibrium model and the strategic or game theoretic model [86].

In the general equilibrium model, it is assumed that each actor in the economy reacts optimally to a market aggregate, namely, the price vector. General equilibrium exists when all prices are in equilibrium; each consumer spends his/her given income in a manner that yields him/her the maximum satisfaction; all producers in each industry are in equilibrium at all prices and output; and the supply and demand for productive resources (factors of production) are equal at equilibrium prices. In comparison with the strategic model, in the general equilibrium model each agent does not take explicitly into account the actions of each other.

On the other hand, in the strategic model, each agent acts taking into consideration the strategies of other agents in the economy. In this case game theory is often used, which is the study of mathematical models of conflict and cooperation among intelligent rational decision-makers. It is clearly specified what each agent knows and how knowledge flows from one to another, and what strategies each agent can take. Explicitly modeling the strategic interaction among agents is of special importance to analysis of imperfectly competitive markets and one of the main recent developments of microeconomics.

3.1.1 Basic Model of a Competitive Economy

The basic model of a competitive economy is given as follows. The basic elements are defined first. All the following items except the last one belong to the exogenous description of the economy. The price vector is endogenous. That means it will be obtained as a part of the solution of the model.

- L is the number of goods.
- N is the number of consumers, and a typical consumer is indexed by n. The set of all consumers is denoted by $\mathbb{N}$.
- $\mathbb{R}^L_+$ is the consumption set for each consumer, which is a set of L-dimensional vectors with nonnegative real elements.
- $\succsim_n$ is the rational preference relation of consumer n on $\mathbb{R}^L_+$.
- u_n is a utility function for consumer n mapping $\mathbb{R}^L_+$ to $\mathbb{R}$ (a set of real numbers). That is, for any consumption bundle, $x = (x_1, \dots, x_L) \in \mathbb{R}^L_+ \rightarrow u_n$ tells us the utility that consumer n associates to that bundle.
- $\omega_n = (\omega_{1n}, \omega_{2n}, \dots, \omega_{\mathrm{Ln}}) \in \mathbb{R}^L_+$ is the endowment of consumer n.
- $p \in \mathbb{R}^L_{++}$ is a strictly positive price vector; $\mathbf{p} = (p_1, \dots, p_\ell, \dots, p_L)$ where $p_\ell > 0$ is the price of the ℓth good.

Definition 3.1 *An allocation*

$$x = ((x_{11}, x_{21}, \dots, x_{L1}), \dots, (x_{1N}, x_{2N}, \dots, x_{\mathrm{LN}}))$$

in $(\mathbb{R}^L_+)^N$ *specifies a consumption bundle for all consumers. A feasible allocation is an allocation such that*

$$\sum_{n \in \mathbb{N}} x_n \leq \sum_{n \in \mathbb{N}} \omega_n$$

or, equivalently, for each ℓ,

$$\sum_{n \in \mathbb{N}} x_{\ell n} \leq \sum_{n \in \mathbb{N}} w_{\ell n}$$

That is, for each piece of goods, the total amount that all consumers consume cannot exceed the total amount that they possess. It should be noted that there is an implicit assumption that the goods are freely disposable. That is, it is not assumed that all the goods should be consumed.

Definition 3.2 *Consumer n's budget set is*

$$B(p, \omega_n) = \{x \in \mathbb{R}_+^L \mid p \cdot x \le p \cdot \omega_n\}$$

Thus the budget set includes all consumption bundles that the consumer can afford to buy at prices $p = (p_1, p_2, \cdots, p_\ell, \cdots, p_L)$, if he/she first sells all of his/her endowment at those prices and then funds his/her purchases with the income.

Definition 3.3 *Consumer n's demand correspondence is*

$$x_n(p, \omega_n) = \{x \in B(p, \omega_n) \mid \textit{there is no } y \in B\,(p, \omega_n) \textit{ with } y \succ_n x\}$$

or, in the form of utility function,

$$x_n(p, \omega_n) = \{x \in B(p, \omega_n) \mid \textit{there is no } y \in B\,(p, \omega_n) \textit{ with } u_n(y) > u_n(x)\}$$

The demand correspondence for consumer n under price p is the set of all affordable consumption bundles without any other affordable consumption bundle more preferable for this consumer under the same price.

For simplicity of analysis, here we assume that for consumer n the preference relation $\succsim_n$ is (a) continuous, (b) strictly increasing (i.e. if $x \succsim_n y$ and $x \ne y$, then $x \succ_n y$), and (c) strictly convex (i.e. if $x \succsim_n y, x \ne y$, and $\alpha \in (0, 1)$, then $\alpha x + (1-\alpha)y \succ_n y$).

In terms of utility function, we assume that the utility function u_n is (a) continuous, (b) strictly increasing (i.e. if $x \ge y$ and $x \ne y$, then $u_n(x) > u_n(y)$), and (c) strictly quasi-convex (i.e. if $u_n(x) \ge u_n(y), x \ne y$, and $\alpha \in (0, 1)$, then $u_n(\alpha x + (1-\alpha)y) > u_n(y)$).

Proposition 3.1 *If $\succsim_n$ is continuous, strictly increasing, and strictly convex (u_n is continuous, strictly increasing, and strictly quasi-convex) then*

1. *$x_n(p, \omega_n) \ne \emptyset$ for any ω_n in $\mathbb{R}_+^L$ and any p in $\mathbb{R}_{++}^L$,*
2. *$x_n(p, \omega_n)$ is a singleton so $x_n(\cdot, \omega_n)$ is a function, and*
3. *$x_n(\cdot, \omega_n)$ is a continuous function.*

3.1.2 Walrasian Equilibrium

We now introduce a key concept of general equilibrium theory, namely the concept of competitive or *Walrasian equilibrium*. In brief, a Walrasian equilibrium is a condition in which the total demand does not exceed the total supply. Indeed, if all goods are desired in the economy, then it is a condition in which the total demand exactly equals the total supply. We can give the following definitions.

Definition 3.4 *The price vector p is a Walrasian (or competitive) equilibrium price if*

$$\sum_{n \in \mathbb{N}} x_n(p, \omega_n) \leq \sum_{n \in \mathbb{N}} \omega_n$$

If we do *not* assume that the demand functions are single-valued, we need a slightly more general form of the definition.

Definition 3.5 *The pair* (p, x) *in* $\mathbb{R}^L_{++} \times (\mathbb{R}^L_{+})^N$ *is a Walrasian equilibrium if* x *is a feasible allocation (that is,* $\sum_{n \in \mathbb{N}} x_n \leq \sum_{n \in \mathbb{N}} \omega_n$*) and, for each* n *in* $\mathbb{N}$,

$$x_n \succsim_n y \text{ for all } y \text{ in } B(p, \omega_n).$$

Since we assume that $\succsim_n$ is strictly increasing, it is easy to see that the only feasible allocations that will be involved in any equilibria are those for which

$$\sum_{n \in \mathbb{N}} x_n = \sum_{n \in \mathbb{N}} \omega_n \tag{3.1}$$

3.1.3 First and Second Fundamental Theorems of Welfare Economics

Efficiency is a key concern of the economist in research, which means the situation in which it is impossible to generate a larger welfare total from the available resources. When speaking about economic efficiency, we often refer to Pareto efficiency, or Pareto optimality, which is a state of allocation of resources in which it is impossible to make any one individual better off without making at least one individual worse off. The strict definition of Pareto efficiency is given as follows.

Definition 3.6 *A feasible allocation* x *is Pareto optimal (or Pareto efficient) if there is no other feasible allocation* y *such that* $y_n \succsim_n x_n$ *for all* n *in* N *and* $y_{n'} \succ_{n'} x_{n'}$ *for at least one* n' *in* N.

This means that a feasible allocation is Pareto optimal if there is no other feasible allocation that makes at least one consumer strictly better off without making any other consumer worse off.

The relationship between economic equilibrium and efficiency is stated by the following theorems.

Theorem 3.1 *(First fundamental theorem of welfare economics) Suppose that for each* n *the preferences* $\succsim_n$ *are strictly increasing and that* (p, x) *is a Walrasian equilibrium, then* x *is Pareto optimal.*

The statement does not hold in the opposite direction; namely, it is not the case that any Pareto optimal allocation is a Walrasian equilibrium. However, under some stronger assumptions, the following statement holds.

Theorem 3.2 *(Second fundamental theorem of welfare economics) Suppose that for each n the preferences* $\succsim_n$ *are strictly increasing, convex, and continuous and that* x *is Pareto optimal with* $x > 0$ *(that is,* $x_{\ell n} > 0$ *for each* ℓ *and each n), then there is some feasible reallocation* ω' *of the endowments (that is,* $\sum_{n \in \mathbb{N}} \omega'_n = \sum_{n \in \mathbb{N}} \omega_n$*) and a price vector p such that* (p, x) *is a Walrasian equilibrium of the economy with preferences* $\succsim_n$ *and initial endowments* ω'*.*

3.2 Noncooperative Game Theory

General equilibrium theory is suitable for the description of competitive markets, which is an idealized model of an economy. In realistic markets, the strategic interaction among agents should be considered. Besides game theory, there are also other ways of modeling such situations: decision theory and mechanism design theory.

Decision theory can be viewed as a theory of one-person games, or a game of a single player against nature. The focus is on preferences and the formation of beliefs. The most widely used form of decision theory argues that preferences among risky alternatives can be described by the maximization of the expected value of a numerical utility function, where utility may depend on a number of things, but in situations of interest to economists often depends on money income. Probability theory is heavily used in order to represent the uncertainty of outcomes, and Bayes' law is frequently used to model the way in which new information is used to revise beliefs. Decision theory is often used in the form of decision analysis, which shows how best to acquire information before making a decision.

Mechanism design theory differs from game theory in that game theory takes the rules of the game as given, while mechanism design theory asks about the consequences of different types of rules. Naturally this relies heavily on game theory. Questions addressed by mechanism design theory include the design of compensation and wage agreements that effectively spread risk while maintaining incentives, and the design of auctions to maximize revenue, or achieve other goals.

3.2.1 Representation of Games

Here we first give the definition of a finite normal-form game and the related concepts. The normal-form (or strategic-form) game is usually represented by a matrix, which shows the players, strategies, and payoffs. More generally it can be represented by any function that associates a payoff for each player with every possible combination of actions.

Suppose that there are a set of players numbered from 1 to N, and the set of these players is denoted by $\mathbb{N}$. Each player has a finite set of actions or strategies, and for player n, it is denoted by S_n. The Cartesian product of these sets is denoted by S. Thus a typical element of S is $s = (s_1, s_2, \dots, s_N)$, where each s_n is a pure strategy of player n and an element of S_n. We call such an s a "pure strategy profile" (sometimes called a "pure strategy combination"). For each player n, a utility function $u_n : S \to R$ is also specified.

Definition 3.7 *A (finite) normal-form game is a triple* (N, S, u), *where* $\mathbb{N} = \{1, 2, \dots, n, \cdots N\}$ *is the set of players,* $S = S_1 \times S_2 \times \cdots \times S_N$ *is the set of profiles of pure strategies, with* S_n *denoting the finite set of pure strategies of player n, and* $u = (u_1, u_2, \dots, u_N)$, *with* $u_n : S \to \mathbb{R}$, *is the utility or payoff function of player n. We call the pair* (N, S) *the "game form". Thus a game is a game form together with a payoff function.*

In more complicated situations, randomized or mixed strategies should be defined, so that each player can form a probabilistic assessment over other players' decisions. Thus when a player makes his/her decision, he/she will choose a lottery over pure strategy profiles. Accordingly, the utility function is interpreted as a measure of the player's preferences over lotteries, namely the von Neumann–Morgenstern utility function.

Definition 3.8 *A mixed strategy of player n is a lottery over his/her pure strategies. One of player n's mixed strategies is denoted by* σ_n *and the set of all mixed strategies of player n is denoted by* $\sum_n$. *Thus* $\sigma_n = (\sigma_n(1), \sigma_n(2), \dots, \sigma_n(K_n))$, *where* $\sigma_n(i)$ *is the probability associated with pure strategy* s_n^i, K_n *is the number of pure strategies of player n,* $\sigma_n(i) \geq 0$ *for* $i = 1, 2, \dots, K_n$, *and* $\sum_{i=1}^{K_n} \sigma_n(i) = 1$. *The Cartesian product* $\Sigma = \Sigma_1 \times \Sigma_2 \times \cdots \times \Sigma_N$ *is the set of all mixed strategy profiles.*

The concept of mixed strategies has aroused a lot of disputes because the idea of randomization, central in mixed strategies, lacks behavioral support.

In 1991, game theorist Ariel Rubinstein described alternative ways of understanding the concept [87]. The first is called purification. A player's mixed strategy is thought of as a plan of action that is dependent on private information which is not specified in the model. Although the player's behavior appears to be random, it is actually deterministic. If we add this information structure to the model, the mixed strategy becomes a pure strategy in which the action depends on the extraneous information. However, it is unsatisfying to have results that hang on unspecified factors. A second interpretation imagines the game players standing for a large population of players. Each of the players chooses a pure strategy, and the payoff depends on the fraction of players choosing each strategy. The mixed strategy hence represents the distribution of pure strategies chosen by each population. However, this does not provide any justification for the case when players are individual agents.

It is more important for us to extend the utility function of a player from that defined on the pure strategy profiles to that defined on mixed strategies. We shall continue to use the same symbol u_n to represent the expected utility of player n as a function of the mixed strategy profile $\sigma = (\sigma_1, \sigma_2, \dots, \sigma_N)$. Intuitively $u_n(\sigma)$ is just the expected value of $u_n(s)$ when s is a random variable with distribution given by σ. Thus

$$u_n(\sigma) = \sum_{s_1 \in S_1} \cdots \sum_{s_N \in S_N} \sigma_1(s_1) \cdots \sigma_N(s_N) u_n(s_1, \dots, s_N)$$

In a similar way, we can define u_n on a more general profile where we have mixed strategies for some players and pure strategies for others.

Let us look now at a canonical example in game theory – the *prisoners' dilemma.*

Two members of a criminal gang are arrested and imprisoned. Each prisoner is in solitary confinement with no means of speaking to or exchanging messages with the other. The prosecutors do not have enough evidence to convict the pair on the principal charge. They hope to get both sentenced to a year in prison on a lesser charge. Simultaneously, the prosecutors offer each prisoner a bargain. Each prisoner is given the opportunity either to: betray the other by testifying that the other committed the crime, or to cooperate with the other by remaining silent. Here is the offer:

- If A and B each betray the other, each of them serves two years in prison.
- If A betrays B but B remains silent, A will be set free and B will serve three years in prison (and vice versa).
- If A and B both remain silent, both of them will only serve one year in prison (on the lesser charge).

It is implied that the prisoners will have no opportunity to reward or punish their partner other than with the prison sentences they get, and that their decision will not affect their reputation in the future. Because betraying a partner offers a greater reward than cooperating with him/her, all purely rational self-interested prisoners would betray the other, and so the only possible outcome for two purely rational prisoners is for them to betray each other.

Such a game is often represented as a labeled matrix as shown in Table 3.1. Here player A's strategies are listed vertically and player B's strategies are listed

Table 3.1 The prisoners' dilemma.

	Prisoner B silent (cooperates)	Prisoner B betrays (defects)
Prisoner A silent (cooperates)	Each serves one year	Prisoner A: three years Prisoner B: goes free
Prisoner A betrays (defects)	Prisoner A: goes free Prisoner B: three years	Each serves two years

horizontally (hence they are sometimes referred to as the row player and the column player respectively).

Here, regardless of what the other decides, each prisoner gets a higher payoff by betraying the other ("defecting"). The reasoning involves an argument by dilemma. Prisoner B will either cooperate or defect. If B cooperates, A should defect, since going free is better than serving one year. If B defects, A should also defect, since serving two years is better than serving three. So either way, A should defect. Parallel reasoning will show that B should defect.

In traditional game theory, some very restrictive assumptions on prisoner behavior are made. It is assumed that both understand the nature of the game, and that, despite being members of the same gang, they have no loyalty to each other and will have no opportunity for retribution or reward outside the game. Most importantly, a very narrow interpretation of "rationality" is applied in defining the decision-making strategies of the prisoners. Given these conditions and the payoffs above, prisoner A will betray prisoner B. The game is symmetric, so prisoner B should act the same way. Since both "rationally" decide to defect, each receives a lower reward than if both were to stay quiet. Traditional game theory results in both players being worse off than if each chose to lessen the sentence of his/her accomplice at the cost of spending more time in jail him/herself.

Here we will elaborate upon some important concepts concerning the solution of a game.

3.2.1.1 Rationality and Common Knowledge

Common knowledge is a special kind of knowledge for a group of agents. There is common knowledge of p in a group of agents G when all the agents in G know p, they all know that they know p, they all know that they all know that they know p, and so on *ad infinitum*. It was first given a mathematical formulation in a set-theoretical framework by Robert Aumann [88].

In game theory, strategic dominance (commonly called simply "dominance") occurs when one strategy is better than another strategy for one player, no matter how that player's opponents may play. Many simple games can be solved using dominance. The opposite, "intransitivity", occurs in games where one strategy may be better or worse than another strategy for one player, depending on how the player's opponents may play.

The iterated elimination (or deletion) of dominated strategies is one common technique for solving games that involves iteratively removing dominated strategies. In the first step, at most one dominated strategy is removed from the strategy space of each of the players since no rational player would ever play these strategies. This results in a new, smaller game. Some strategies – that were not dominated before – may be dominated in the smaller game. The first step is repeated, creating a new even smaller game, and so on. The process stops when no dominated strategy is found for any player. We call the strategies that remain *iteratively undominated* or *correlatedly rationalizable*. This process is valid since it is assumed that rationality among players is common knowledge, that is, each player knows that the rest of the players are rational, and each player knows that the rest of the players know that he/she knows that the rest of the players are rational, and so on *ad infinitum*.

3.2.1.2 Nash Equilibrium

A more widely used concept is that of Nash equilibrium or strategic equilibrium, first defined by John F. Nash in the early 1950s [89, 90].

Nash equilibrium is a solution concept of a noncooperative game involving two or more players, in which each player is assumed to know the equilibrium strategies of the other players, and no player has anything to gain by changing only their own strategy. If each player has chosen a strategy, and no player can benefit by changing strategies while the other players keep theirs unchanged, then the current set of strategy choices and the corresponding payoffs constitutes a Nash equilibrium.

An equilibrium is a profile of mixed strategies, one for each player, with the property that, if each player's uncertainty about what the others will do is represented by the profile of mixed strategies, then his/her mixed strategy has positive weight only on those pure strategies that bring him/her the maximum expected utility. We give the formal definition of Nash equilibrium first.

Definition 3.9 *A strategic equilibrium (or Nash equilibrium) of a game (N, S, u) is a profile of mixed strategies $\sigma = (\sigma_1, \sigma_2, \ldots, \sigma_N)$ such that, for each $n = 1, 2, \ldots, N$ for each $s_n, t_n \in S_n$, if $\sigma_n(s_n) > 0$, then*

$$u_n(\sigma_1, \ldots, \sigma_{n-1}, s_n, \sigma_{n+1}, \ldots, \sigma_N) \geq u_n(\sigma_1, \ldots, \sigma_{n-1}, t_n, \sigma_{n+1}, \ldots, \sigma_N)$$

Remember how we extended the definition of u_n from the pure strategies to the mixed strategies at the beginning of this section.

We can relate this concept to undominated strategy.

Proposition 3.2 *Any strategic equilibrium profile consists of iteratively undominated strategies.*

Let us look back to the example of the prisoners' dilemma and calculate the equilibrium. The strategy of "do not confess" is dominated. That is, "confess" is better whatever the other player is doing. Thus for each player "confess" is the only undominated strategy and hence the only iteratively undominated strategy. Thus by the proposition "(confess, confess)" is the only equilibrium.

3.2.2 Existence of Equilibrium

The existence of equilibrium is important in game theory. We begin our analysis with a simple example (Table 3.2). Player 1 chooses the row strategies and player 2 (simultaneously) chooses the column strategies. The resulting payoffs are indicated in the appropriate box of the matrix, with player 1's payoff appearing first.

A (mixed) strategy for player 1 (that is, an assessment by player 2 of how player 1 would play) can be expressed by a vector $(x, 1 - x)$, where x lies between 0 and 1 and denotes the probability of playing T. Similarly, a strategy for player 2 is a vector $(y, 1 - y)$.

Now, given x, the payoff-maximizing value of y is depicted by Figure 3.1(a); and vice versa, given y, the payoff-maximizing value of x is depicted by Figure 3.1(b). When the two figures are drawn together, we get Figure 3.1(c). It is evident that this game possesses a single equilibrium, namely $x = \frac{1}{2}$, $y = \frac{1}{3}$. Thus, in a self-enforcing assessment, player 1 must assign a probability of $\frac{1}{3}$ to player 2 playing L, and player 2 must assign a probability of $\frac{1}{2}$ to player 1 playing T.

The game of Figure 3.1 is an example in which a unique equilibrium completely determines the solution. However, we cannot reach such a conclusion in general.

In Figure 3.1 we see that, although there is no pure strategy equilibrium, a mixed strategy equilibrium exists. The main result of noncooperative game theory tells us that this statement holds quite generally. In the landmark work of Nash [89, 90], he defined what we now call Nash equilibrium: "an equilibrium point is an n-tuple such that each player's mixed strategy maximizes his payoff if the strategies of the others

Table 3.2 Payoffs from a simple game.

	Player 2	
Player 1	L	R
T	2.0	0.1
B	0.1	2.0

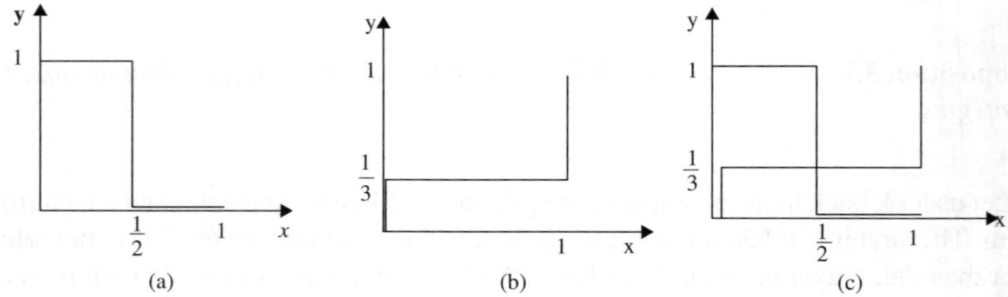

Figure 3.1 Existence of equilibrium.

are held fixed. Thus each player's strategy is optimal against those of the others." By putting the problem in this framework, Nash employed the Kakutani fixed point theorem in his 1950 paper. In his 1951 paper he used a variant of it – the Brouwer fixed point theorem – to prove that there had to exist at least one set of mixed strategies that mapped back into themselves for nonzero-sum games, namely, a set of strategies that did not call for a shift in strategies that could improve payoffs. Nash's existence theorem is given as follows [89, 90].

Theorem 3.3 *If we allow mixed strategies, then every game with a finite number of players in which each player can choose from finitely many pure strategies has at least one Nash equilibrium.*

3.3 Mechanism Design

3.3.1 Principles of Mechanism Design

Mechanism design is an important concept in modern microeconomics, aiming at optimal allocation of resources when the information needed to determine an optimal allocation is privately held by the agents who will get the resources. An agent's reveal of his/her information will influence the allocation results, which in turn will affect his/her utility. In such a situation, how can the allocator simultaneously elicit the information that is privately held and choose the optimal allocation? This should be solved by the so-called mechanism design theory [91].

In economy, people have information about their resources and desires, and people take actions for producing, redistributing, and consuming resources. In markets, the agents' actions may depend on others' information as it has been communicated and employed in the economic institution. Then we should view economic institutions as mechanisms for communicating people's information and coordinating people's actions. To evaluate whether the economic institution is good or not, we should study how it performs in this communication and coordination role.

So the crucial question is, what kinds of economic coordination schemes are feasible? A feasible economic coordination scheme could be implemented by many different economic institutions, but it is helpful to begin by considering a very centralized institution where every individual communicates separately and confidentially with a trustworthy central allocator. Suppose, first, that each individual confidentially submits all his/her private information to the allocator, and then, based on all these submissions, the allocator recommends to each individual what actions he/she should take under the economic institution. However, if we allow that individuals can be dishonest or disobedient to the allocator, then the economic mechanism must give people incentives to share information and to act appropriately according to the original objective of the economic system.

First, for the reason that our economic scheme depends on the individuals' private information, which is hard for others to observe, we need to give individuals an incentive to share their true information. This problem of having people share information honestly is called "adverse selection". On the other hand, because the economic scheme requires people to choose hidden actions and exert efforts that are hard for others to monitor, we need to give people the incentive to act obediently according to the scheme. This problem of getting people to act obediently to an economic scheme is called "moral hazard". If it is a rational equilibrium for everyone to be honest and obedient to the central allocator who is implementing the economic coordination scheme, then the scheme is incentive-compatible.

There are two important things concerned with incentive-compatible coordination schemes. First, they can be modeled mathematically by a set of inequalities called incentive constraints. Second, although we defined incentive compatibility by thinking about honesty and obedience in communication with a central allocator, in fact these incentive-compatible schemes can point to everything that can be implemented by rational equilibrium behavior in any economic institution or mechanism. This assertion of generality is called the "revelation principle".

The revelation principle asserts that any rational equilibrium of individual behavior in any economic institution must be equivalent to an incentive-compatible coordination scheme. Thus, without loss of generality, a trustworthy allocator can plan to make honesty and obedience the best policy for everyone, as illustrated in Figure 3.2. In a single object auction problem, Myerson states the revelation principle as follows [92].

Proposition 3.3 *Given any feasible auction mechanism, there exists an equivalent feasible direct revelation mechanism which gives to the seller and all bidders the same expected utilities as in the given mechanism.*

To prove the revelation principle, suppose that there is a general coordination mechanism and an equilibrium that describes rational individual strategies with the

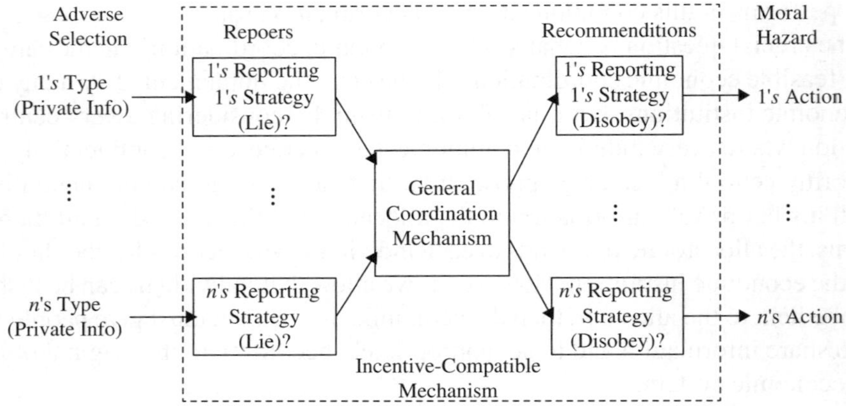

Figure 3.2 The revelation principle.

possibility of dishonest report and dishonest action in this mechanism. We need to describe how an allocator would implement the equivalent incentive-compatible allocation scheme where honesty and obedience is an equilibrium. When the allocator has received a confidential submission of every individual's private information, the equivalent incentive-compatible allocation scheme will first compute the information that everyone has sent in the given equilibrium. Then it will compute the behavior that the given mechanism will indicate for each individual based on these submissions. Then it will compute the action that each individual should choose in the given equilibrium. Finally, it will confidentially send to each individual the computed action he/she should choose. If any individual could gain by being dishonest or disobedient to the allocator under this scheme, then he/she could have gained by being dishonest or disobedient to him/herself under his/her given equilibrium strategy in the mechanism. However, in a rational equilibrium, nobody can gain by lying to him/herself or disobeying his/her own optimal strategy [93].

To explore the relationship between mathematical programming and mechanism design theory, this section illustrates how standard results in linear programming play a role in the optimal auction of a single good [94].

3.3.2 *Optimization of a Single Commodity Auction*

Suppose that a single risk-neutral seller has a single commodity, which he/she wishes to sell to n risk-neutral agents. The opportunity cost of the commodity to the seller is normalized to zero. The value that agent i assigns to the commodity is t^i and is private to him/her. The value assigned to the commodity by an agent is called his/her "type".

The uncertainty that the seller has about each agent's type and that each agent has about the types of other agents is extracted independently from the set $T = \{1, 2, \ldots, m\}$. Let F be the distribution and f its density. For convenience, assume $f_t > 0$ for all $t \in T$. Set T is called the "type space" and is assumed to be discrete. The assumption of a discrete type space allows the problem to be solved by linear programming.

3.3.2.1 Expected Allocations and Payments

Here we call $\mathbf{t}$ a profile of types, which is an n-tuple of types, and denote the set of profiles of length n by T^n. Let $\mathbf{a}$ be an allocation rule. For simplicity of notation, we focus on allocation rules that are anonymous, i.e. independent of the names of agents. This involves no loss of generality, since each agent's type is extracted from the same distribution and there is symmetry in the set of feasible allocations. The symmetry assumption allows us to focus on one agent, such as agent 1.

If $\mathbf{t}$ is the profile of types, then $\mathbf{a}[i, \mathbf{t}^{-1}]$ is the allocation to agent 1 when he/she is of type $i \in T$ and all other agents have the profile $\mathbf{t}^{-1}$. By symmetry, this must be the same as the allocation to any other agent in the same profile that reports type i.

It will sometimes be convenient to write this as $\mathbf{a}_i(t)$. Here we do not restrict $\mathbf{a}$ to be deterministic. Thus, randomized rules are allowed. Let $\pi(\mathbf{t}^{-1})$ be the probability of the agents having types that give rise to the profile $\mathbf{t}^{-1}$. We further denote the number of agents with type i in profile $\mathbf{t}$ by $n_i(\mathbf{t})$.

The *expected allocation* (also called the *interim allocation*) that agent 1 with type i receives is denoted by $\mathcal{A}_i$, i.e.

$$\mathcal{A}_i = \sum_{\mathbf{t}^{-1} \in T^{n-1}} \mathbf{a}[i, \mathbf{t}^{-1}]\pi(\mathbf{t}^{-1}). \tag{3.2}$$

It should be noted that $\mathcal{A}$ is defined under the assumption that all other agents truthfully report their type.

Similarly we define a payment rule $\mathbf{p}$ and an expected payment $\mathbf{P}$. Specifically, P_i is the expected payment of agent 1 if he/she reports that he/she is type i. Let $\upsilon(\mathcal{A}_i \mid i)$ be the expected value that agent 1 of type i receives with a lottery over allocations with expected value $\mathcal{A}_i$. With the assumption that the type of an agent is their value, we have

$$\upsilon(\mathcal{A}_i \mid i) = \sum_{\mathbf{t}^{-1} \in T^{n-1}} i\mathbf{a}_i[i, \mathbf{t}^{-1}]\pi(\mathbf{t}^{-1}) = i\mathcal{A}_i \tag{3.3}$$

and $\upsilon(\mathcal{A}_j \mid i) = i\mathcal{A}_j$.

3.3.2.2 Formulation

Based on the above problem description and notation, the following model based on linear programming can be formulated to determine the maximum expected revenue for the seller:

$$Z = \max_{(\mathbf{p},\mathbf{a})} \quad n\sum_{i=1}^{m} f_i P_i \tag{3.4}$$

$$\text{s.t.} \quad \upsilon(\mathcal{A}_i \mid i) - P_i \geq \upsilon(\mathcal{A}_j \mid i) - P_j, \quad \forall i, j \in T \tag{3.5}$$

$$\upsilon(\mathcal{A}_i \mid i) - P_i \geq 0, \quad \forall i \in T \tag{3.6}$$

$$\mathcal{A}_i = \sum_{\mathbf{t}^{-1} \in T^{n-1}} \mathbf{a}_i[i, \mathbf{t}^{-1}]\pi(\mathbf{t}^{-1}) \tag{3.7}$$

$$\sum_{i \in T} n_i(\mathbf{t})\mathbf{a}_i(\mathbf{t}) \leq 1, \quad \forall \mathbf{t} \in T^n \tag{3.8}$$

$$\mathbf{a}_i(\mathbf{t}) \geq 0, \quad \forall i \in T, \forall \quad \mathbf{t} \in T^n \tag{3.9}$$

Let us examine the model and constraints in order.

- The objective function (3.4) is to maximize the total expected revenue.
- Constraint (3.5) is the constraint of incentive compatibility. The left-hand side is the expected surplus of an agent from truthfully reporting their type. The right-hand side is the expected surplus from misreporting their type. This constraint is not optional. It is forced upon the seller by equilibrium considerations. The revelation principle asserts that, for any mechanism and any Bayes–Nash equilibrium of it, there is a corresponding direct mechanism where truth telling is an equilibrium. Thus constraint (3.5) implies the assumption that the agents will play a particular equilibrium of the game induced by the mechanism.
- Constraint (3.6) ensures that each agent receives a nonnegative surplus from participation. This prevents the seller from extracting unlimited amounts of money from any one agent. The right-hand side of the constraint implicates that the agents have no outside options.
- Constraint (3.7) ensures that the expected allocation rule is consistent with the underlying allocation rule.
- Constraint (3.8) means that no more than one unit of the commodity can be allocated either as a whole or in part.
- Constraint (3.9) means that no agent can receive a negative amount or probability share of the commodity.

Now we can discuss a little further constraints (3.5) and (3.6), which are related to the revelation principle. They can be rewritten as

$$P_i - P_j \leq v(\mathcal{A}_i \mid i) - v(\mathcal{A}_j \mid i).$$

We can illustrate these incentive compatibility constraints with a network, as shown in Figure 3.3. The vertex i of the network represents type i. Every ordered pair of vertices (j, i) are connected by a directed edge of length $v(\mathcal{A}_i \mid i) - v(\mathcal{A}_j \mid i)$. The constraints (3.6) can be incorporated by assuming a dummy type of value 0.

For fixed $\mathcal{A}_i$, system (3.5) is feasible if and only if this network has no negative-length cycle. If this network is free of negative-length cycles, each P_i is upper bounded by the length of the shortest path from an arbitrarily chosen root

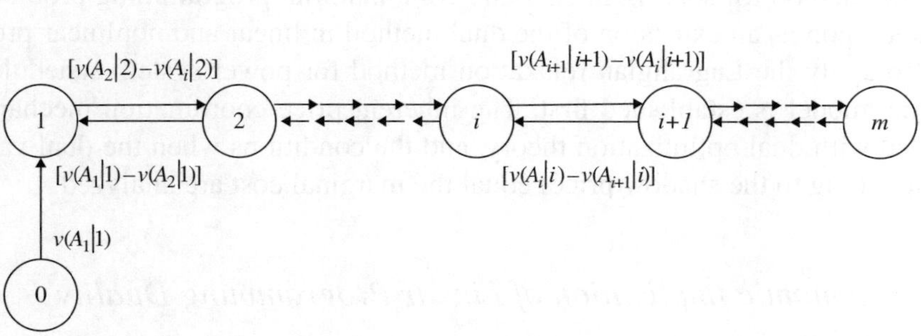

Figure 3.3 Network of types.

vertex, r, to vertex i. To ensure that (3.6) holds, choose the root r to be the vertex corresponding to the dummy type 0.

Considering the nonnegative cycle condition applied to cycle $i \to i+1 \to i$ in Figure 3.3, it implies

$$[v(\mathcal{A}_{i+1} \mid i+1) - v(\mathcal{A}_i \mid i+1)] + [v(\mathcal{A}_i \mid i) - v(\mathcal{A}_{i+1} \mid i)] \geq 0$$
$$\Rightarrow (i+1)[\mathcal{A}_{i+1} - \mathcal{A}_i] + i[\mathcal{A}_i - \mathcal{A}_{i+1}] \geq 0$$
$$\Rightarrow \mathcal{A}_{i+1} - \mathcal{A}_i \geq 0$$

Hence, (3.5) implies that $\mathcal{A}$ is monotonic. The inverse statement is also true as follows [94].

Proposition 3.4 *System (3.5) is feasible if and only if the corresponding expected allocations are monotonic. That is, if $r \leq s$, then $\mathcal{A}_r \leq \mathcal{A}_s$.*

It turns out that the problem of deciding whether a given interim allocation rule is feasible can be solved as a transportation problem [94].

3.4 Duality Principle and Its Economic Implications

In power system optimal planning and operation, various mathematical optimization methods have been used, such as linear programming, nonlinear programming, integer and combinatorial programming, etc. In addition, graph theory and artificial intelligence approaches have emerged. Different methods are employed depending on the specific problems.

The basic mathematical foundation of microeconomics is optimization theories. Currently in economics, the most widely used method is linear and nonlinear programming. A particularly noteworthy aspect is the relationship between duality and shadow prices in economic systems with limited resources. This chapter introduces the economic implication of duality in linear and nonlinear programming. The Lagrangian relaxation method for solving integer and combinatorial programming problems can be looked upon as an extension of the dual method in linear and nonlinear programming. To apply the Lagrangian relaxation method for power system scheduling, its economic model is established first. The inherent price coordination mechanism is explained with dual optimization theory, and the conditions when the dual variables (corresponding to the shadow price) equal the marginal cost are analyzed.

3.4.1 Economic Implication of Linear Programming Duality

An example of linear programming is given first. Suppose that a firm manufactures n kinds of different products under m production capacity or scarce resource constraints.

The manager wants to set the output x_j of product j to be such that the total income z_x can be maximized. If c_j is the market price of product j, then

$$z_x = \sum_{j=1}^{n} c_j x_j$$

Suppose also that a_{ij} is the quantity of resource i needed in manufacturing one product j, and b_i is the total quantity of resource i that the firm has. Then we have

$$a_{i1}x_1 + a_{i2}x_2 + \cdots + a_{in}x_n \leq b_i, \quad i = 1, \ldots, m$$

namely, the quantity of resource i needed in manufacturing product n cannot exceed b_i. Then the mathematical model becomes

$$\begin{aligned} \max\ z_x &= c_1x_1 + c_2x_2 + \cdots + c_nx_n \\ \text{s.t. } a_{11}x_1 &+ a_{12}x_2 + \cdots + a_{1n}x_n \leq b_1 \\ a_{21}x_1 &+ a_{22}x_2 + \cdots + a_{2n}x_n \leq b_2 \\ &\vdots \\ a_{m1}x_1 &+ a_{m2}x_2 + \cdots + a_{mn}x_n \leq b_m \\ x_j &\geq 0, \quad j = 1, 2, \ldots, n \end{aligned} \tag{3.10}$$

Now let us consider a related question. Suppose that y_i is the shadow price of resource i, and the manager wants to pass all revenue of the company on to its scarce resources, namely y_i should be chosen such that the total book value of resource m in manufacturing per unit of product j is not less than the price of product j. Then we have

$$a_{1j}y_1 + a_{2j}y_2 + \cdots + a_{mj}y_m \geq c_j, \quad j = 1, \ldots, n$$

The total value of resources that the company can obtain is

$$z_y = b_1y_1 + \cdots + b_my_m$$

Now the problem is that the company wants to price the resources to minimize the total cost. Then the mathematical model of the problem becomes

$$\begin{aligned} \min\ z_y &= b_1y_1 + b_2y_2 + \cdots + b_my_m \\ \text{s.t.}\quad a_{11}y_1 &+ a_{21}y_2 + \cdots + a_{m1}y_m \geq c_1 \\ a_{12}y_1 &+ a_{22}y_2 + \cdots + a_{m2}y_m \geq c_2 \\ &\vdots \\ a_{1n}y_1 &+ a_{2n}y_2 + \cdots + a_{mn}y_m \geq c_n \\ y_i &\geq 0, \ i = 1, 2, \ldots, m \end{aligned} \tag{3.11}$$

If (3.10) is called the primal problem, then (3.11) is called the dual problem of (3.10). In fact, the dual of a dual problem is the primal problem.

For the primal and dual problems, the following theorems hold.

Theorem 3.4 *For any feasible solutions of primal and dual problems, we have*

$$z_x \leq z_y \tag{3.12}$$

Theorem 3.5 *(Duality theorem)*
(a) Suppose both the primal problem and the dual problem have finite feasible solutions. Then the two problems have finite optimal solutions x^* and y^*, and the two optimal values of the objective functions are equal, i.e.

$$z_x^* = z_y^* \tag{3.13}$$

(b) If the primal problem or the dual problem has a feasible solution, and it gives a finite optimal objective function value, then the other problem has an optimal feasible solution, and it gives the same optimal objective function value.

Theorem 3.6 *If the derivatives exist, it can be proved that*

$$y_i^* = \frac{\partial z_x^*}{\partial b_i} \quad (i = 1, 2, \ldots, m) \tag{3.14}$$

Theorem 3.4 shows that the value passed on to the resources is not less than the total revenue generated. The duality theorem shows that, under the optimal condition, the book value passed on to the resources does not exceed the total revenue obtained from the sale of the products, and thus they are equal. The total book value of resources can be looked upon as the opportunity cost. The duality theorem shows that, under the optimal condition, the sales revenue is equal to the opportunity cost. Theorem 3.6 shows that, under the optimal condition, the book value of resource i is equal to its marginal contribution to the total revenue.

Therefore, in the primary problem, the manager directly determines the quantity of n products to maximize his/her revenue under the constraints of m resources. In the dual problem, the manager solves the same problem indirectly. He/she allocates all scarce resources to the manufacturing of n products until the minimum total value of the resources (or opportunity cost) is equal to the maximum total revenue in utilization of the resources.

3.4.2 Economic Implication of Duality in Nonlinear Programming

The duality of nonlinear programming is closely related to the Kuhn–Tucker (KT) conditions and the saddle-point problem of the Lagrangian function.

3.4.2.1 Kuhn–Tucker Condition

For a general nonlinear programming problem

$$\begin{aligned} \max \ & z = f(x_1, x_2, \dots, x_n) \\ \text{s.t.} \ & g_i(x_1, x_2, \dots, x_n) \le b_i, \quad i = 1, \dots, u \\ & g_i(x_1, x_2, \dots, x_n) \ge b_i, \quad i = u+1, \dots, v \\ & g_i(x_1, x_2, \dots, x_n) = b_i, \quad i = v+1, \dots, m \\ & x_j \ge 0, \quad j = 1, 2, \dots, n \end{aligned} \tag{3.15}$$

where the functions f and g_i are not necessarily linear, but they have continuous first derivatives and satisfy certain regularity conditions. Below the superscript "*" indicates the optimal. The Kuhn–Tucker theorem can be stated as follows.

Theorem 3.7 *The necessary condition for $x^* = (x_1^*, x_2^*, \dots, x_n^*)^{\mathrm{T}}$ to be the optimal solution is that, under the feasibility conditions in (3.15), there are m numbers $\lambda_1^*, \lambda_2^*, \dots, \lambda_m^*$ such that the following conditions hold.*

1. *For the positive components x_j^* of x^*,*

$$\frac{\partial f(\mathbf{x}^*)}{\partial x_j} - \sum_{i=1}^{m} \lambda_i^* \left(\frac{\partial g_i(\mathbf{x}^*)}{\partial x_j} \right) = 0 \tag{3.16}$$

For the zero components $x_j^* = 0$ of $\mathbf{x}^*$,

$$\frac{\partial f(\mathbf{x}^*)}{\partial x_j} - \sum_{i=1}^{m} \lambda_i^* \left(\frac{\partial g_i(\mathbf{x}^*)}{\partial x_j} \right) \le 0 \tag{3.17}$$

where $\partial f(\mathbf{x}^*)/\partial x_j$ and $\partial f(\mathbf{x}^*)/\partial x_j$ are the partial derivatives at $x_j = x_j^*$, $j = 1, 2, \dots, n$.

2. *One has*

$$\lambda_i^* \ge 0 \ (i = 1, \dots, u) \quad \textit{and} \quad \lambda_i^* \le 0 \ (i = u+1, \dots, v)$$

The sign (positive or negative) of λ_i^* is not restricted $(i = v+1, \dots, m)$. When the ith constraint at the optimal solution $x = x^*$ is not active (i.e. the strict inequality holds)

$$\lambda_i^* = 0 \tag{3.18}$$

3. *If $\lambda_i^* \ne 0$, then $b_i - g_i(x^*) = 0$; if $\lambda_i^* = 0$, then $b_i - g_i(x^*) \ge 0$ $(i = 1, \dots, u)$ and*

$$b_i - g_i(x^*) \le 0 \quad (i = u+1, \dots, v) \tag{3.19}$$

3.4.2.2 Saddle-point Problem

For function $F(x, \lambda)$, where $x = (x_1, \dots, x_n)^T$ and $\lambda = (\lambda_1, \dots, \lambda_m)^T$, if there is a neighborhood ϵ of (x^*, λ^*) in which, for all (x, λ), one has

$$(x, \lambda^*) \leq F(x^*, \lambda^*) \leq F(x^*, \lambda) \tag{3.20}$$

then $F(x, \lambda)$ is called a saddle-point of (x^*, λ^*).

That is, the function F of vector x and λ is maximized with regard to x and minimized with regard to λ at the point (x^*, λ^*).

The Lagrangian function of the above nonlinear programming problem is defined as

$$L(x, \lambda) = f(x) + \sum_{i=1}^{m} \lambda_i[b_i - g_i(x)] \tag{3.21}$$

Then the following saddle-point optimality holds.

Theorem 3.8 *(Saddle-point theorem) If the Lagrangian function $L(x, \lambda)$ has a saddle-point at (x^*, λ^*), namely the inequality*

$$L(x, \lambda^*) \leq L(x^*, \lambda^*) \leq L(x^*, \lambda) \tag{3.22}$$

holds, then x^* is the optimal solution of the primal problem, and λ^* is the optimal solution of the problem $\min_\lambda \max_x L(x, \lambda)$.

The problem $\min_\lambda \max_x L(x, \lambda)$ is called the dual problem, and the following relation holds:

$$\max f(x) = \min_{\lambda} \max_{x} L(x, \lambda) \tag{3.23}$$

The following theorem holds about saddle-point optimality conditions and KT conditions.

Theorem 3.9 *If (x^*, λ^*) is the saddle-point of Lagrangian function $L(x, \lambda)$, then (x^*, λ^*) satisfies the KT condition (3.16)–(3.19), and so the multipliers in the saddle-point conditions are also those in KT conditions.*

3.4.2.3 Economic Implication of KT Conditions

Here we explain this problem with peak load pricing of a public utility.

With rising fuel costs, many public utilities may find it necessary to adopt the following pricing mechanism: charging higher price at peak load periods, and lower price at non-peak load periods. Assuming that the issue of energy storage is not considered, the utility is interested in whether such a pricing mechanism can maximize its profits.

Take $q_1, q_2, \dots, q_{24}$ to be the loads (e.g. electricity) in 24 hours of the day, $p_1, p_2, \dots, p_{24}$ to be the corresponding prices, $C(q_1, q_2, \dots, q_{24})$ to be the total daily production costs, and $g(y)$ to be the daily capacity costs.

The problem is to maximize the total revenue in a day $z = f(q_1, q_2, \ldots, q_{24})$. Here,

$$\begin{aligned} z &= \sum_{i=1}^{24} p_i q_i - C(q_1, \ldots, q_{24}) - g(y) \\ \text{s.t.} \quad & 0 < q_i \leq y, \quad i = 1, 2, \ldots, 24 \end{aligned} \tag{3.24}$$

Assume that all $q_i > 0$, i.e. there is some load in every hour of the day, and $y > 0$, and also assume that the price p_i is independent of load q_i, namely $\partial p_i / \partial q_i = 0$.

From (3.17) and (3.18) in KT conditions, there are 24 nonnegative numbers $\lambda_1, \lambda_2, \ldots, \lambda_{24}$, such that

$$\frac{\partial z}{\partial q_i} = p_i - \frac{\partial C}{\partial q_i} - \lambda_i = 0, \quad i = 1, 2, \ldots, 24 \tag{3.25}$$

$$\frac{\partial z}{\partial y} = -\frac{\partial g(y)}{dy} + \sum_{i=1}^{24} \lambda_i = 0 \tag{3.26}$$

Moreover, the KT condition (3.19) requires $\partial z / \partial \lambda_i = y - q_i \geq 0$ and

$$\lambda_i (y - q_i) = 0, \quad i = 1, \ldots, 24 \tag{3.27}$$

At non-peak load periods, $y > q_i$, and thus from (3.27) we can get $\lambda_i = 0$. From (3.25) we can conclude that at non-peak load periods, $p_i = \partial C / \partial q_i$, namely the optimal prices are equal to the marginal costs.

On the other hand, at peak load periods t, it is possible that $y = q_t$, and $\lambda_t > 0$, and then from (3.25) we have

$$p_t = \frac{\partial C}{\partial q_t} + \lambda_t \tag{3.28}$$

That is to say, at any peak load period t, the optimal price should exceed the marginal operation cost; the price markup is λ_t. From (3.26), we have

$$\frac{\mathrm{dg}}{\mathrm{dy}} = \sum_{i=1}^{24} \lambda_i$$

That is, the total price markup at all peak periods should be equal to the marginal cost of capacity. The above analysis shows that, from the viewpoint of profit maximization, the fees charged should be higher at the peak load periods.

3.4.3 *Economic Implication of Lagrangian Relaxation Method*

Now let us discuss the economic implication of the Lagrangian relaxation method, the framework of which is shown in Figure 3.4. For nonlinear programming, the Lagrange multipliers can be generally considered as the marginal costs or shadow prices, but for the Lagrangian relaxation method integer programming, because the

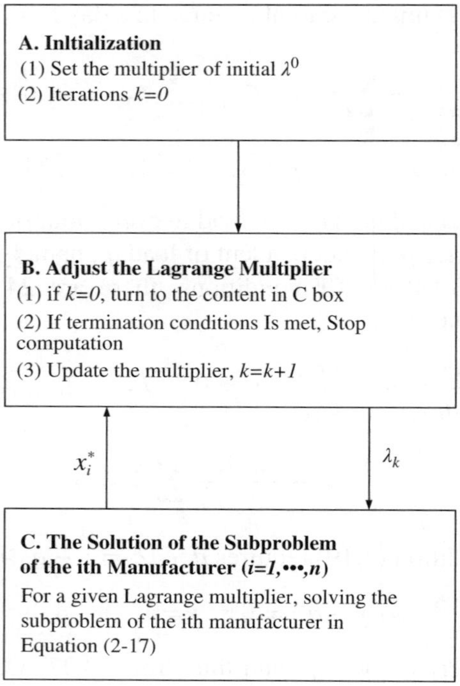

Figure 3.4 Framework of Lagrangian relaxation algorithm.

objective function is not differentiable and the corresponding KT conditions do not exist, it is difficult to explain directly with marginal costs. However, we can still give an explanation in terms of price mechanism.

When the marked demand constraints of the primal problem (2.3) are relaxed, the Lagrangian problem (2.23) can be formed. The Lagrange problem can be seen as such an economic model: the market demand either can be supplied by the firms within the market, or can be purchased from (or sold to) the outside market. Namely, the first item of objective function (2.23)

$$\sum_{i=1}^{n}\sum_{j=1}^{m} c_{ij}x_{ij} \tag{3.29}$$

is the total production cost of firms with the market, and the second item

$$\sum_{j=1}^{m} \lambda_j \left(d_j - \sum_{i=1}^{n} x_{ij} \right) \tag{3.30}$$

is the total expense of buying ($d_j > \sum_{i=1}^{n} x_{ij}$) or revenue of selling ($d_j \le \sum_{i=1}^{n} x_{ij}$) product j with the price λ_j incurred by the difference between market demand and supply.

The sum of these two items is the total cost required to meet the market demand. This problem can also be looked at in this way. The market purchases product j with price $lambda_j$ to meet market demand, and the total cost of all products is $\sum_{j=1}^{m} \lambda_j d_j$. A part of the products is purchased from firms within the market, and for this part of the products the profit yielded within the market (sale revenue minus production cost) should be deducted when calculating the actual cost. Summing up the profit of all firms, we get

$$\sum_{i=1}^{n}\left(\sum_{j=1}^{m} \lambda_j x_{ij} - \sum_{j=1}^{m} c_{ij} x_{ij}\right) = -\sum_{i=1}^{n}\sum_{j=1}^{m}(c_{ij} - \lambda_j)x_{ij} \tag{3.31}$$

The difference between the two items is the total cost required to meet the market demand, namely the expression after the second equals sign of (2.23).

The subproblem of an individual firm is that, after the prices of various products are given by the market, each firm determines the yield of various products according to its production capacity and per unit production cost, and the objective is to maximize its profit. The Lagrangian problem is to calculate the total cost by subtracting the total profit incurred within the market from the expense of purchasing m products.

For the economic model of the primal problem (2.3), the outside market does not exist, namely the market demand for all products must be met by the firms within the market. For the economic model of the dual problem (2.23), because of the existence of the outside market, its production costs must be less than the production cost economic model described by the primal problem. This is because when the market demand can be completely met by firms within the market, namely $\sum_{i=}^{n} x_{ij} \geq d_j,\ j = 1, \ldots, m$, the solution is feasible for the primal problem, and when the outside market exists, for any given nonnegative market prices $\lambda_j, j = 1, \ldots, m$, the firms will be able to sell $\sum_{i=1}^{n} x_{ij} - d_j,\ j = 1, \ldots, m$ product to the outside market, and thereby lower the total production cost.

Now let us discuss the dual problem (2.28), which is to determine the optimal Lagrange multipliers, i.e. the best prices of various products. The market coordinates the production of different firms through the price mechanism. When the price of a product is low, such as $\lambda_j = 0$ for product j, for the subproblem (2.25) of firm $i = 1, \ldots, n$, the optimal solution is $x_{ij}^* = 0$, so the yield of product j in the whole market $\sum_{i=1}^{n} x_{ij}^* = 0$. On the other hand, if the price of a product is high, such as $\lambda_j \to \infty$ for product j, for the subproblem (2.25) of firm $i = 1, \ldots, n$, under the constraints of production capacity, the optimal solution must be that having x_{ij} as large as possible. Let $x_{ij}^* = \bar{x}_{ij}$. Then for the whole market, one must have

$$\sum_{i=1}^{n} x_{ij}^* = \sum_{i=1}^{n} \bar{x}_{ij} \gg d_j$$

namely, the yield of product j is much more than the market demand. Therefore the economic implication of the subgradient method (2.33) is to adjust the market prices according to the difference between market demand and firms' supply. When the market demand for product j exceeds firms' supply, namely $d_j > \sum_{i=1}^{n} x_{ij}^*$, the price of this product λ_j will rise. On the contrary, when the supply exceeds market demand, namely $\sum_{i=1}^{n} x_{ij}^* \geq d_j$, the price of this product λ_j will go down. In the ideal situation (i.e. without considering the duality gap), the final goal of price adjustment is to achieve supply–demand balance in the market, namely $\sum_{i=1}^{n} x_{ij}^* \geq d_j$. At this time $\lambda_j = 0$, the objective function value of the primal problem is equal to that of the dual problem, and the total production costs when the outside market exists and does not exist (the actual situation) are equal. The primal problem reaches the minimum, and the dual problem reaches the maximum.

In solving the dual problem, when λ is given, the problem is to solve the Lagrangian problem (2.23), where x is the decision variable. When there is a change of production Δx_{ij} of product j for firm i, the change of objective function value will be $\Delta z_d(\lambda)$. At the optimal solution x^*, for any Δx_{ij}, one has $\Delta z_d(\lambda) \geq 0$. When c_{ij} is a constant, this is possible only when

$$x_{ij}^* = \begin{cases} 0 & c_{ij} \geq \lambda_j \\ \overline{x}_{ij} & c_{ij} < \lambda_j \end{cases} \tag{3.32}$$

If c_{ij} is a function of x_{ij}, supposing x_{ij} is a continuous variable and $c_{ij} = c_{ij}(x_{ij})$ is differentiable, when the production capacity constraints are not active, at the optimal solution x_{ij}^*, one has

$$\left.\frac{\partial z_d(\lambda)}{\partial x_{ij}}\right|_{x_{ij}=x_{ij}^*} = 0 \tag{3.33}$$

$$\frac{\mathrm{d}(c_{ij}(x_{ij})x_{ij} - \lambda_j x_{ij})}{\mathrm{d}x_{ij}} = 0 \tag{3.34}$$

$$\lambda_j = \frac{\mathrm{d}(c_{ij}(x_{ij})x_{ij})}{\mathrm{d}x_{ij}} = \frac{\mathrm{d}z}{\mathrm{d}x_{ij}} \tag{3.35}$$

where λ_j is the marginal cost of product j.

If x_{ij} is an integer, the derivatives do not exist. For an increment Δx_{ij} of x_{ij}, the marginal cost can be calculated as

$$MC^+(x_{ij}) = \frac{c_{ij}(x_{ij} + \Delta x_{ij})(x_{ij} + \Delta x_{ij}) - c_{ij}(x_{ij})x_{ij}}{\Delta x_{ij}} \tag{3.36}$$

If $MC^+(x_{ij}) < \lambda_j$, namely,

$$\begin{aligned}\Delta z_d(\lambda) &= [c_{ij}(x_{ij} + \Delta x_{ij})(x_{ij} + \Delta x_{ij}) - \lambda_j(x_{ij} + \Delta x_{ij})] \\ &\quad - [c_{ij}(x_{ij})x_{ij} - \lambda_j x_{ij}] \\ &= c_{ij}(x_{ij} + \Delta x_{ij})(x_{ij} + \Delta x_{ij}) - c_{ij}(x_{ij})x_{ij} - \lambda_j \Delta x_{ij} \\ &< 0\end{aligned} \tag{3.37}$$

then apparently $z_d(\lambda)$ can be decreased by increasing x_{ij}. On the contrary, for x_{ij}, using the negative increment $-\Delta x_{ij}$, the marginal cost can be calculated as

$$MC^{-}(x_{ij}) = \frac{c_{ij}(x_{ij} - \Delta x_{ij})(x_{ij} - \Delta x_{ij}) - c_{ij}(x_{ij})x_{ij}}{-\Delta x_{ij}} \tag{3.38}$$

If $MC^{-}(x_{ij}) > \lambda_j$, namely,

$$\begin{aligned}\Delta z_d(\lambda) &= [c_{ij}(x_{ij} - \Delta x_{ij})(x_{ij} - \Delta x_{ij}) - \lambda_j(x_{ij} - \Delta x_{ij})] \\ &\quad - [c_{ij}(x_{ij})x_{ij} - \lambda_j x_{ij}] \\ &= c_{ij}(x_{ij} - \Delta x_{ij})(x_{ij} - \Delta x_{ij}) - c_{ij}(x_{ij})x_{ij} + \lambda_j \Delta x_{ij} \\ &< 0\end{aligned} \tag{3.39}$$

then $z_d(\lambda)$ can be decreased by decreasing x_{ij}.

Hence, at the optimal solution x^*_{ij}, we have

$$MC^{-}(x_{ij}) \leq \lambda_j \leq MC^{+}(x_{ij}) \tag{3.40}$$

In the ideal situation, we have

$$MC^{+}(x^*_{ij}) = MC^{-}(x^*_{ij}) = MC(x^*_{ij}) = \lambda_j \tag{3.41}$$

namely λ_j is the marginal production cost of product j and it is the same for all firms. Therefore, in the ideal situation, the optimal solution of the dual problem λ^* is the marginal cost of the economic system described by the primal problem.

In optimal scheduling/operation of electric power systems, the linear programming, nonlinear programming, and integer and combinatorial programming methods are extensively used. The duality theory of mathematical programming is worthy of particular note. Because the dual variables correspond to the shadow prices of limited resources, in linear and nonlinear programming, they are the marginal values or marginal costs under the optimal situation. In mathematics, the duality theory of linear programming, nonlinear programming and Kuhn–Tucker conditions have an intrinsic relationship. Power system optimal scheduling is an integer programming problem, and thus appropriate integer programming methods should be used for solution. The Lagrangian relaxation method makes use of duality in integer and combinatorial programming, which can be viewed as an extension of the duality in linear and nonlinear programming. There is an intrinsic relation between the dual optimization principle and the market price mechanism. The economic model and price coordination mechanism of the Lagrangian relaxation method is established. The dual variables (or Lagrange multipliers) of the Lagrangian relaxation method correspond to market prices inducing the market supply and demand balance, and they are the marginal cost of the products in the ideal situation.

4

Power System Planning

Power system planning is often divided into generation planning, transmission planning, and distribution planning, all of which will be introduced in the following sections.

4.1 Generation Planning Based on Lagrangian Relaxation Method

The aim of generation planning is to seek the most economical generation expansion scheme that achieves a certain reliability level according to the forecast of demand growth in a certain period of time. The mathematical model of generation planning generally consists of two parts: the power plant investment decision, and power system operation planning. The mathematical models of power system optimal planning and operation are always high-dimensional, nonconvex, discrete, and nonlinear, and thus solution is difficult. The approaches of integer programming, nonlinear programming, network flow optimization, and stochastic optimization have been successfully applied to them [95–105].

A number of successful generation planning models have been presented, which include the Wien Automatic System Planning Package (WASP) of the International Atomic Energy Agency [106] and the Jiaotong Automatic System Planning Package (JASP) of Xi'an Jiaotong University, P.R. China [107].

WASP optimizes the generation planning according to generating unit categories, which neglects the effect of the geographical distribution of system load and power plants. These types of models only take into account the influence of the difference in generating units' categories in the power system. In a developing country such as China, with a vast territory, diverse geographical conditions, and huge hydro power

Power System Optimization – Large-scale Complex Systems Approaches, First Edition.
Haoyong Chen, Honwing Ngan and Yongjun Zhang.

resources, the investment and operational costs in power system generation planning are closely related to the locations of the newly built power plants. Furthermore, errors are introduced in generation planning when WASP combines hydroelectric stations and each hydroelectric station's features are not adequately taken into account. In order to take these features into account, hydroelectric power stations should be treated separately in power plant investment decisions and probabilistic production simulations.

For the above reasons, generation planning can be satisfactorily optimized not according to categories of generating units, but rather according to power plants. JASP uses the generation planning model optimized according to power plants [107]. The following are some of the special features of this model.

1. It takes account of the influence of plant location on the investment and operation costs.
2. It considers features arising from the hydrological and geographical conditions.
3. It is a flexible model that is flexible enough to conform to the deregulation environment of the electricity market.

The mathematical model of JASP consists of two parts: the power plant investment decision, and the probabilistic production simulation. Two different methods are applied to the power plant investment decision: one is a heuristic optimization method, and the other is the Lagrangian relaxation (LR) technique. The equivalent energy function (EEF) method is used for probabilistic production simulation. The two parts are solved by a decomposition–coordination method. The framework of JASP is shown in Figure 4.1. Here X and Y are investment decision variables of the thermal power plants and hydroelectric stations, respectively, and Z denotes the vector of operation decision variables. The transmission capacity is assumed to be adequate and transmission planning has not been considered in the model.

Simulation results show that JASP can not only overcome the "curse of dimensionality" but also find an economical and technically sound generation planning scheme.

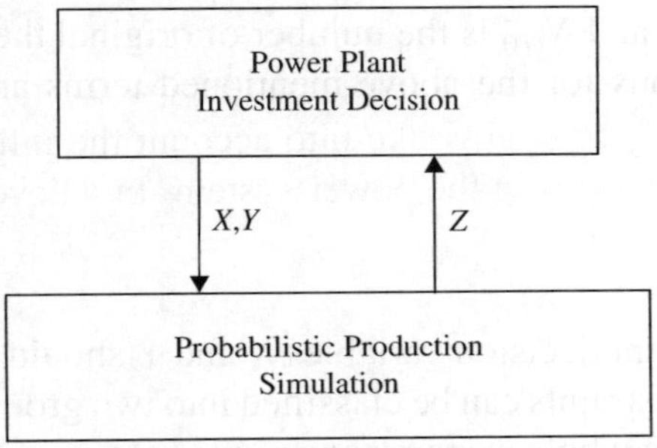

Figure 4.1 The framework of JASP. *Source*: Chen 2004. Reproduced with permission from Elsevier.

4.1.1 *Problem Formulation*

The decision variables of the thermal power plants and hydroelectric stations are denoted respectively as X and Y in the generation investment decision model. Suppose that the planning period is N_t years and there are N_{gf} thermal power plants and N_{gh} hydroelectric stations for candidacy. Then the dimensionality of X and Y is $N_g f \cdot N_t$ and $N_g h \cdot N_t$, respectively. However, the definitions of these two groups of decision variables are completely different. For thermal plants, X_{ti} denotes the unit number of thermal power plants i going into operation in year t in the planning period. For hydroelectric stations, the units should be continuously constructed according to the plan in the feasibility report after the dam has been built. Therefore, $Y_{tj} = 1$ denotes the first batch of generating units in hydroelectric station j to go into operation in year t.

4.1.1.1 Objective Function

A number of generating engineering projects are generally involved in generation planning. These engineering projects have different operation lives and different operation starting times. We use the equivalent annual value method to deal with these problems. The equivalent annual value means the average annual expenditure in power systems over the planning period, which includes fixed annual cost and the average annual operation cost of the engineering project.

The objective function is

$$\min F = \sum_{t=1}^{N_t} \left[\sum_{i=1}^{N_{gf}} a_{ti} X_{ti} + \sum_{j=1}^{N_{gh}} b_{tj} Y_{tj} \right] + \mathrm{CRF} \left(\sum_{i=1}^{N_{gf}} c_{ti} X_{ti} + \sum_{j=1}^{N_{gh}} d_{tj} Y_{tj} + \sum_{k=1}^{N_{gf0}} e_{tk} \right) \tag{4.1}$$

Here a_{ti} is the fixed cost of every unit for thermal power plant i to commit in year t; b_{tj} is the fixed cost of unit investment of hydroelectric station j in year t; CRF is the capital recovery factor; c_{ti} is the sum of the present values of the annual operation costs of the units committed in year t in thermal power plant i; d_{tj} is the sum of the present value of the annual operation costs in the planning period when the hydroelectric station j is put into operation in year t; e_{tk} is the annual operation cost of the original thermal power plant k in year t; and N_{gf0} is the number of original thermal power plants. The mathematical expressions for the above-mentioned terms are defined in Wang and McDonald [107].

4.1.1.2 Constraints

First of all, the investment decision variables X and Y should be integers and not less than zero. The other constraints can be classified into two groups: the individual power plant constraints and the whole system constraints.

The individual power plant constraints are as follows. The number of units that are committed each year in a thermal power plant, X_{ti}, should be limited by the permissible installation number M_{ti}, viz.

$$X_{ti} \le M_{ti} \quad (t = 1, 2, \ldots, N_t;\ i = 1, 2, \ldots, N_{gf}) \tag{4.2}$$

The total number of committed units of thermal power plant i should not exceed the given ultimate number of committed units T_{gi},

$$\sum_{t=1}^{N_t} X_{ti} \le T_{gi} \quad (i = 1, 2, \ldots, N_{gf}) \tag{4.3}$$

The decision variable Y_{tj} for a hydroelectric station should satisfy the following mutual exclusion constraints:

$$\sum_{t=1}^{N_t} Y_{tj} \le 1 \quad (j = 1, 2, \ldots, N_{gh}) \tag{4.4}$$

To obtain the solution easily, the variable X_{tj} related to hydroelectric stations is introduced. It means the number of units to be committed in year t in hydroelectric station j. However, it is not an independent variable.

The whole system constraints include electric power balance condition and electric energy balance condition.

The electric power balance condition is

$$\sum_{\tau=1}^{t} \left(\sum_{i=1}^{N_{gf}} X_{\tau i} W_i + \sum_{j=1}^{N_{gh}} X_{tj} W_j \right) \ge C_t \quad (t = 1, 2, \ldots, N_t) \tag{4.5}$$

where W_i is the single unit capacity of thermal plant i; W_j is the expected output power of hydroelectric station j in a dry year; and C_t is the total output capacity of the new units that need to be committed in the system in year t.

The electric energy balance condition is

$$\sum_{\tau=1}^{t} \left(\sum_{i=1}^{N_{gf}} X_{\tau i} W_i H_i^t + \sum_{j=1}^{N_{gh}} X_{tj} W_j H_j^t \right) \ge E_t \quad (t = 1, 2, \ldots, N_t) \tag{4.6}$$

where H_i^t and H_j^t are respectively the hours of utility of thermal power plant i and hydroelectric station j in year t; and E_t is the electric energy output that has to be supplied by the newly built power plants in the system in year t. The generation energy of a hydroelectric station is calculated through its average power output in a normal year.

The uncertainty of load has not been included in the model, but the necessary generation capacity can be guaranteed by satisfying the least required electric power and energy output in (4.5) and (4.6).

The decomposition–coordination technique for large-scale system optimization is adopted to solve the above generation planning model. The generation investment decision model solves for the optimal investment decision variables X and Y, while the probabilistic production model solves for the optimal operation decision variables Z, especially H_i^t and H_j^t, when X and Y are fixed at some values. These two parts are solved iteratively and in a coordinated way until some convergent criterion is reached.

4.1.2 Lagrangian Relaxation for Generation Investment Decision

The above generation investment decision model is a nonlinear integer programming problem. It is almost impossible to find the mathematically optimal solution. On the other hand, heuristic algorithms can offer an approximation to the optimal solution with relatively less computational time. The previous version of JASP uses a heuristic algorithm [107]. There are some flaws concerning this heuristic algorithm, such as a lack of mathematical theory foundation and suboptimal performance.

The new version of JASP adopts the Lagrangian relaxation [62] approach to optimize the generation investment decision. LR is an integer programming algorithm that has been applied to power system unit commitment and generation scheduling problems with great success [17, 108]. The basic idea of LR is to relax the systemwide constraints such as the power and spinning reserve requirements by using Lagrange multipliers, then to decompose the problem into a number of commitment or generation scheduling subproblems of individual units, which are much easier to solve. LR not only has a relatively rigorous mathematical foundation, but also can overcome the dimensional obstacle and get quite good suboptimal solutions.

Based on a similar idea, JASP uses LR to solve the generation investment decision model. First, relax the systemwide constraints such as the electric power balance condition and electric energy balance condition by using Lagrange multipliers, then decompose the problem into a number of individual power plant planning subproblems, which are much easier to solve. The subgradient optimization technique is used to solve the dual problem, and different optimization methods such as dynamic programming are used to solve the subproblems. The framework of LR is shown in Figure 4.2.

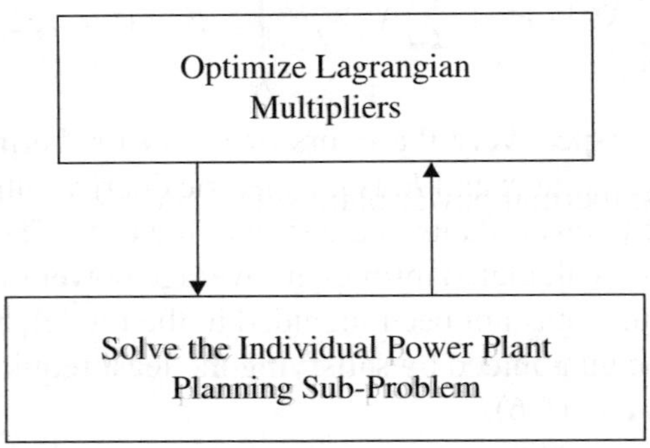

Figure 4.2 The framework of the Lagrangian relaxation method. *Source*: Chen 2004. Reproduced with permission from Elsevier.

4.1.2.1 Dual Problem

The mathematical model of (4.1)–(4.6) is a mixed integer programming problem. To decompose the problem, we first relax systemwide constraints (4.5) and (4.6) and formulate the dual problem by the principle of duality:

$$
\begin{aligned}
L = \sum_{t=1}^{N_t} \Bigg[& \sum_{i=1}^{N_{gf}} a_{ti} X_{ti} + \sum_{j=1}^{N_{gh}} b_{tj} Y_{tj} \\
& + \mathrm{CRF} \left(\sum_{i=1}^{N_{gf}} c_{ti} X_{ti} + \sum_{j=1}^{N_{gh}} d_{tj} Y_{tj} + \sum_{k=1}^{N_{gf}} e_{tk} \right) \Bigg] \\
& + \sum_{t=1}^{N_t} \Bigg\{ \lambda_t \left[C_t - \sum_{\tau=1}^{t} \left(\sum_{i=1}^{N_{gf}} X_{\tau i} W_i + \sum_{j=1}^{N_{gh}} X_{\tau j} W_j \right) \right] \\
& + \mu_t \left[E_t - \sum_{\tau=1}^{t} \left(\sum_{i=1}^{N_{gf}} X_{\tau i} W_i H_i^t + \sum_{j=1}^{N_{gh}} X_{\tau j} W_j H_j^t \right) \right] \Bigg\}
\end{aligned} \tag{4.7}
$$

where λ_t and μ_t are Lagrange multipliers associated with the electric power balance condition and electric energy balance condition at year t respectively. For notational convenience, we define

$$
\begin{aligned}
\lambda &= (\lambda_1, \lambda_2, \ldots, \lambda_{N_t})^{\mathrm{T}} \\
\mu &= (\mu_1, \mu_2, \ldots, \mu_{N_t})^{\mathrm{T}}
\end{aligned} \tag{4.8}
$$

Using the duality theorem and exploiting the decomposable structure of (4.7), we can form a two-level maximum–minimum optimization problem.

Given multipliers λ and μ, the low-level problem consists of individual power plant planning subproblems: thermal power plant subproblem (P_i), $i = 1, 2, \ldots, N_{gf}$,

$$
\begin{aligned}
\min_{(X_{1i}, X_{2i}, \ldots, X_{N_t i})} L_i = \sum_{t=1}^{N_t} \Bigg[& (a_{ti} + \mathrm{CRF} \cdot c_{ti}) X_{ti} \\
& - \lambda_t \left(\sum_{\tau=1}^{t} X_{\tau i} \right) W_i - \mu_t \left(\sum_{\tau=1}^{t} X_{\tau i} \right) W_i H_i^t \Bigg]
\end{aligned} \tag{4.9}
$$

subject to (4.2) and (4.3), and hydroelectric station subproblem (P_j), $j = 1, 2, \ldots, N_{gh}$,

$$\min_{(X_{1j}, X_{2j}, \ldots, X_{N_t j})} L_j = \sum_{t=1}^{N_t} \left[(b_{tj} + \text{CRF} \cdot d_{tj}) Y_{ti} - \lambda_t \left(\sum_{\tau=1}^{t} X_{\tau j} \right) W_j - \mu_t \left(\sum_{\tau=1}^{t} X_{\tau j} \right) W_j H_j^t \right] \tag{4.10}$$

subject to (4.4).

The high-level dual problem (primal–dual) is to optimize Lagrange multipliers, viz.

$$\max_{\lambda,\mu} = \sum_{i=1}^{N_{gf}} L_i^*(\lambda, \mu) + \sum_{i=1}^{N_{gh}} L_j^*(\lambda, \mu) + \sum_{t=1}^{N_t} \left(\text{CRF} \sum_{k=1}^{N_{gf}} e_{tk} + \lambda_t C_t + \mu_t E_t \right) \tag{4.11}$$

subject to $\lambda \geq 0$ and $\mu \geq 0$, where $L_i^*(\lambda, \mu)$ is the optimal objective function value of problem (P_i), and $L_j^*(\lambda, \mu)$ is the optimal objective function value of problem (P_j).

The high-level dual problem and the low-level subproblems are solved alternately and iteratively until the optimal dual solution is found.

4.1.2.2 Solving Individual Power Plant Subproblems

For thermal power plant subproblem (4.9), we define $X_{\sum ti} = \sum_{\tau=1}^{t} X_{\tau i}$ and regard it as the state variable. We can see that $X_{\sum ti}$ has the characteristic that, given the current value, the optimal decision for each of the remaining values does not depend on the previous values. Then the forward dynamic programming algorithm is suitable for solving this problem. Using $X_{\sum ti}$ as the state variable, the state space for one year or one stage is the integers from 0 to T_{gi}, and the state transition must satisfy constraint (4.3).

For the hydroelectric station subproblem (4.10), we only need to enumerate N_t schemes to determine the optimal time for the first batch of generating units to go into operation.

4.1.2.3 Solving Dual Problem by Subgradient Optimization Method

Although the primal problem is nonconvex, the dual problem can be easily proved concave. Therefore, the dual problem has a unique maximum solution. Since the objective function of the dual problem is not differentiable, a subgradient algorithm is used to update λ and μ as follows [63]:

$$\begin{aligned}\lambda^{j+1} &= \lambda^j + \alpha_j g_\lambda, \quad \text{for } j = 0, 1, 2, \ldots \\ \mu^{j+1} &= \mu^j + \alpha_j g_\mu\end{aligned} \tag{4.12}$$

subject to $\lambda \geq 0$ and $\mu \geq 0$, where α_j is a scalar step size and g_λ is the subgradient of $L(\lambda, \mu)$ with respect to λ and g_μ is the subgradient of $L(\lambda, \mu)$ with respect to μ:

$$\begin{aligned} g_\lambda &= \left[C_t - \sum_{\tau=1}^{t} \left(\sum_{i=1}^{N_{gh}} X^*_{\tau i} W_i + \sum_{j=1}^{N_{gh}} X^*_{\tau j} W_j \right) \Bigg|_{t=1,2,\ldots,Nt} \right]^{\mathrm{T}} \\ g_\mu &= \left[E_t - \sum_{\tau=1}^{t} \left(\sum_{i=1}^{N_{gh}} X^*_{\tau i} W_i H_i^t + \sum_{j=1}^{N_{gh}} X^*_{\tau j} W_j H_j^t \right) \Bigg|_{t=1,2,\ldots,Nt} \right]^{\mathrm{T}} \end{aligned} \tag{4.13}$$

Here $X^*_{\tau i}$ $(i = 1, 2, \ldots, N_{gf},\ \tau = 1, 2, \ldots, N_t)$ and $X^*_{\tau j}$ $(j = 1, 2, \ldots, N_{gh},\ \tau = 1, 2, \ldots, N_t)$ are respectively the optimal solutions of problems (P_i) and (P_j).

A practical choice of step size α_j is

$$\alpha_j = \gamma_j \frac{\overline{L} - L}{(g_\lambda^{\mathrm{T}}, g_\mu^{\mathrm{T}})^{\mathrm{T}} \cdot (g_\lambda^{\mathrm{T}}, g_\mu^{\mathrm{T}})^{\mathrm{T}}} \tag{4.14}$$

where $\overline{L}$ is an overestimate of the maximum of L, which can be estimated by the heuristic optimization method presented in Wang and McDonald [107], and the sequence λ_j starts from a number between 0 and 2 and is halved every threshold number of iterations until the resulting α_j is sufficiently small.

Since the dual solution is generally infeasible for the primal problem due to its nonconvexity, an optimization method based on merit order is used to generate a good feasible solution for the primal problem based on the dual results.

4.1.2.4 Extension of Generation Investment Decision Model to the Deregulation Environment of the Electric Power Industry

The equivalent annual value of the total investment and operation cost of the power system expressed by (4.1) is under the hypothesis that the system total load and energy demand in the planning horizon can be forecasted. That hypothesis is embodied in electric power balance condition (4.5) and electric energy balance condition (4.6). In the deregulated power market condition, however, the demand is determined by market equilibrium, and the optimizing object is to maximize the social welfare. The objective function is:

$$
\begin{aligned}
\max W &= B - F \\
&= \mathrm{CRF}\sum_{t=1}^{N_t}\left[\left(\sum_{i=1}^{N_l} p_{ti}W_{ti} + \sum_{i=1}^{N_l} q_{ti}E_{ti}\right)\right] \\
&\quad - \mathrm{CRF}\sum_{t=1}^{N_t}\left[\left(\sum_{i=1}^{N_{gf}} \pi_{ti}W_{ti} + \sum_{j=1}^{N_{gh}} \pi_{tj}W_{tj}\right)\right] \\
&\quad - \mathrm{CRF}\sum_{t=1}^{N_t}\left[\left(\sum_{i=1}^{N_{gf}} \sigma_{ti}W_{ti} + \sum_{j=1}^{N_{gh}} \sigma_{tj}E_{tj} + \sum_{k=1}^{N_{gf0}} \sigma_{tk}E_{tk} + \sum_{l=1}^{N_{gh0}} \sigma_{tl}E_{tl}\right)\right]
\end{aligned}
\tag{4.15}
$$

where W is the social welfare; B is the benefits to the consumers from consuming the electric power energy; p_{ti} is the benefit of per unit generation capacity of consumer i in year t; q_{ti} is the benefit of per unit generation energy of consumer i in year t; π_{ti} and π_{tj} are respectively the investment costs of per unit generation capacity of thermal power plant i and hydroelectric station j in year t; and σ_{ti} and σ_{tj} are respectively the operation costs of per unit generation energy of thermal power plant i and hydroelectric station j in year t. All benefits and costs are converted to their present values first, and then converted to the annual equivalent value by multiplying the capital recovery factor (CRF). The new mathematical model can also be solved by the LR method.

If the demand side of the competitive electricity market is regulated, as is being done in many developing countries, the mathematical model of (4.1)–(4.6) can remain unchanged. We still solve this problem by the LR method. In thermal power plant subproblem (4.9) and hydroelectric station subproblem (4.10), if we view λ_t and μ_t as the prices of per unit generation capacity and per unit generation energy, respectively, then L_i equals the negative value of the net benefits, namely,

$$
\begin{aligned}
-L_i &= NB_i = B_i - F_i \\
&= \lambda_t\left(\sum_{\tau=1}^{t} X_{\tau i}W_i + \mu_t\left(\sum_{\tau=1}^{t} X_{\tau i}\right) W_i H_i^t\right) - \sum_{t=1}^{N_t}[(a_{ti} + \mathrm{CRF}\cdot c_{ti})X_{ti}]
\end{aligned}
\tag{4.16}
$$

Minimization of L_i is equivalent to maximization of the net benefits, just as suppliers often do.

When the system total load and energy demand are forecasted and fixed at given values, the solutions of dual problem (4.7), namely λ and μ, are the system marginal generation capacity costs and the system marginal generation energy costs, respectively.

4.1.3 Probabilistic Production Simulation

The main purpose of probabilistic production simulation is to simulate the dispatch of generating units, and to estimate the production cost. Because the effect of unit forced outage and maintenance is considered, we can get system reliability indices such as expected energy not served (EENS) and loss of load probability (LOLP).

In power system operation and planning, EENS is more meaningful than LOLP, and that means electric energy is the key variable. This chapter presents an EEF (equivalent energy function) approach for probabilistic production simulation. The EEF approach calculates electric energy consumed in different load level segments and modifies it directly when unit failure effects are taken into consideration [109].

A load duration curve can be described by

$$t = F(x) \tag{4.17}$$

where x is the load level, and t is the time interval during which the load is larger than or equal to x. Assume the simulation time horizon is T. From (4.17) we can get

$$p = F(x)/T = f(x) \tag{4.18}$$

where p is considered as the probability at which the load is larger than or equal to x. From (4.18) we can see that load uncertainty is easy to include in the above probabilistic production simulation model. From $F(x)$ we can define a discrete energy function as

$$E(J) = \int_x^{x+\Delta x} F(x)\,\mathrm{d}x = T\int_x^{x+\Delta x} f(x)\,\mathrm{d}x \tag{4.19}$$

with

$$J = [X/\Delta x] + 1 \tag{4.20}$$

where $[X/\Delta x]$ means the biggest integer not greater than $X/\Delta x$, and $E(J)$ corresponds to the area under a section of the load curve from x to Δx, or the energy that corresponds to this section of the load.

Suppose that the probability distribution of the primary load duration curve is $f^{(0)}(x)$, the corresponding energy function is $E^{(0)}(J)$, the probability distribution of the load duration curve when generating unit $i-1$ has been committed is $f^{i-1}(x)$, and the corresponding energy function is $E^{i-1}(J)$. The generating unit i is assumed to have a capacity of C_i and a forced outage rate of q_i. We have

$$\begin{aligned} E^{(i)}(J) &= T\int_x^{x+\Delta x} [p_i f^{(i-1)}(x) + q_i f^{(i-1)}(x - C_i)]\,\mathrm{d}x \\ &= p_i E^{(i-1)}(J) + q_i E^{(i-1)}(J - K_i) \end{aligned} \tag{4.21}$$

where $p_i = 1 - q_i$ and $K_i = C_i/\Delta x$

Generating unit i's energy output E_{gi} is

$$E_{gi} = Tp_i \int_{x_{i-1}}^{x_{i-1}+C_i} f^{(i-1)}(x)\,\mathrm{d}x = p_i \sum_{J=J_{i-1}+1}^{J_i} E^{(i-1)}(J) \tag{4.22}$$

where $J_{i-1} = x_{i-1}/\Delta x$ and $J_i = (x_{i-1} + C_i)/\Delta x = J_{i-1} + K_i$.

The load energy not served by the system is

$$E_{Di} = \sum_{J>J_i} E^{(i)}(J) \tag{4.23}$$

Assuming that there are n generating units in the power system, then

$$\text{EENS} = E_{Dn} = \sum_{J>J_n} E^{(n)}(J) \tag{4.24}$$

The EEF method is not only more efficient than any other method available in the area, but also more flexible in handling assigned energy units, such as the hydroelectric generating units [110].

If there are hydroelectric generating units in the power system, they should substitute for the thermal generating units, sharing the peak load as far as possible. Maximum fuel consumption savings can be achieved as these thermal generating units have the highest coal consumption rate.

Figure 4.3 shows how the hydroelectric generating unit shares the peak load. The shaded portion is the load shared by the hydroelectric generating unit. The curve gc is derived by shifting the primary load duration curve toward the left by the capacity C_H

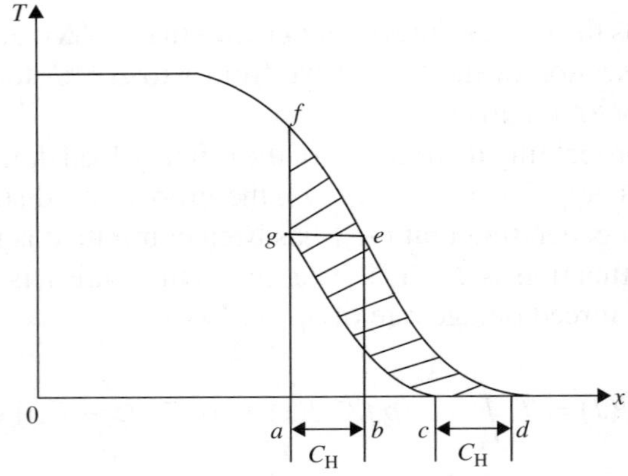

Figure 4.3 A single hydroelectric generator unit under peak load condition. *Source*: Chen 2004. Reproduced with permission from Elsevier.

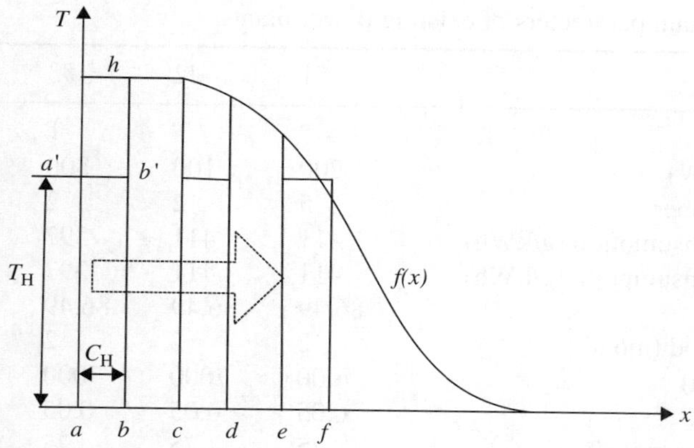

Figure 4.4 Procedure to determine the loading position of a hydroelectric generating unit. *Source*: Chen 2004. Reproduced with permission from Elsevier.

of the hydroelectric generating unit. The area of the shaded portion gcdf is equal to the hydroelectric generating unit's given energy E_A. In this way, the hydro energy E_A is most economically and fully utilized. The rest of the load shared by the thermal generating units is the area surrounded by 0acgfh. In Figure 4.3, it is not difficult to prove that the areas acg and bde are identical, so the effect is the same whether the hydroelectric generating unit shares the load abef or the shaded portion gcdf. Therefore the principle for load sharing by a hydroelectric generating unit is to locate a section of the equivalent load duration curve which corresponds to the unit's operation capacity C_H and has an area equal to the unit's given energy E_A. The process of determining a hydroelectric generating unit's operation position is shown in Figure 4.4, where $T_H = E_A/C_H$. When there are two or more hydroelectric generating units in the power system, the procedure for determining the hydroelectric generating units' operation positions is based on a similar methodology but becomes much more complicated.

4.1.4 Example

JASP has been successfully applied to generation planning of some practical power grids in China. To make things clearer, we illustrate the algorithm features with the optimal planning problem of a relatively small example system with the data taking from a real-world generation system.

4.1.4.1 Problem Data

The example system has 10 power plants with a total capacity of 3.5 GW. The main parameters are listed in Table 4.1.

The example system has 10 new power plants for candidacy. The main parameters are listed in Table 4.2. The system load growth in the planning period is shown in

Table 4.1 The main parameters of existing power plants.

	#1	#2	#3	#4	#5
Plant type[a]	T	T	T	H	H
Unit capacity (MW)	200	100	300	50	150
Installed unit number	5	2	2	2	2
Base load coal consumption (g/kWh)	411	411	397	0	0
Peak load coal consumption (g/kWh)	411	411	397	0	0
Coal price ($/ton)	86.49	86.49	86.49	0	0
Maintenance period (month)	2	2	2	1	1
Hours of utility (h)	7000	7000	7000	3600	3600
Forced outage rate	0.05	0.05	0.05	0.03	0.03
Minimum power output (%)	75	75	75	20	20
Number of retired units[b]	0	0	0	0	0
Cost of maintenance and operation (%)	7	7	7	2	2
	#6	**#7**	**#8**	**#9**	**#10**
Plant type[a]	T	H	T	T	H
Unit capacity (MW)	100	100	300	200	200
Installed unit number	2	4	1	0	2
Base load coal consumption (g/kWh)	411	0	397	400	0
Peak load coal consumption (g/kWh)	411	0	397	400	0
Coal price ($/ton)	110.15	0	86.49	86.49	0
Maintenance period (month)	2	1	2	2	1
Hours of utility (h)	7000	4430	7000	7000	3400
Forced outage rate	0.05	0.03	0.05	0.05	0.03
Minimum power output (%)	75	20	75	75	20
Number of retired units[b,c]	0	0	−1	−2	0
Cost of maintenance and operation (%)	7	2	7	7	2

Notes:
[a]Type T denotes thermal or nuclear power plant.
[b]Type H denotes hydroelectric stations.
[c]The negative retired unit number denotes number of units mandatorily committed.

Table 4.3. In addition to the above data, the necessary problem data include information on the load curve, the expected power output and average power output of each hydroelectric station in a dry year and a normal year, etc. Although hydro power generation is stochastic, a 95% or above guarantee rate of average power output data can be achieved in long-term generation planning.

4.1.4.2 Results

The discount rate of capital is assumed to be 10%, and the capacity reserve is assumed to be 25% of the yearly maximum load. Comparison is made between optimization

Table 4.2 The main parameters of candidate power plants.

	#11	#12	#13	#14	#15
Plant type	T	T	T	T	T
Unit capacity (MW)	200	300	300	200	600
Candidate unit number	4	4	4	4	4
Base load coal consumption (g/kWh)	325	320	320	325	315
Peak load coal consumption (g/kWh)	325	320	320	325	315
Coal price ($/ton)	86.49	86.49	86.49	86.49	86.49
Maintenance period (month)	2	2	2	2	2
Hours of utility (h)	7000	7000	7000	7000	7000
Forced outage rate	0.05	0.05	0.05	0.05	0.05
Minimum power output (%)	75	75	75	75	75
Economic life (y)	25	25	25	25	25
Cost of maintenance and operation (%)	7	7	7	7	7
Investment ($/kW)	850	925	925	850	1200
Construction time (y)	2	2	2	2	2
	#16	#17	#18	#19	#20
Plant type	T	H	H	H	H
Unit capacity (MW)	600	100	100	50	100
Candidate unit number	4	2	4	2	3
Base load coal consumption (g/kWh)	315	0	0	0	0
Peak load coal consumption (g/kWh)	315	0	0	0	0
Coal price ($/ton)	110.15	0	0	0	0
Maintenance period (month)	2	1	1	1	1
Hours of utility (h)	7000	3443	3885	3600	3200
Forced outage rate	0.05	0.03	0.03	0.03	0.03
Minimum power output (%)	75	0	0	0	0
Economic life (y)	25	30	30	30	30
Cost of maintenance and operation (%)	7	2	2	2	2
Investment ($/kW)	1200	1388	1540	2300	1836
Construction time (y)	2	4	5	4	4

Table 4.3 The annual growth of system total load and energy.

Year	1	2	3	4	5
Maximum load (MW)	2683	3088	3513	3978	4553
Energy (10^9 kWh)	17.48	20.12	22.89	25.92	29.67
Year	6	7	8	9	10
Maximum load (MW)	4987	5510	6102	6743	7307
Energy (10^9 kWh)	32.5	35.91	39.77	43.95	47.62

Table 4.4 System economic and reliability indices.

	Annualized cost	EENS	LOLP
Heuristic	6.576	0.8143	0.0089
LR	6.185	0.4053	0.0048

results of two different methods, viz. heuristic algorithm and the Lagrangian relaxation method for solving the generation investment decision model. The system economic and reliability indices are listed in Table 4.4. The annualized cost indicates the equivalent annual value of total investment and operation cost in the planning period. From Table 4.4 we can see that the LR method can find a much better generation planning scheme in economical and technical indices.

The construction scheme of the new plants is shown in Table 4.5. The heuristic algorithm results in a generation expansion of 5.20 GW, not including 0.70 GW of

Table 4.5 Construction scheme of the new plants.

Old version	Year									
	1	2	3	4	5	6	7	8	9	10
#11	0	0	0	1	0	1	1	1	0	0
#12	0	0	0	0	0	0	0	1	1	1
#13	0	0	0	1	0	0	1	1	1	0
#14	0	0	1	1	1	1	0	0	0	0
#15	0	0	0	0	0	0	0	0	0	1
#16	0	0	0	0	0	0	0	0	0	0
#17	0	0	0	0	1	1	0	0	0	0
#18	0	0	0	0	1	1	1	0	0	0
#19	0	0	0	0	0	0	0	0	0	0
#20	0	0	0	1	1	1	0	0	0	0
New version	Year									
	1	2	3	4	5	6	7	8	9	10
#11	0	1	1	1	1	0	0	0	0	0
#12	0	0	0	0	0	0	1	1	1	1
#13	0	0	0	0	0	0	1	1	1	1
#14	0	0	0	1	1	1	0	0	1	0
#15	0	0	0	0	0	0	0	0	0	0
#16	0	0	0	0	0	0	0	0	0	0
#17	0	0	0	1	1	0	0	0	0	0
#18	0	0	0	0	1	1	1	1	0	0
#19	0	0	0	0	0	0	0	0	0	1
#20	0	0	0	1	1	1	0	0	0	0

mandatory expansion in Table 4.1. The total capacity of the system is 9.40 GW at the end of the planning period, which is the sum of the newly installed capacity of 5.90 GW (optimal capacity expansion plus mandatory capacity expansion) and the original capacity of 3.50 GW. The LR method results in a generation expansion of 4.95 GW. The total capacity of the system is 9.15 GW with the newly installed capacity of 5.65 GW. Compared with the heuristic algorithm, the optimization result of the LR method diminishes the installation of one 600 MW thermal unit and enlarges the installation of one 300 MW thermal unit and one 50 MW hydroelectric unit. So the LR method acquires a generation planning scheme with better peak load adjustment characteristics, more rational generation mix, and a lower investment and operation cost by increasing the installation of relatively small thermal units and hydroelectric units and decreasing the installation of large thermal units.

4.1.5 *Summary*

The mathematical model of the generation planning program JASP consists of two parts: the power plant investment decision and the probabilistic production simulation. Two different optimization methods, viz. the heuristic algorithm and the LR method, have been applied to the power plant investment decision. The equivalent energy function method is used for probabilistic production simulation.

The following are some of the special features of JASP.

1. The generation planning model is optimized according to power plants, which is particularly suitable for a developing country with a vast territory, diverse geographical conditions, and huge hydro power resources.
2. The generation planning model treats the hydroelectric power stations separately, each hydroelectric station's features being adequately taken into account.
3. The generation planning model treats the hydroelectric power stations separately, each hydroelectric station's features being adequately taken into account.
4. The mathematical model and solution method are flexible enough to conform to the deregulation environment of the electricity market.

Simulation results show that JASP not only can overcome the "curse of dimensionality" but also can find a much better generation planning scheme in economical and technical indices.

4.2 Transmission Planning Based on Improved Genetic Algorithm

The object of transmission system planning is to determine when, where, and what type of transmission line should be constructed, under the condition that the generation planning and load forecasting in the planning stage are known. The allocation of transmission costs in a competitive environment requires determination of optimal transmission network planning. As a result, the need for methods that are able to synthesize optimal plans becomes more important than ever. The optimal plan is to give the best planning scheme that has minimum construction and operating costs and can satisfy the need of reliable transmission of the power flow.

The long-term transmission expansion planning problem is a large, complex, mixed integer nonlinear programming problem that has presented a major challenge to all known heuristic, optimization approaches. Many mathematical models have been proposed in the past for optimal transmission network planning [111–116]. Most of them are static models, giving an optimal solution for a fixed horizon year [8, 112–116]. Others are dynamic planning, giving an optimal solution for multi-stage development, considering the evolution in power demand through time [111, 112].

The horizon year planning problem is also known as the single-stage problem. It involves determining the network connection scheme for a particular generation and load pattern in a future horizon year, and does not deal with the timing problem. For the single-stage planning problem, the available information consists of the original network structure and parameters, the candidate lines (rights of way and parameters), the construction investment cost of each candidate line, and the transmission capacities (ratings) of each existing line and each candidate line. The objective of this problem is to determine the most economical planning scheme(s) to meet the load demand in the horizon year subject to the security or reliability constraints. The security or reliability constraints may consist of the overload constraints in the basic planning network and the $N-1$ security constraints [113–116].

The multi-stage planning of a transmission system has to consider not only how to form the network schemes of every stage but also how to coordinate the network schemes of every stage. Because of the combinatorial nature of the transmission network planning, it has not been well solved by conventional optimization techniques.

The genetic algorithm (GA) approach is a global optimization method based on the mechanism of evolution and natural genetics, which has been successfully used for solving single-stage transmission network planning [113, 115].

This chapter presents an improved GA approach to optimal multi-stage transmission network planning. A fitness function including investment and overload constraint is constructed. The overload is checked by DC load flow. A concise codification model called the "redundant binary-coded technique" is proposed. By this technique, the crossover operation can be executed inside the gene so that the recombinatorial and search functions of the crossover operator are well utilized. The simulated annealing (SA) selector is used to adjust the fitness function in the evolution process, which can avoid GA's premature convergence to a certain extent and improve GA's ability to find the global optimal solution. Some improvements are employed to speed up the algorithm's convergence, such as keeping excellent seeds, mutation in pairs, etc.

Computational programs have been developed based on the mathematical model and algorithm proposed. Three case studies are applied to demonstrate the usefulness and effectiveness of the suggested GA-based multi-stage transmission network planning.

4.2.1 Mathematical Model

In multi-stage network planning, besides the questions of what and where the new transmission facilities should be installed, one should also answer when to install new transmission facilities stage by stage. In this section, a mathematical model using GA for multi-stage transmission system planning is proposed.

The objective of multi-stage transmission system planning is to minimize the sum of present values in the whole planning period:

$$\min \sum_{k=0}^{N_p-1} \frac{l_1 u(k) + l_2(x(k+1), k+1)}{(1+r)^{m(k-1)}} \tag{4.25}$$

$$\text{s.t.} \quad \begin{cases} x(k) \in X(k) \\ u(k) \in U(k) \end{cases} \tag{4.26}$$

$$x(k+1) = x(k) + u(k) \tag{4.27}$$

$$x(0) = C \tag{4.28}$$

where f is the total present cost; r is the annual interest rate; N_p is the number of stages in the planning period; $m(k) = \sum_{i=0}^{k} y(i)$ is the number of years from original planning year to the end year of stage k, with $y(i)$ the number of years in stage i; $u(k)$ is the expanding scheme from stage k to stage $k+1$; $U(k)$ is the set of admissible expanding schemes of stage k; $l_1(u(k))$ is the investment of expanding scheme $u(k)$; $x(k)$ is the network state of stage k; $X(k)$ is the set of admissible network states in stage k; and $l_2(x(k+1), k+1)$ is the operational cost of network state $x(k+1)$ in stage k, which consists of the cost of line loss and the discount cost of transmission facilities.

The multi-stage transmission system planning should search a series of feasible expanding schemes $u(k) \quad (k = 0, 1, \ldots, N_p - 1)$ such that (4.25) is minimized, subject to the system equations and the constraints (4.26) and (4.27), and the initial condition (4.28).

In (4.25), each candidate line is an independent integral decision variable. It represents when or where the line is to be built during the planning period. The problem is a complex large discrete optimization problem. The genetic algorithm is very suitable to solve such problems.

Equation (4.26) represents the feasible constraints. The constraints may include overload, $N-1$ criteria, node voltage limits, etc. In this chapter, only the overload constraint is considered in every stage. The set of admissible network states $x(k)$ should satisfy

$$|F_j| \leq \overline{F_j} \tag{4.29}$$

where $|F_j|$ and $\overline{F_j}$ are the transmission power flow and transmission limitation of line j, respectively.

4.2.1.1 Encoding

Encoding multi-stage transmission network planning needs several binary codes consisting of a gene to express the building state of a candidate line.

Assuming that a candidate line i can be built since stage N_s, then an integral variable Z_i in interval $[N_s, N_p + 1]$ will express the stage at which the line i will be built. When $Z_i = N_p + 1$, the line is not built in the whole planning period. The variable Z_i represents $N_p + 2 - N_{s,i}$ states of candidate line i. If it needs N_b binary codes to completely express these states, the length of the gene for this line N_b satisfies

$$2^{N_{b,i}} \geq N_p + 2 - N_{s,i} \tag{4.30}$$

The chromosome string is a composition of genes according to the order of all candidate lines:

$$C = \{G_1, G_2, \ldots, G_i, \ldots, G_m\} \tag{4.31}$$

where G_i is a gene of line i which has N_{bi} binary codes, and m is the number of total candidate lines. A chromosome string is a planning scheme. It contains the messages of the building stage for each candidate line.

When decoding a gene of a chromosome string, it should be noted that the states of the gene might be greater than the number stages in which the line can be built. The number of the gene's states expressed by N_b binary codes is 2^{N_b}; the number of stages in which the line can be built is $N_p + 2 - N_s$. The decoding of the gene may have two cases. The reasonable decoding equation is given as follows:

$$Z_i = \begin{cases} N_{s,i} + N & 0 \leq N \leq N_p + 1 - N_{s,i} \\ N_{s,i} + N - N_{E,i} & N_p + 1 - N_{s,i} \leq N \leq 2^{N_{b,i}} - 1 \end{cases} \tag{4.32}$$

where

$$N_{E,i} = N_p + 2 - N_{s,i}$$

4.2.1.2 Fitness Function

In the multi-stage network planning model, the overload constraints are treated by penalty factors in its fitness function. The fitness function of the problem has the following form:

$$f = \sum_{k=0}^{N_p-1} \frac{l_1(u(k)) + l_2(x(k+1), k+1)}{(1+r)^{m(k-1)}} + \sum_{k=0}^{N_p-1} P(k)S(k) \tag{4.33}$$

where $P(k)$ is the overload penalty coefficient in stage k, and $S(k)$ is the amount of overload. In the evolution procedure, through crossover and mutation operation, some new generated chromosome string may have islands or the network is not connective. Thus $S(k)$ can be calculated as

$$S(k) = \begin{cases} W_1(k) & \text{connecting} \\ W_2(k) & \text{nonconnecting} \end{cases} \tag{4.34}$$

Here $W_1(k)$ is the total overload amount calculated by a DC load flow model if the scheme $x(k)$ is a connective network; and $W_2(k)$ is a given large number if the scheme $x(k)$ is not a connective network for the reason that its overload amount cannot be calculated directly.

4.2.2 *Improvements of Genetic Algorithm*

As is well known, the simple genetic algorithm (SGA) is a robust optimization technique that works above a set of candidate solutions named "population" and performs a number of operations (selection, crossover, mutation) based on a genetic mechanism. The main problems of SGA are concerned with two aspects. The first one is its slower convergence speed, and the second one is that it cannot give an explicit convergence condition.

To overcome these disadvantages, some improvement measures have been adopted as follows.

4.2.2.1 Keeping Excellent Seeds

Keeping excellent seeds means that a few promising schemes in every generation are directly copied to the next generation. This improvement measure can avoid losing excellent seeds in the evolution procedure. However, they will be replaced if schemes better than them are produced in the later generations. In our experience, maintaining the percentage of excellent seeds at 10–15% of the total population can give satisfactory results. When the evolution procedure has finished, a group optimal and suboptimal solutions are obtained simultaneously.

4.2.2.2 Mutation in Pairs

Mutation in pairs means that, when natural evolution cannot give any better solution in several continued generations, we use an idea similar to a tabu search to check the excellent seeds being kept. For one of the excellent seeds, we try to force one of its genes to mutate from 1 to 0, then check if the scheme appears to show overload or nonconnectivity. If these phenomena do not appear, the mutation is successful. If these phenomena appear, the system is either divided into two parts or connected through a minimum cut with a lower transmission capacity. So we try to add a shorter candidate line across the two parts to replace the original one. If a shorter (or economic) candidate line exists and it makes the new scheme feasible, then the system has an efficient mutation. The gene of the candidate line is changed from 0 to 1. Through this mutation in pairs procedure, the best solution may be quickly obtained.

4.2.2.3 Simulated Annealing

There are quite a few papers using simulated annealing to improve GA convergence [114, 116]. In this chapter, the fitness function (4.33) is scaled according to the simulated annealing method as

$$F_i = \exp\left(\frac{f_{avg} - f_i}{T_n}\right), \quad i = 1, 2, \ldots, N_{pop} \tag{4.35}$$

where f_i is the fitness of the ith chromosome; F_i is the scaled fitness of f_i; f_{avg} is the average value of total f_i in the whole population size N_{pop}; and T_n is given by

$$T_n = \ln\left(\frac{T_0}{M+1} + 1\right)$$

in which $T_0 = 100$, and M is the total number of generation iterations.

Combined with the simulated annealing technique, the improved GA can increase the ability to find the global optimal solution.

4.2.3 Example

Based on the mathematical model and algorithm proposed above, the computational program has been coded on a personal computer. Three example systems (case 1, six-bus single-stage system; case 2, 19-bus four-stage system; and case 3, 71-bus two-stage system) are used to demonstrate the usefulness and effectiveness of the suggested multi-stage transmission network planning approach based on improved GA.

4.2.3.1 Case 1

The single-stage transmission network planning of the six-bus system [8] is tested using the developed program based on the above model and improved GA. The number of candidate lines is 22. By keeping excellent seeds, a group of solutions are obtained simultaneously to provide the planner with further choice. Figure 4.5 shows four suboptimal schemes as well as two optimal schemes.

When the population size is 60, Table 4.6 gives the relationship between the generations at which the optimal scheme first appears and the number of excellent seeds (N_k) to be kept.

From Table 4.6 we can see that when N_K is in the interval between 6 and 10, i.e. when N_K/N_{pop} = 10–15%, the appearance of the optimal solutions is earlier than in other cases. Especially, if $N_K = 0$, i.e. we do not use the measure of keeping excellent

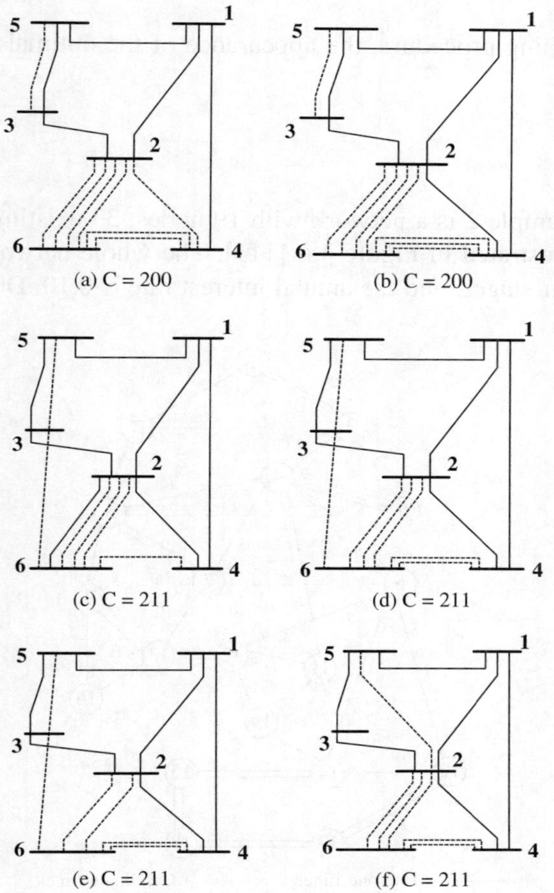

Figure 4.5 Planning solutions of six-bus system, in case 1. *Source*: Wang 2001. Reproduced with permission from IEEE.

Table 4.6 Results of keeping excellent seeds changing.

N_k	First generation for scheme (a) to appear[a]	First generation for schemes (a–d) to appear[a]
0	42	>50
2	17	21
4	21	33
6	12	16
8	13	14
10	15	19
12	21	>50

Note:
[a]Schemes refer to those in Figure 4.5.

seeds in the searching procedure, the appearance of the optimal solutions becomes very slow.

4.2.3.2 Case 2

The numerical example 2 is a network with 19 nodes, 33 existing lines and 21 candidate lines, as illustrated in Figure 4.6 [111]. The whole network planning period is divided into four stages, and the annual interest rate is 0.10. Detailed data for this

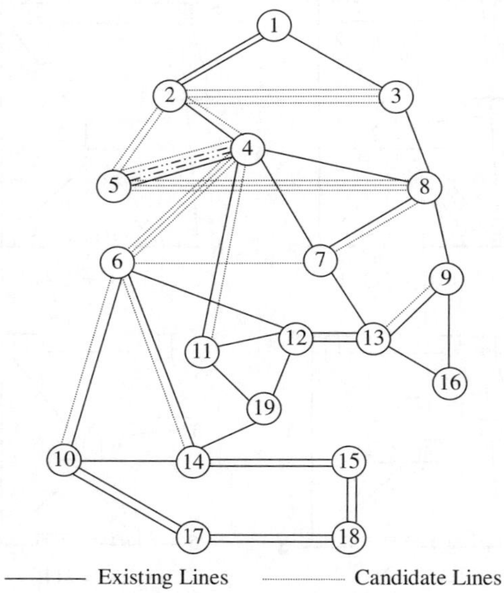

Figure 4.6 System structure of case 2. *Source*: Wang 2001. Reproduced with permission from IEEE.

example system are given in Tong *et al.* [111]. An optimal result with present value of investment 662.62 is obtained therein based on a critical feasible configuration matching method.

In this chapter, one optimal and four suboptimal schemes are obtained by the proposed IGA as shown in Table 4.7. We can see that the present values of investment of these planning schemes are 649.92, 650.92, 653.23, 653.53, and 653.89. Each of these planning results is better than the result given by Tong *et al.* [111]. The four stages of the expanding procedure of the optimal scheme in Table 4.7 is shown in Figure 4.7.

In order to illustrate the convergence of the proposed GA model, we give the following two indices. The first index is the convergence rate. It represents the percentage of times the GA can get the optimal solution when it began from different original states. For this example, we started from 50 different original states and, through GA searching, the optimal solution with the cost 649.92 appeared 37 times, so the convergence rate is about 74%. That means the GA can converge to an optimal solution with a larger probability. By the way, in the 50 searching procedures, the optimal solution with highest cost is 659.47. It is still better than the solution (662.62) given by Tong *et al.* [111].

The second index is the average value of the optimal solution in each generation for the 50 different original states searching procedures. Figure 4.8 shows the average values when the total number of generations is 400. From this we can see that almost all searches are convergent before 100 generations and it shows satisfactory convergence.

Table 4.7 Optimal and suboptimal schemes of case 2.

Scheme	Optimal	Subopt. 1	Subopt. 2	Subopt. 3	Subopt. 4
Stage 1	4–5 4–6 5–8 5–8	2–4 4–5 4–6 5–8 5–8	4–5 4–6 5–8 5–8	4–5 4–6 5–8 5–8	4–5 4–6 5–8 5–8
Stage 2	5–8 2–4	5–8	2–3 2–3 2–5	2–4 5–8	2–4 4–6 5–8
Stage 3	2–3 2–3 2–5 4–6	2–3 2–3 2–5 4–6	2–4 4–6 5–8	2–3 2–3 2–5 4–5 4–6	2–3 2–3 2–5
Stage 4	2–3 4–5	2–3 4–5	2–3 4–5	2–3	2–3 4–5
Cost	649.92	650.92	653.23	653.53	653.89

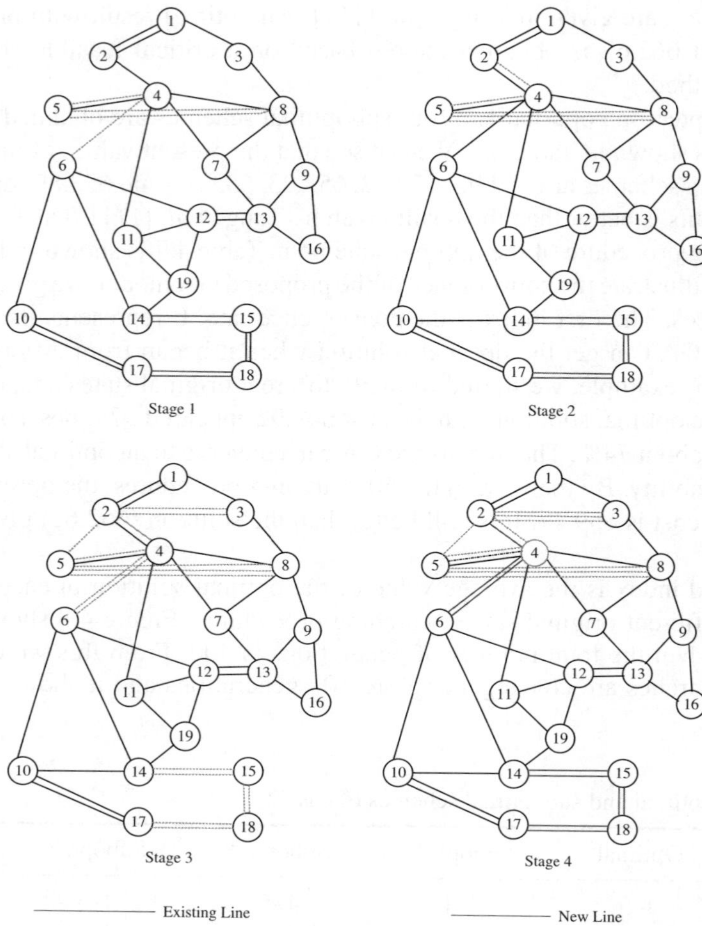

Figure 4.7 Optimal expanding procedure of case 2. *Source*: Wang 2001. Reproduced with permission from IEEE.

4.2.3.3 Case 3

The data of example 3 is a Brazilian system supplied by Dr. R. Romero [114, 115]. With a little simplification, the system has 71 buses, 104 branches, and 159 candidate lines. There are two planning stages with terminal years 2002 and 2008, and the annual interest rate is 0.10.

Some optimal and suboptimal solutions are obtained by the author's work. Their present values of investment are 2493.17, 2516.11, 2525.18, 2525.88, and 2525.98, respectively.

Table 4.8 gives the expanding scheme of the optimal solution in the two stages. Its total investment is 2889.56. The value of the static cost 2889.56 is larger than the value 2600.59 reported by Gallego *et al.* [115], which is obtained by 2008

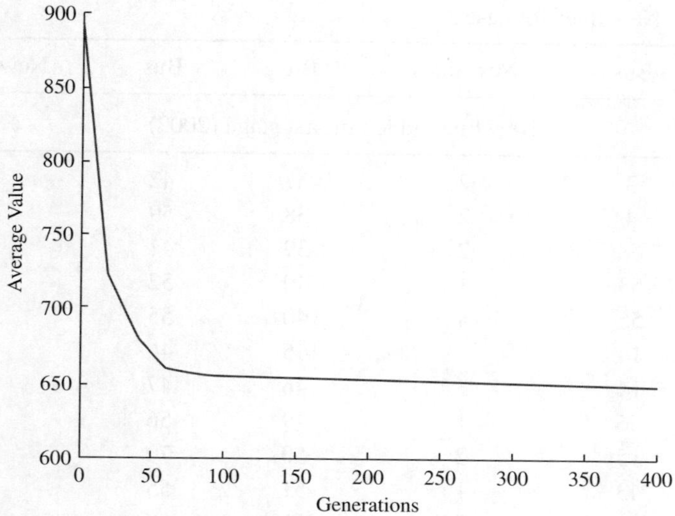

Figure 4.8 Average convergence curve of case 2. *Source*: Wang 2001. Reproduced with permission from IEEE.

horizon year planning. Through DC flow calculation, we see that the scheme given by Gallego *et al.* [115] does not satisfy the transmission demand of year 2002 because the load distribution in year 2002 is quite different from that in year 2008. The result verifies the necessity for multi-stage planning to solve practical problems.

4.2.4 Summary

The improved genetic algorithm approach is presented for multi-stage transmission network planning. The overload constraint in normal conditions is considered based on a DC power flow model. Some efficient improvements are applied to speed up convergence. The simulated annealing technique to scale fitness is also applied in GA to further speed up its convergence and improve the GA's ability to find the global optimal solution.

The test results on three example systems show that the proposed approach is very suitable for solving transmission network planning problems. For the example of the well-known six-node system, almost all of the optimal and suboptimal solutions have been found. For the example of the 19-node system, the algorithm finds one optimal and four suboptimal solutions. All of these planning solutions are better than the solution given by the critical feasible configuration matching method. Example 3 is a real system, the result of which verifies the necessity for multi-stage planning to solve practical problems. A lot of experimental calculations illustrate that the proposed GA model can converge to an optimal solution with a very large probability.

Table 4.8 New lines for case 3.

Bus	Bus	New lines	Bus	Bus	New lines
		New lines added in first stage (2002)			
2	3	2	37	42	1
3	4	2	38	59	1
3	68	2	39	41	1
4	53	1	40	52	2
4	55	4	40	55	3
11	14	1	45	46	1
12	14	3	46	47	4
13	56	1	49	56	1
14	15	2	50	70	1
14	43	1	51	55	1
15	41	5	51	58	1
17	47	7	53	54	2
17	66	4	57	69	2
20	54	2	57	70	1
21	55	1	58	59	1
23	40	2	60	61	1
24	52	2	60	62	1
29	30	1	60	63	3
29	58	2	63	64	1
32	48	1	64	65	1
33	43	1	65	66	1
36	70	1	67	68	1
		New lines added in second stage (2008)			
1	2	1	29	30	1
3	4	3	33	43	1
3	68	1	36	70	1
4	53	1	37	43	1
12	14	1	38	59	2
14	15	1	40	52	1
14	43	1	49	56	1
15	41	1	50	70	1
17	47	5	57	69	1
17	66	1	60	62	1
20	54	1	62	71	1
21	55	1	65	66	1
24	52	3	65	67	1
25	28	1			

4.3 Transmission Planning Based on Ordinal Optimization

4.3.1 Introduction

Transmission planning aims at strengthening an existing transmission network to serve power producers and customers in an optimal way. Owing to the large-scale nature of a transmission system and its inherent complexities, transmission planning has always been a complex problem. It has been formulated as a large-scale mixed integer nonlinear optimization problem. Various optimization techniques have been used to solve the problem, such as linear programming [8, 117], dynamic programming, nonlinear programming [112, 118], and mixed integer programming [119–122]. At the same time, various divide-and-conquer strategies such as Benders' decomposition [123–125], hierarchical decomposition, and the branch-and-bound algorithm have also been applied to solve large-scale transmission planning problems. A more complete literature survey can be found in Lee *et al.* [126].

Industry restructuring in recent years has resulted in the separation of generation and transmission systems and the introduction of competitive electricity markets. A comprehensive methodology, called the transmission economic assessment methodology (TEAM), was developed by the California Independent System Operator and London Economics International to evaluate the economic benefits of transmission expansion [127]. TEAM advocates the use of simulations for a large number of scenarios.

The economic impacts of alternative transmission enhancement schemes are different for different market participants. The participants of an electricity market include independent power producers, large customers, transmission network owners, and the independent system operator. The interests of different parties vary a great deal. Furthermore, other factors, such as uncertainties in generation patterns, transmission congestions, and regulatory policy changes, need to be considered in the evaluation. Transmission expansion planning in the restructured industry has become much more complicated than before [128]. No simple mathematical model can capture all the major factors in the transmission expansion planning. Computer simulation, in particular Monte Carlo simulation, has become the only viable approach for assessing alternative plans for transmission expansion.

A large number of scenarios and operating conditions, as well as uncertainties regarding contracts and bidding, must be considered in planning. Simulation-based transmission planning of such a stochastic system for practical large networks will require computational resources that are commonly unavailable.

If, on the other hand, analytical approaches can be used to complement the simulation-based search methods so that the search for optimum performance can be narrowed down to a set of *good enough solutions*, then the computational requirements may be manageable. Ordinal optimization (OO) [129] is a method that provides a theoretical foundation for such an approach.

4.3.2 *Transmission Expansion Planning Problem*

We assume that there is an entity, which may be the transmission company, the independent system operator (ISO), or the regional transmission organization (RTO), that is responsible for planning the expansion of the transmission network (i.e. when and where to install new lines, capacities, and types, etc.). The economic effects of transmission planning on various market players are different. Transmission owners are concerned about their investment returns, generation companies about congestion rents affecting their profits, the system operator about congestion revenues, and consumers about electricity prices after network enhancement. On the other hand, societal outage costs may very well be reduced after adding line capacities in the network. The magnitudes of the economic effects on generators, consumers, the system operator, and society depend on system operating conditions, which change from moment to moment. The benefit of an expansion scheme to an individual participant may take several years to realize. To simulate the effects of multiple years of an expansion plan on market players, an hourly based dispatch model is more or less necessary for the whole expansion time horizon.

Industry restructuring is an ongoing process in many parts of the world. Its initial focus has been on the competitive market for generation. Transmission systems remain largely regulated, and the rules and regulations for transmission expansion are mostly unsettled. The obligations and responsibilities of the participants are not fully defined, except that some general characteristics can be detected. Transmission planning has become an optimization problem whose variables are strongly stochastic and lumpy (discrete), with multiple participants having different objectives. Traditional optimization formulations and techniques are no longer appropriate. In this chapter, we propose the application of ordinal optimization to the transmission planning problem. The development of OO is motivated by the complexities of large-scale, stochastic, discrete-event nonlinear dynamic systems, such as manufacturing systems, whose performance can only be evaluated by way of computer simulations.

Because the transmission planning problem is not fully standardized and specified, we will not attempt to give a definitive algorithmic solution to the problem. Instead, our goal is to demonstrate that the OO approach is viable. Therefore, our formulation of the transmission planning problem and its solution algorithm are for illustration purposes only. For that purpose, and for the ease of exposition and understanding, the classical transmission planning formulation that has been commonly used in the past decades is used in this chapter. The OO approach can be adopted by planners to a formulation that incorporates the issues and considerations relevant to individual systems and to the development of models and performance indices that the OO approach requires.

4.3.2.1 Classical Transmission Planning Model

For simplicity, we use the classical optimization model in this section to represent the multi-year transmission planning problem. We assume that the objective function of the exact transmission planning model is to minimize the total cost, which is formulated as

$$\min T = I + G + L \tag{4.36}$$

where T represents total cost; I represents the investment cost; G represents production cost under the optimal dispatching condition; and L represents the cost of loss of load as a result of contingencies.

We assume that the expanded line capacities are added yearly in a Y-year span and that the rights of way planned and authorized for building new lines are already specified. The transmission planner has the option to decide on which rights of way to use in order to build new lines and their capacities. Each combination of the lines built in one year is called a "transmission expansion scheme" for that year. The annual planning schemes over the span of Y years are illustrated in Table 4.9. Assume that there are M rights of way authorized for building new lines, and that the number of lines built on the right of way m ($m = 1, \dots, M$) in year y ($y = 1, \dots, Y$) is Line_{my}. Each column in the table represents the expansion scheme of the year. The whole table represents the expansion scheme of the whole planning span.

1. *Investment cost.* The investment cost is calculated as the total expansion investment over the planning span. It is formulated as

$$I = \sum_{y=1}^{Y} \frac{I^y}{(1+\gamma)^{y-1}}$$

$$I^y = \sum_{m=1}^{M} I_m^y \tag{4.37}$$

where symbol I_m^y represents the expansion investment on the mth right of way in year y; I^y is the investment cost in year y; γ is the discount rate; and I is the investment value corresponding to the beginning year of the investment. The net present value (NPV) approach is used here.

Table 4.9 Planning schemes in Y-year transmission expansion planning.

Right of way m	Year 1	Year 2	…	Year Y
Right of way 1	Line_{11}	Line_{12}	…	Line_{1Y}
Right of way 2	Line_{21}	Line_{22}	…	Line_{2Y}
⋮	⋮	⋮	⋱	⋮
Right of way M	Line_{M1}	Line_{M2}	…	Line_{MY}

2. *Production cost.* The commonly used quadratic function (4.38) is used to represent the production cost function of a generator i. The optimal hourly production cost G_h can be obtained by minimizing the total generation cost for each hour $h = 1, \dots, 8760$, subject to the power balance (DC load flow) and other operating constraints:

$$\begin{aligned} G_h &= \min \sum_i (a_i g_i^2 + b_i g_i + c_i) \\ \text{s.t.}\quad g_i - d_i &= \sum_{j=1}^{N} B_{ij}\theta_i \\ &\sum_i g_i = D_h \\ &\underline{l_{ij}} \le B_{ij}(\theta_i - \theta_j) \le \overline{l_{ij}} \\ &\underline{g_i} \le g_i \le \overline{g_i} \\ &i, j = 1, \dots\ N \end{aligned} \tag{4.38}$$

Here g_i represents the real power generation of the generator on bus i; a_i, b_i, and c_i are the constant coefficients of power generation; D_h is the total load in hour H; B_{ij} represents the susceptance between node i and j; θ_i represents the phase angle of node i; $\underline{l_{ij}}$ and $\overline{l_{ij}}$ represent the upper and lower limits of line ij; $\overline{g_i}$ and $\underline{g_i}$ represent the upper and lower limits of g_i; and N is the total number of nodes. For year y, the total production cost G^y is the sum of G^h for 8760 h:

$$G^h = \sum_{h=1}^{8760} G_h \tag{4.39}$$

The NPV of the production cost for the whole planning span can be represented as

$$G = \sum_{y=1}^{Y} \left(\frac{G^y}{(1+\gamma)^{y-1}} \right) \tag{4.40}$$

3. *Cost of loss of load.* In abnormal operation conditions, a good network configuration may help to avoid load shedding. Providing a reliable network under system contingencies is one of the goals of transmission expansion planning. The dispatch modes are different for different transmission network configurations.

   The loss of load cost (LOLC) depends on the cost of load shedding due to a system contingency. The LOLC can be calculated using one of the three loss-of-cost functions: exponential, quadratic, and hyperbolic functions [130]. In this chapter, the quadratic form $\varphi(\cdot)$ as in (4.41) is used to represent LOLC,

$$\varphi_{n,h}(\Delta) = \left(\frac{4\xi}{D_h^n} \right)^2 \Delta^2 \tag{4.41}$$

where $\varphi_{n,h}$ represents LOLC at node n in hour h; Δ represents the amount of lost load, $\Delta = (D_h^n - d^n)$, where D_h^n is the load at node n in hour h; d^n is the load at node n after the contingency; and ξ is a rough estimated cost assuming that half of the total load is lost. In Europe, ξ is usually assumed to be 30–60 times the regular electricity price [130]. The value of LOLC caused by the contingency in hour h, L_h, can be formulated as

$$L_h = \sum_{n=1}^{N} \varphi_{n,h}(D_h^n - d^n) \tag{4.42}$$

For the given probabilities of all contingencies, we can obtain the expected hourly LOLC for all the contingencies considered. The expected hourly LOLC, EL_h, is formulated as

$$EL_h = \sum_{k=1}^{K} p_k L_{h,k} \tag{4.43}$$

where K is the size of the contingency set considered; and p_k is the probability of contingency k. The yearly LOLC can be formulated as the summation of EL_h, as

$$L^y = \sum_{k=1}^{8760} EL_h. \tag{4.44}$$

The total LOLC over the whole planning horizon can be obtained from

$$L = \sum_{y=1}^{Y} \left(\frac{L^y}{(1+\gamma)^{y-1}} \right). \tag{4.45}$$

To solve the proposed transmission planning model, (4.36)–(4.45), both the normal and abnormal operation modes of all hours of the whole planning span need to be calculated. Obtaining the exact solution for a multi-year transmission planning problem will result in a huge computational burden.

4.3.3 Ordinal Optimization

The ordinal optimization theory developed by Ho *et al.* [128, 131–139] is for solving simulation-based complex optimization problems. It has recently been applied to many areas in power systems, such as optimal power flow (OPF) with discrete control [140] and bidding strategies of power suppliers in markets [141]. In this chapter, the theory is applied to solve the multi-year transmission expansion planning problem.

The goal of ordinal optimization (OO) is to find good enough solutions for a complex optimization problem. A good enough subset $G(\Theta)$ is the subset consisting of the top best solutions, say, the top 5% in the solution space. However, it is difficult to find the subset G for a simulation-based problem unless all the solutions in the solution space are calculated and compared. The OO method uses rough estimates from a crude model to rank the solutions. However, even with the use of a crude model, estimation of performance values for all solutions of a large solution space Θ may not be computationally feasible. Ordinal optimization theory uses a representative set with N samples, Θ^N, to represent the original solution space. If the elements of the representative set are randomly selected, the probability of an event where at least one of the N samples will fall within the top 5% of the whole solution space is

$$P = 1 - (1 - 0.05)^N.$$

If $N = 1000$, then the probability that the top 5% good enough solutions are not in the $N = 1000$ samples is $P = 0.95^{1000} = 5.29 \times 10^{-23} \approx 0$. In this chapter, we will use $N = 1000$ samples to represent the solution space Θ. However, calculating all $N = 1000$ accurate solutions by computer simulations is still a formidable task. The goal of ordinal optimization is to reduce the number of necessary but computationally costly simulations.

4.3.3.1 Good Enough Subset and Selected Subset

Within the defined finite solution space ($\Theta^N = 1000$), the number of top $n\%$ solutions is $g = N \times n\%$. The top g solutions compose the "good enough" subset G of Θ^N. In the later parts of this section, g is selected to be $1000 \times 5\% = 50$.

The multi-year transmission planning problem can symbolically be represented as finding the minimum performance value T among all possible expansion schemes,

$$\min T = T(x) \tag{4.46}$$

where the variable x corresponds to a set of line expansion schemes. For example, building certain lines on the approved rights of way in each year of the Y-year span is referred to as an expansion scheme. Assuming there are 1000 potential expansion schemes $x_1, \ldots, x_{1000}$, then a high-quality solution could be found by calculating the performance values for 1000 schemes, $T_1, \ldots, T_{1000}$, and choosing the best one. Such an exhaustive search method is what we want to avoid. Instead, we reduce the search space to a small selected subset $S(\Theta^N)$ and perform simulations within the selected subset.

The key to the determination of a selected subset is to be sure that the selected subset intersects with the "good enough" subset. The selected subset may be determined by using certain fast evaluation methods, such as mathematical algorithms, heuristics, etc., based on the crude approximate models of the system.

For example, we may first use a crude model to obtain a rough estimate of the performance values $T'_1, \ldots, T'_{1000}$ for the potential expansion schemes $x_1, \ldots, x_{1000}$ in problem (4.46). The rough evaluation should take much less time than the accurate calculation. Thus, the top 50 schemes found by the crude model may not be the same as (i.e. not aligned with) the top 50 schemes (subset G) obtained from the accurate model. However, if we select enough schemes (subset S) from the rough estimates, the selected subset S will have a high probability of overlapping with the elements in G. The number of overlapped elements with G is called the *alignment level k*. Our goal is to find a subset S including at least elements of G. The probability of this event is called the *alignment probability p*.

For a given alignment probability p and the alignment level k, the size of the selected subset S is determined by the requirement that the probability that S overlaps the good enough subset G with at least k elements greater than p,

$$\text{Prob } |G \cap S| \geq k \geq p \tag{4.47}$$

Obviously, the determination of the size of the selected subset is dependent on the nature of the underlying optimization problem. The ordinal optimization theory broadly divides the optimization problems into several classes. The classification is accomplished by way of constructing an *ordered performance curve* to be introduced below.

4.3.3.2 Ordered Performance Curve

The ordered performance curve (OPC) may be constructed based on the estimated performance values obtained by the crude model. The 1000 estimated performance values are arranged in ascending order (for minimization problem). The X axis of the resulting plot is the scheme labels; whereas the Y axis represents the (estimated) performance values. The shape of an OPC determines the nature of the underlying optimization problem. The shapes of OPC curves can be broadly categorized into five classes, i.e. flat, U-shape, neutral, bell, and steep, as shown in Figure 4.9 [128, 131].

For a minimization problem, a smaller performance value means a good scheme, and a higher performance value means a bad scheme. For a particular problem, if more small-value schemes are found, then the problem has more good schemes. In Figure 4.9, the problem with a flat OPC has more good schemes. The five OPC curves in the figure represent five classes of optimization problems: (1) flat—many good schemes; (2) U-shape—many good and bad schemes; (3) neutral—good and bad schemes equally distributed; (4) bell—many intermediate schemes; and (5) steep—many bad schemes.

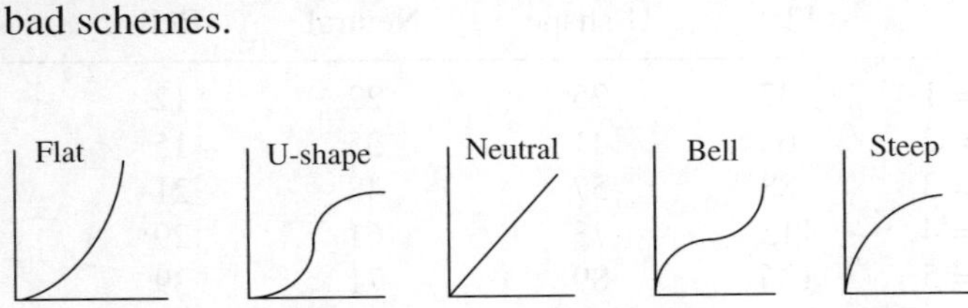

Figure 4.9 Five types of OPCs. *Source*: Xie 2007. Reproduced with permission from IEEE.

4.3.3.3 Size of Selected Subset

In Lau and Ho [131], through extensive simulations and statistical analysis, a formula is derived to relate the size of the selected subset S to (i) the shape of the OPC, (ii) the size of the good enough subset G, (iii) the alignment level, (iv) the alignment probability, and (v) the error bound w between the performance values from the crude model and the accurate model.

Assume the requirements for our transmission planning optimization problem are:

1. size of representative set $N = 1000$;
2. "good enough" subset G is defined as the top 5% solutions of Θ^N, thus $g = N \times n\% = 1000 \times 5\% = 50$;
3. alignment level of $G \cap S$ is $k = 1, 2, 3, 4$, or 5; and
4. alignment probability of $G \cap S$ is $p = 95\%$.

Assuming the error bound is 0.5, the size of the selected subset for the optimization problems with the five different OPC shapes are calculated based on the formula provided in Lau and Ho [131], and the results are as tabulated in Table 4.10. The size of the selected subset for the case $g = 10$ and $k = 1$ is also given in Table 4.10 for comparison.

From Table 4.10, we find that, for an optimization problem with a bell shape OPC curve, the size of the selected subset is $s = 12$ (for $k = 1$). This means that, after a rough estimation, if we pick the best 12 schemes from the rough estimation to run the exact evaluations, there is a 95% probability that at least one scheme ($k = 1$) out of the 12 will fall in the "good enough" subset G. If alignment level k is set to be 2, then 15 schemes need to be calculated for exact evaluations to guarantee that at least two schemes will fall in the good enough subset G with a probability of 95%.

Note from the rows of Table 4.10 that the size of the selected subset decreases from a flat shape to a steep shape, that is, an optimization problem with a flat OPC shape requires more selected schemes than a problem with a steep OPC curve. This is because, for a flat shape OPC, there are many schemes whose performance values

Table 4.10 Size of selected subset for five OPC-based problems.

		Ordered performance curve (OPC) shape				
		Flat	U-shape	Neutral	Bell	Steep
$g = 50$	$k = 1$	37	25	22	12	11
	$k = 2$	63	41	35	15	13
	$k = 3$	88	57	48	21	14
	$k = 4$	113	73	61	29	16
	$k = 5$	136	89	71	39	19
$g = 10$	$k = 1$	219	153	125	45	31

based on the crude model are more or less equally good (small). The exact performance value of each individual scheme is the estimated value plus the error in the estimation. The error terms are unknown. Adding such error terms will change the order of the ranking of the schemes that are flat (more or less the same performance) significantly. Therefore, more schemes need to be selected in order to capture sufficient good enough schemes.

On the other hand, if the OPC is steep, although fewer schemes out of the 1000 are needed for an exact calculation, the quality of the 1000 samples may be lower. In this case, it may be prudent to increase the initial sample from 1000 to a larger number in order to capture more "good enough" schemes.

4.3.4 Crude Model for Transmission Planning Problem

In this section, we propose crude models for the quick estimation of investment cost, production cost, and loss of load cost to be used in the ordinal optimization approach.

4.3.4.1 Investment Cost

The exact formula (4.37) for the investment cost does not require much computation effort; therefore, it can be used directly.

4.3.4.2 Crude Model for Production Cost

The evaluation of production cost given in (4.40) for a span of Y years requires simulations (solving OPF) for $Y \times 8760$ periods to be carried out, which constitutes a considerable computational burden. A crude model is needed for the rough estimation of production cost. For example, a possible crude model may use a seasonal typical-day dispatch mode and production cost to represent all the days in the season. The simulation times can thus be reduced from $Y \times 8760$ to $Y \times 4 \times 24 = Y \times 96$, almost a 100-fold reduction. In this section, the hat symbol $\hat{}$ is used to represent the rough estimate of the value from the crude model. The estimated production cost from the crude model can be expressed as

$$\hat{G} = \sum_{y=1}^{Y} \left(\frac{\hat{G}^y}{(1+\gamma)^{(y-1)}} \right) \tag{4.48}$$

4.3.4.3 Crude Model for Loss of Load Cost

A piecewise linear approximation can be used to give a rough estimate of the cost of loss of load, as expressed in

$$\hat{L} = \sum_{y=1}^{Y} \left(\frac{\hat{L}^y}{(1+\gamma)^{(y-1)}} \right) \tag{4.49}$$

In the approximation, a few load points between peak load and off-peak load are chosen. For example, we can select the base load level and the medium load level. The selected load points divide the load curve into several pieces. By calculating the LOLC of the selected load levels, we can use the linear approximation to roughly calculate the LOLC for all load levels.

4.3.4.4 Crude Model of Transmission Planning Cost

By combining (4.37), (4.48), and (4.49), a rough estimate of the total cost of the transmission planning problem (4.36) can be obtained as

$$\min \hat{T} = I + \hat{G} + \hat{L}. \tag{4.50}$$

The rough estimates obtained from (4.50), of course, are different from the accurate values obtained from the accurate simulation models (4.36)–(4.45). The difference is the error, which can be regarded as a random variable [129]

$$T = \hat{T} + \text{error} \tag{4.51}$$

The probability distribution of the random variable error is usually assumed to follow a uniform distribution $\mu(-\omega, \omega)$ [131]. An estimate of the error bound ω can be obtained by performing a calculation of a small number of schemes for both the exact and rough estimates of the performance values; ω will be used in the calculation of the size of the selected subset.

4.3.5 Example

The six-node Garver system is used to demonstrate the proposed ordinal optimization approach to the multi-year transmission planning problem. There are 11 rights of way available for transmission line expansions. The time span for the expansion is assumed to be five years.

The original six-node Garver system is shown in Figure 4.10. Table 4.11 has the basic data for the modified Garver system, Table 4.12 has the quadratic generation cost function ($), and Table 4.13 has the quadratic loss of load cost function ($).

The network configuration of the original Garver system and the numbers of the target expansion lines are given in Table 4.14.

Assume that the annual growth rate of the system peak load is 20% and the discount rate is 8%. The $N-1$ security criterion is used for the contingency analysis to calculate the loss of load cost. The outage probabilities of existing lines are set to be 0.01, while that of the new lines are assumed to be 0.005. It is forecasted that, at the end of the planning horizon, the system peak load will reach 1530 MW. The load curve of the seasonal typical day is chosen as the daily load curves for all days in the season.

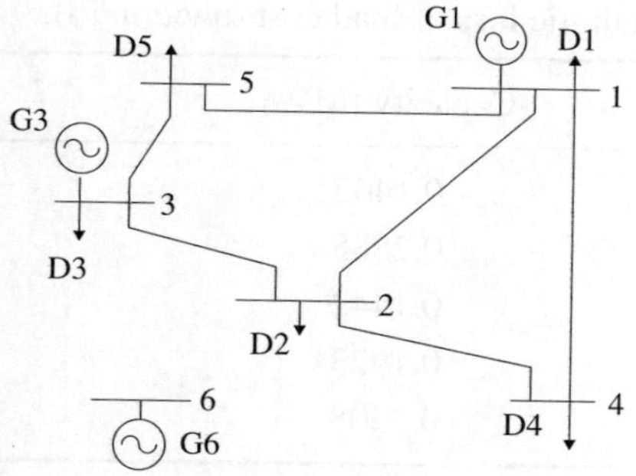

Figure 4.10 Original six-node Garver system. *Source*: Xie 2007. Reproduced with permission from IEEE.

Table 4.11 Basic data for the modified Garver system.

Right of way (from–to)	Reactance of each line in the right of way	Original number of lines in the right of way	Capacity of each line (MW)	Investment for new line ($\$10^4$/line)
1 (1–2)	0.40	1	100	40
2 (1–4)	0.20	1	100	60
3 (1–5)	0.20	1	100	20
4 (1–6)	0.38	0	100	68
5 (2–3)	0.20	1	100	20
6 (2–4)	0.30	1	100	40
7 (2–6)	0.30	0	100	30
8 (3–5)	0.20	1	100	20
9 (3–6)	0.38	0	100	48
10 (4–6)	0.30	0	100	30
11 (5–6)	0.21	0	100	61

Table 4.12 Quadratic generation cost function ($).

Generator	Capacity (MW)	Cost function
G1	600	$0.01g^2 + 20g + 150$
G3	400	$0.03g^2 + 30g + 180$
G6	800	$0.02g^2 + 25g + 100$

4.3.5.1 Create the Representative Set Θ_N

There are 746 496 possible ways to add new lines to the 11 rights of way. For five years, the number comes out to be 3 732 480. After eliminating infeasible schemes, the number will still be staggering. As suggested by the ordinal optimization theory,

Table 4.13 Quadratic loss of load cost function ($).

Load	Capacity (MW)	Cost function
D1	0.1442	$11.71\Delta^2$
D2	0.2885	$3.9\Delta^2$
D3	0.1442	$10.41\Delta^2$
D4	0.1923	$7.32\Delta^2$
D5	0.2308	$6.01\Delta^2$

Table 4.14 Original and maximum target number of lines.

Right of way	1	2	3	4	5	6	7	8	9	10	11
Original no. of lines	1	1	1	0	1	1	0	1	0	0	0
Max. target no. of lines	4	4	4	4	4	4	4	4	4	4	4

a representative set of 1000 samples may be used to represent the original solution space. The 1000 samples are selected randomly. However, each expansion scheme is tested for each year for feasibility (i.e. to see if it satisfies the yearly peak load requirement without violating operating constraints). If the requirement is not satisfied, the scheme will not be selected as a sample. The procedure is repeated until 1000 feasible samples are selected.

4.3.5.2 Ordered Performance Curve

Based on the crude model given in (4.50), the total cost of each sample scheme is calculated roughly. The ordered performance curve can be drawn based on the rough estimates, as shown in Figure 4.11. In comparison to the five OPC shapes in Figure 4.9, we find that the OPC curve of the proposed transmission planning problem is a bell shape curve.

To obtain the error bound ω, 30 samples are selected for accurate calculations. The standard deviation of the normalized error of the 30 samples is found to be $\sigma = 0.1701$. We use $2\sigma = 0.3402$ as the error bound and ω as one of the parameters in the determination of the size of the selected subset.

4.3.5.3 Size of Selected Subset

Based on the above-mentioned requirement, suppose we want to have at least k "good enough" solutions in the selected subset. We find that the size of the selected subset for bell shape OPC is $s = 12$ according to Table 4.10.

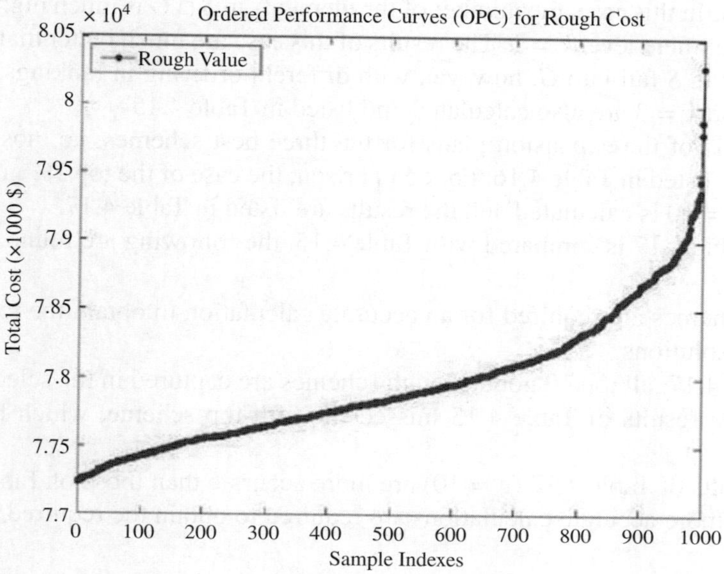

Figure 4.11 Ordered performance curve for rough estimation. *Source*: Xie 2007. Reproduced with permission from IEEE.

4.3.5.4 "Good Enough" Solutions

The top 12 schemes obtained by (4.50) are selected as belonging to the selected subset S. In Table 4.15, the index numbers of the schemes in S are listed according to the ascending order of the objective values $\hat{T}$. They are selected from $N = 1000$ samples. After calculating the 12 schemes accurately, it is found that the best solution in the selected set is scheme no. 865, which is ranked as the eighth by the rough estimation. The "good enough" subset G (top 5% of Θ^N) is calculated and listed in Table 4.15. The underlined index numbers are the schemes that fall in both S and G. As described in (4.47), there is a 95% probability that at least one ($k = 1$) element will fall in both G

Table 4.15 Planning schemes in the selected subset and "good enough" subset (subset G is top 1% of Θ^N).

Selected subset S	257	994	62	42	10	539	854	865	291	
($k = 1, S = 45$)	896	38	973	361	944	99	768	30	551	111
	166	470	572	860	809	247	656	934	189	
	621	450	256	643	233	377	198	569	919	
	442	332	683	878	13	322	565	476		
Good enough set G	865	854	291	42	189	994	257	62	539	
($g = 10$)	973									

and S, $S \cap G$. In this case, the number of the elements in $S \cap G$ is much higher than the required alignment level $k = 1$. The results of this case are much better than expected; all elements in S fall into G, however, with different ordering in rankings. The cases for $k = 2$ and $k = 3$ are also calculated and listed in Table 4.15.

The results of the expansion plans for the three best schemes, i.e. nos. 865, 854, and 291, are listed in Table 4.16. For comparison, the case of the top 1% good enough solutions $g = 10$ is calculated and the results are listed in Table 4.17.

When Table 4.17 is compared with Table 4.15, the following are found.

1. More schemes are required for an accurate calculation to obtain the top 1% good enough solutions.
2. In Table 4.17, all top 10 good enough schemes are captured in the selected subset; while the results of Table 4.15 missed the fifth top scheme, which has scheme no. 189.
3. The results of Table 4.17 ($g = 10$) are more accurate than those of Table 4.15($g = 50$) and more accurate calculations are required to obtain the required accuracy.

4.3.5.5 Comparison of Rough and Accurate Evaluations

For each of the 1000 sample expansion schemes, we carry out the accurate evaluation of (4.36)–(4.45) based on the exact simulation model. The rough estimated values $\hat{T}$ and the exact values T are plotted in Figure 4.12. The sample indices are ranked according to the ascending order of accurate simulation results (the smooth curve).

The errors between the rough estimates and the exact values are calculated according to the following:

$$\text{error (in \$10 000)} = \text{rough total cost} - \text{exact total cost}$$

$$\text{normalized error (\%)} = \frac{\text{error (in \$10 000)}}{\text{exact total cost}}$$

Figure 4.12 shows that (1) the errors arising from the calculations based on the crude model seem to be biased downward, and (2) the errors get larger when the cost is higher. The error bias is attributed to the way the crude model is constructed in which the congestion costs that occur mainly at peak loading were largely not captured. The high cost cases are mostly due to the contributions from the loss-of-load costs, which are more sensitive to whether exact operating constraints can be satisfied.

It should be pointed out that, though the OO theory assumes that errors are random, biased errors do not render the theory invalid. This is because the assumption concerning the error terms is used on two occasions. The first occasion is in the proof that the OPCs with or without errors (i.e. based on a crude or the exact models) belong to the same type (relative to the five types in Figure 4.9). Clearly, biased errors would

Table 4.16 Line expansions of scheme nos. (a) 865, (b) 854, and (c) 291 ($10 000).

(a) Right of way	Year 1	Year 2	Year 3	Year 4	Year 5	Total
1 (1–2)	1	1	0	1	1	4
2 (1–4)	1	1	0	0	0	2
3 (1–5)	1	1	0	0	1	3
4 (1–6)	0	0	1	0	0	1
5 (2–3)	1	0	0	0	1	2
6 (2–4)	1	0	0	1	1	3
7 (2–6)	1	1	1	0	1	4
8 (3–5)	1	1	0	1	0	3
9 (3–6)	1	0	1	0	1	3
10 (4–6)	1	1	0	0	1	3
11 (4–5)	1	1	1	0	0	3
I	36.900	24.167	17.747	7.938	16.759	103.51
G	11 529	13 354	15 606	17 207	19 479	77 175
L	46.99	1.241	5.6148	1.7241	3.4018	58.972
T	11 613	13 380	15 629	17 216	19 499	77 337

(b) Right of way	Year 1	Year 2	Year 3	Year 4	Year 5	Total
1 (1–2)	1	1	0	1	0	3
2 (1–4)	1	1	1	0	1	4
3 (1–5)	1	1	0	1	1	4
4 (1–6)	1	1	1	0	0	3
5 (2–3)	1	0	0	1	0	2
6 (2–4)	0	1	0	1	1	3
7 (2–6)	1	1	1	0	1	4
8 (3–5)	1	1	1	0	0	3
9 (3–6)	1	0	0	0	0	1
10 (4–6)	1	1	1	0	0	3
11 (4–5)	1	1	0	1	0	3
I	39.700	34.167	17.837	14.368	11.025	117.09
G	11 529	13 354	15 606	17 207	19 479	77 174
L	58.003	0.0000	0.0000	0.0000	0.0000	58.003
T	11 627	13 388	15 624	17 221	19 490	77 349

(c) Right of way	Year 1	Year 2	Year 3	Year 4	Year 5	Total
1 (1–2)	1	1	0	0	0	2
2 (1–4)	1	0	0	1	0	2
3 (1–5)	1	0	1	0	0	2
4 (1–6)	1	1	0	1	1	4
5 (2–3)	1	1	1	1	0	4
6 (2–4)	0	0	1	1	0	2
7 (2–6)	1	0	1	1	1	4

(*continued overleaf*)

Table 4.16 (*continued*)

(c) Right of way	Year 1	Year 2	Year 3	Year 4	Year 5	Total
8 (3–5)	1	0	1	0	1	3
9 (3–6)	1	0	1	1	1	4
10 (4–6)	1	1	0	0	0	2
11 (4–5)	1	0	1	0	1	3
I	39.700	14.630	20.490	21.116	16.685	112.62
G	11 529	13 354	15 606	17 207	19 479	77 174
L	58.003	9.4612	0.3151	0.0000	0.0000	67.779
T	11 627	13 378	15 627	17 228	19 495	77 355

Table 4.17 Planning schemes in the selected subset and "good enough" subset (subset G is top 5% of Θ^N).

Selected subset S	257	994	62	42	10	539	854	865	291	896
($k = 1, S = 12$)	38	973								
Good enough	865	854	291	42	189	994	257	62	539	973
set G	198	10	683	656	572	944	768	38	247	896
($g = 50$)	99	860	256	450	809	361	111	30	551	166
	621	383	440	233	934	442	345	710	643	332
	470	246	333	284	351	878	377	476	254	919
Selected subset S	257	994	62	42	10	539	854	865	291	896
($k = 2, S = 15$)	38	973	361	944	99					
Good enough	865	854	291	42	189	994	257	62	539	973
set G	198	10	683	656	572	944	768	38	247	896
($g = 50$)	99	860	256	450	809	361	111	30	551	166
	621	383	440	233	934	442	345	710	643	332
	470	246	333	284	351	878	377	476	254	919
Selected subset S	257	994	62	42	10	539	854	865	291	896
($k = 3, S = 21$)	38	973	361	944	99	768	30	551	111	166
	470									
Good enough	865	854	291	42	189	994	257	62	539	973
set G	198	10	683	656	572	944	768	38	247	896
($g = 50$)	99	860	256	450	809	361	111	30	551	166
	621	383	440	233	934	442	345	710	643	332
	470	246	333	284	351	878	377	476	254	919

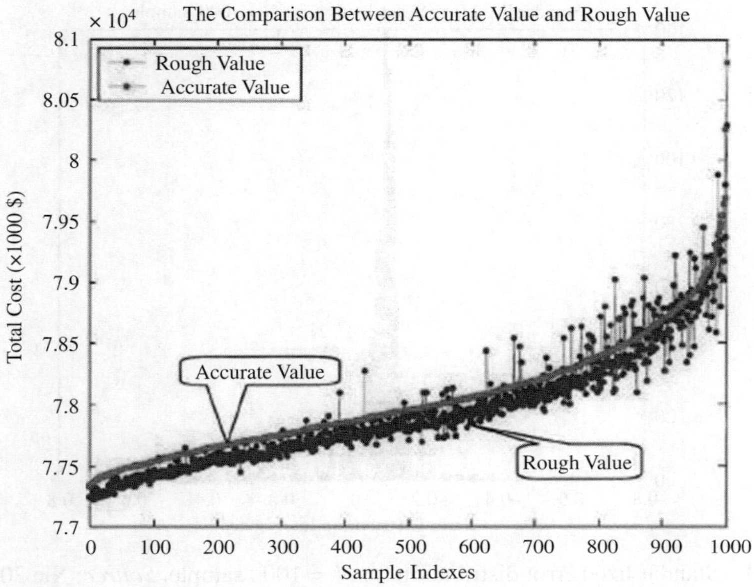

Figure 4.12 Ordered performance curve for the Garver system. *Source*: Xie 2007. Reproduced with permission from IEEE.

not change the shape of OPC either. The second occasion is when the error term is used in the selection of the size of the "selected subset S", where the "error bound" is one of the parameters in the selection formula. The error bound used in the theory was assumed to be "two-sided", whereas the biased error may be "one-sided". The use of a two-sided error bound in the formula in this case may be a little conservative (the size of S is unnecessarily large) when the errors are one-sided. In our example, errors seem to be biased. This may be part of the reason why our results are better (they capture all the good enough solutions, rather than just some of them) than the OO theory predicted.

The distribution of errors of the rough estimation is illustrated in Figure 4.13. Comparing the exact results and rough estimations of the 1000 samples, we found that scheme no. 75 has the maximum error. The error of this scheme is listed in Table 4.18. The maximum error is only 0.94%. The small errors between the crude model and the exact model may explain the reason why all elements in the selected subset in Table 4.15 are from the good enough subset.

The normalized errors for $N = 1000$ samples have a standard deviation of $\sigma = 0.0867$, which is in fact smaller than that of the 30 samples we used to determine subset size s.

The average time of exact computation of a sample scheme is about 1.04 h. We used three computers to calculate the 1000 samples. It took about 15 days to obtain the exact results for all 1000 samples on three computers. However, the ordinal optimization approach takes only 2.12 h on one computer. The computation time has been reduced

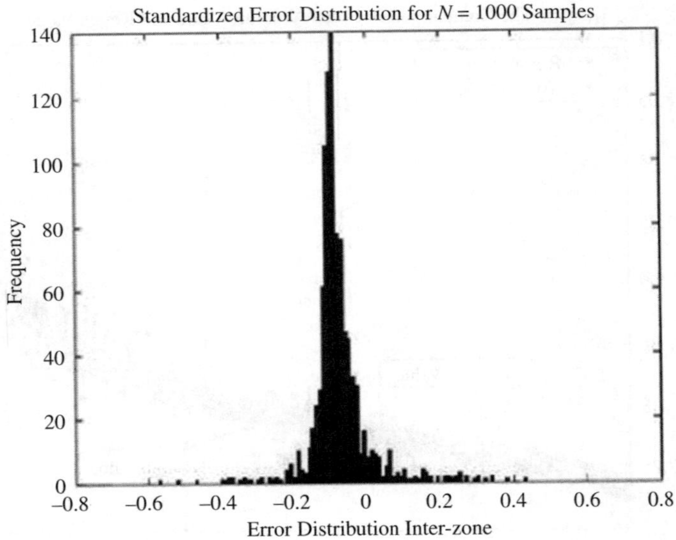

Figure 4.13 Standardized error distribution for $N = 1000$ sample. *Source*: Xie 2007. Reproduced with permission from IEEE.

Table 4.18 Error of scheme no. 75 ($10 000).

Scheme no.	Accurate cost	Rough cost	Error	Error (%)
75	79327.4	78579.5	–747.9	0.94

to about 0.5%. The computers used have a Pentium 4 microprocessor, a 2.4 GHz CPU, and 512 MB of memory.

4.3.6 Summary

Transmission expansion planning in the restructured industry is not standardized. However, its complexity makes simulation the only viable approach to evaluate the performance of alternative planning schemes. We submit that ordinal optimization can be effectively applied for the simulation-based multi-year transmission planning problem. The goal of this chapter is not to document how to apply the technique to a particular system. Rather, it is to demonstrate that the OO approach is effective in the sense that it is able to pick a "selected subset" in which sufficient "good enough" solutions can be found, and therefore simulations may be confined to the elements of a small selected subset. This was done based on the calculations made on a crude model. The classical transmission planning problem is used for demonstration. Furthermore, a brief and self-contained exposition of the ordinal optimization theory is presented. The approach can be adopted by planners with a proper selection of the crude and exact models of the specific transmission planning problem.

4.4 Integrated Planning of Distribution Systems Based on Hybrid Intelligent Algorithm

The problems of traditional energy shortage and environmental pollution are becoming more and more serious, and people are paying more attention to renewable energy generation and demand-side response (DSR) mechanism. DSR is very important for integrated planning of the power distribution network, especially the current large-scale access of distributed generation (DG). At present, the main research on DSR is on different pricing strategies of demand response technologies [142–149]. The price of the interruptible load is an important content of DSR. The power company and users signed the agreed price of interruptible load contract, and the power company can cut off part of the user's power load by making appropriate compensations in price. The load characteristics will be improved through determining the interruptible load and the interruption duration correctly.

Distribution network planning is a multi-objective integer programming problem, and with the addition of DSR and DG the planning dimension and the difficulty are increased. According to the planning problem considering DSR [150], the DSR model is solved by using a greedy randomized adaptive search procedure in transmission network planning. The game model of DSR based on a smart grid platform is researched by using equilibrium theory [151], and the demand response game market rule is given. The concept of a new energy power system and the overall planning scheme based on power, grid, and load response have been proposed [152]. On the DG location and capacity problem, the improved genetic algorithm is used to solve this optimization problem, but the computing speed of the genetic algorithm is slow and will converge to a local optimal solution sometimes. Three different generator models including wind power generation, photovoltaic power generation, and gas turbine power generation have been established [153], and the particle swarm optimization method has been adopted to determine the optimal capacity allocation for various different forms of power generation. However, power distribution network planning considering DG and DSR needs to be further studied.

In order to reflect the impact of DG and DSR on distribution network planning, the comprehensive planning framework based on DG and DSR (which here mainly adopts interruptible load management) is proposed in this section. The planning objective function is formed by taking into consideration the investment and operation cost of DG, transmission loss cost, compensation cost for interruptible loads, purchasing electricity cost, and the environmental benefits after integration of DG. The hybrid intelligent algorithm based on support vector machine (SVM) and particle swarm optimization (PSO) algorithm is used to get the best DSR plan and the position and capacity of DG. Finally, based on an actual 42-node distribution system in China, the economy and reliability of three schemes are compared to verify the effectiveness of integrated planning and the necessity of DSR.

4.4.1 Mathematical Model of Integrated Planning Based on DG and DSR

Stochastic chance-constrained programming (SCCP) has been proposed by Charnes and Cooper [154], with the remarkable characteristic that the constraints are satisfied at a certain confidence level. This mathematical model offers a powerful means of modeling stochastic decision systems, with the assumption that the stochastic constraints will hold at least α of the time, where α is referred to as the confidence level provided as an appropriate safety margin by the decision-maker.

The SCCP model is usually expressed as follows [142]:

$$\begin{cases} \min \bar{f} \\ \text{s.t.} \quad p_r\{f(x,\xi) \leq \bar{f}\} \geq \beta \\ \qquad p_r\{g_i(x,\xi) \leq 0,\ j = 1,2,\ldots,v\} \geq \beta \end{cases} \tag{4.52}$$

where x and ξ respectively present the decision vector and the random vector, $p_r\{\cdot\}$ presents the probability of an event, α and β are the confidence levels which decision-makers give in advance, and $\bar{f}$ is the minimum value of the objective function at the confidence level β.

4.4.1.1 Objective Function

The integrated planning for a distribution network covers network planning, DG planning, and DSR, and it is easy to fall into the "dimensionality". The planning objective function considers the investment and operation annual cost of DG, transmission loss cost, compensation cost for interruptible loads, purchasing electricity cost, and the environmental benefits after integration of DG.

The optimization objective [143] is

$$\min C = C_{loss} + C_{DSR} + C_{DG} - C_b - C_e \tag{4.53}$$

where C_{loss} is the annual loss cost, C_{DG} is the investment and operation annual cost of DG, C_{DSR} is the compensation cost for interruptible loads, C_b is the saving cost for purchasing electricity, and C_e is the environmental benefits.

1. *Annual loss cost*

$$C_{loss} = C_{ps} \times \sum_{i=1}^{k} (P_{loss\_i} \times \tau_{\max\_i}) \tag{4.54}$$

where C_{ps} is the unit electricity selling price, k is the number of branches in the distribution system, P_{loss_i} is the active power loss of the ith branch, and $\tau_{\max\_i}$ is the annual maximum load loss hours of the ith branch.

2. *Compensation cost for interruptible loads*

$$C_{DSR} = \sum_{i=1}^{n_{NSR}} P_{DSR\_i} \times T_{DSR\_i} \times (C_{ps} + C_{pi}) \tag{4.55}$$

where n_{NSR} is the number of interruptible loads, P_{DSR_i} and T_{DSR_i} respectively present the interruptible load and interrupt time of the ith interruptible load, and C_{pi} is the unit compensation cost for interruptible load.

3. *Investment and operation annual cost of DG*

$$C_{DG} = \sum_{i=1}^{N_{DC}} \left(\frac{a(1+a)^m}{(1+a)^m - 1} \times r_i \times P_{DG\_i} + W_{DG\_i} \right) \tag{4.56}$$

where N_{DC} is the number of DGs, a is the discount rate, m is the durable years of DG, r_i is the unit capacity cost of DG in the ith node, P_{DG_i} is the power capacity of DG in the ith node, and W_{DG_i} is the annual operation cost of DG in the ith node.

4. *Saving cost for purchasing electricity*

$$C_b = \left(\sum_{i=1}^{N_{DG}} P_{DG\_i} \times T_{DG\_i} + \sum_{i=1}^{N_{DSR}} P_{DSR\_i} \times T_{DSR\_i} \right) \times C_{pb} \tag{4.57}$$

where T_{DG_i} is the annual utilization hours of the ith DG, and C_{pb} is the unit feed-in tariff.

5. *Environmental benefits*

$$C_e = \left(\sum_{i=1}^{N_{DG}} P_{DG\_i} \times T_{DG\_i} + \sum_{i=1}^{N_{DSR}} P_{DSR\_i} \times T_{DSR\_i} \right) \times C_{pe} \tag{4.58}$$

where C_{pe} is the unit environmental cost of the thermal power plant—in general, the main pollutants of thermal power generation include sulfur dioxide, nitrogen dioxide, carbon monoxide, oxygen compounds, total suspended particulates, fly ash, and slag, etc.

4.4.1.2 Constraints

1. *Flow constraints*

$$\begin{cases} P_{DGi} - P_{Li} = V_i \sum_{j=1} V_j (G_{ij} \cos\theta_{ij} + B_{ij} \sin\theta_{ij}) \\ Q_{DGi} - Q_{Li} = V_i \sum_{j=1} V_j (G_{ij} \sin\theta_{ij} - B_{ij} \cos\theta_{ij}) \end{cases} \tag{4.59}$$

where P_{DGi} and Q_{DGi} are the active and reactive power injections in the ith node by DGs, P_{Li} and Q_{Li} are the active and reactive power loads in the ith node, V_i is the voltage amplitude of the ith node, G_{ij} is the conductance of the branch $i-j$, B_{ij} is the susceptance of the branch $i-j$, and θ_{ij} is the phase angle difference between the ith node and the jth node.

2. *Voltage constraints*

$$V_{\min} \le V_i \le V_{\max} \tag{4.60}$$

where $V_{\min}$ and $V_{\min}$ are the upper and lower limits of the voltage amplitude.

3. *Branch power flow constraints*

$$S_j \le S_j^{\max} \tag{4.61}$$

where $S_j^{\max}$ is the upper limit of the jth line capacity.

4. *Interruptible load constraints*

$$P_{Dmin\_i} \le P_{DSR\_i} \le P_{Dmax\_i} \tag{4.62}$$

$$T_{Dmin\_i} \le T_{DSR\_i} \le T_{Dmax\_i} \tag{4.63}$$

where P_{Dmax_i} and P_{Dmin_i} are the upper and lower limits of the interruptible load, and T_{Dmax_i} and T_{Dmin_i} are the upper and lower limits of the interruption duration annually.

5. *Installed capacity constraints of DG*

$$0 \le S_{DGi} \le S_{Li} \tag{4.64}$$

where S_{DGi} is the installed capacity of the ith DG, and S_{Li} is the maximum permissible installed capacity of the ith DG. DG will cause a power flow change, and it is necessary to limit the installed capacity of DGs in order to make this power flow change in a controllable range. The installed capacity of DG is not allowed to be larger than 20% of the maximum load of the node that DG accesses.

4.4.2 Hybrid Intelligent Algorithm

In recent years, the particle swarm optimization (PSO) algorithm has been widely used in power system optimization problems. But by using PSO for solving the chance-constrained programming problem, one needs to estimate the random function through the random simulation [155], so the estimated value of power loss under a variety of wind scenarios should be calculated through many different instances of power flow calculation. To make the estimates as accurate as possible, the number of simulations must increase, which makes the solution process very time-consuming. In addition, stochastic simulation is used to test the feasibility of the solution and calculate the fitness of each particle of the swarm in the process of iteration. When the number of particles and the number of iterations are large, the estimation process by using stochastic simulation will be very time-consuming. In practical application, the efficiency of the algorithm is a very important problem. In fact, what we need is the relative comparison of each solution. Calculating the exact value of the objective function is not necessary in the solution process, so we can seek an approximate estimation algorithm. As a new intelligent algorithm, support vector machine (SVM) has achieved good performance in function approximation [156]. In order to accelerate the solution process, the SVM model can be trained to replace the traditional power loss calculation.

Therefore, a hybrid intelligent algorithm is used to solve the integrated planning of distribution systems with DG and DSR, and the process is shown in Figure 4.14. The steps of the hybrid intelligent algorithm can be listed as follows:

1. Input the original data and get information about nodes, branches, and equipment parameters, then determine the limit of voltage, current-carrying capacity, DG installed capacity, and interruptible load.
2. Establish the mathematical model of integrated planning.
3. Under a series of wind scenarios based on the Weibull distribution [157, 158], generate some planning randomly and calculate the power transmission loss. Then these data form the training set for the approximation function.
4. Use the genetic algorithm (GA) to optimize the parameters of SVM kernel function, then train the SVM model and get the SVM approximation function.
5. Generate the initial particle swarm and test the feasibility of these initial particles.
6. Get the target value (power loss) of each particle by the SVM approximation function.
7. Calculate the fitness (overall cost of the planning scheme) of each particle based on the target value according to the objective function of integrated planning.
8. Update the velocity and position of the initial particle based on PSO.
9. Update the particle and test its feasibility.
10. Go back to step 6, and repeat until termination condition is satisfied.

4.4.3 Example

A practical 42-node distribution network is researched in this section, the wiring diagram for which is shown in Figure 4.15. The total length of all the lines is 26 km, the total number of users is 5974, including seven industrial users, two power plant electricity loads, and 5938 agricultural users or business users. A chemical plant whose annual peak load reaches 1737 kW is the largest industrial user in this distribution network, and the rest of the industrial electricity loads are less than 400 kW. Thus the chemical plant is chosen as an interruptible load for DSR.

Here assume that the unit compensation cost for interruptible load is 0.6 Yuan/kWh, the unit electricity selling price is 0.74 Yuan/kWh, the unit feed-in tariff of the thermal power plant is 0.48 Yuan/kWh, the unit environmental cost of the thermal power plant is 0.1171 Yuan/kWh [159], the unit capacity investment cost of a wind turbine

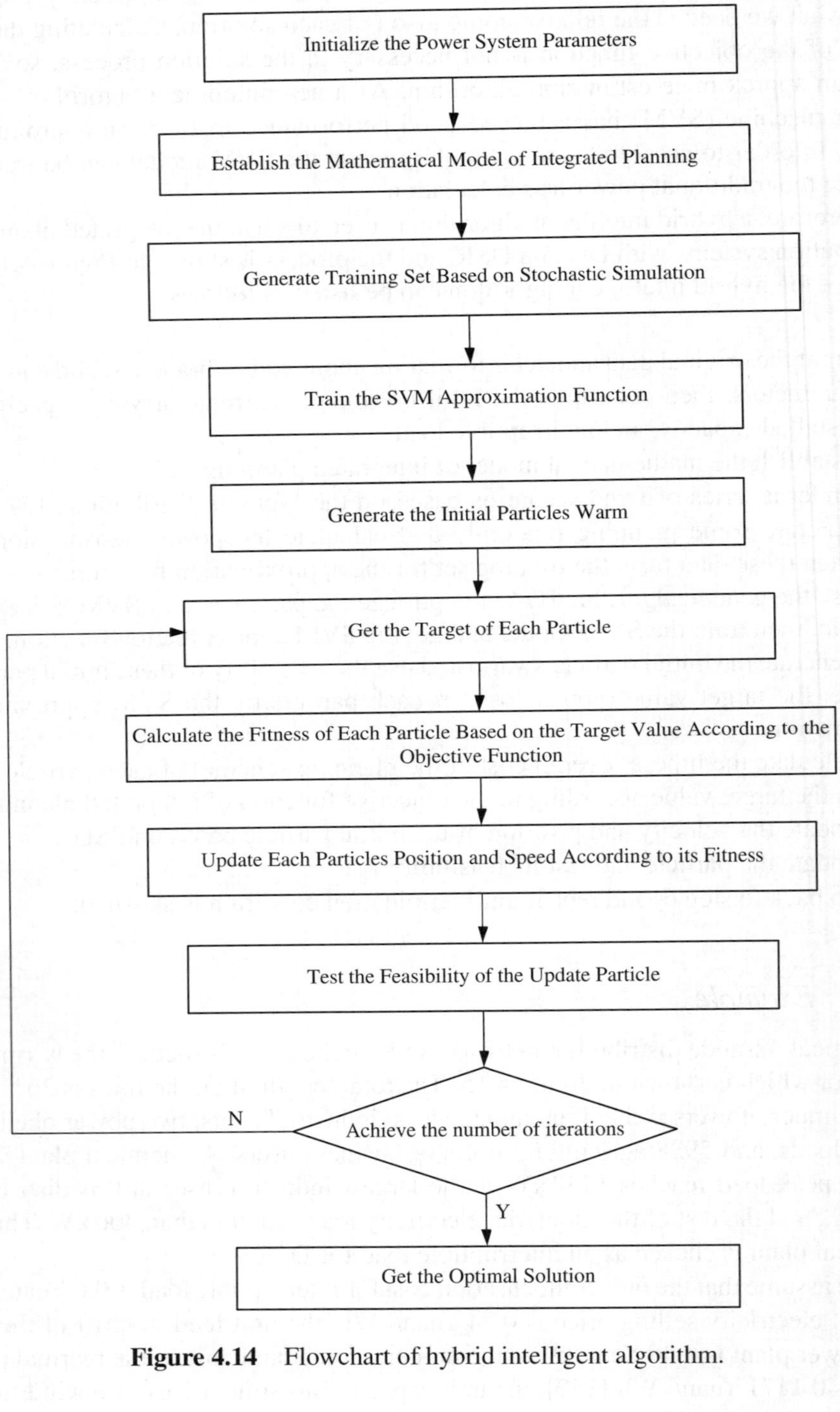

Figure 4.14 Flowchart of hybrid intelligent algorithm.

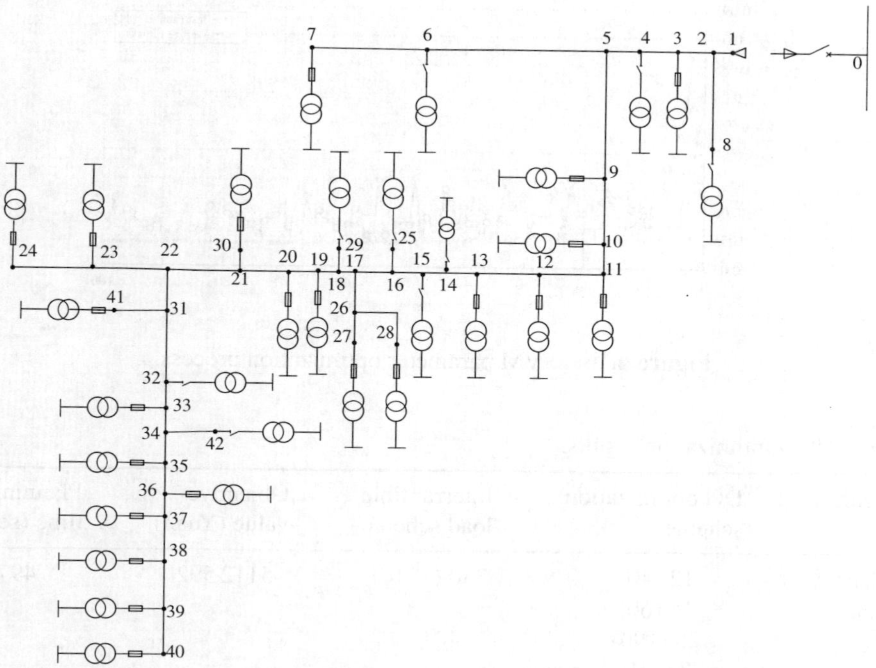

Figure 4.15 Diagram of the example distribution system.

is 6300 Yuan/kWh, the durable years of a wind turbine is 25 years, the discount rate is 0.1, the population size of PSO is 20, the iteration number is 500, the wind scenario simulation number of each training sample is 200, and the number of training samples for SVM is 2500.

The penalty parameter c and the kernel function parameter g have a great influence on the accuracy of the approximation function [160], so the parameters are optimized by GA. The population is 20, the iteration number is 200, the range of c is [0.1, 100], and the range of g is [0.01, 1000]. The parameter optimization process is shown in Figure 4.16. The optimal values of c and g are 1.0907 and 14.0419.

In order to test the accuracy and efficiency of the hybrid intelligent algorithm, the traditional method combined with stochastic simulation and PSO is used to solve the SCCP of integrated planning first. The number of random wind samples is 200, the confidence level of the objective function is 0.5, and the confidence level of voltage constraints is 0.95. In addition, according to the flowchart shown in Figure 4.14, the SCCP of integrated planning is solved with the hybrid intelligent algorithm. The optimization results of these two algorithms are shown in Table 4.19. In the DG optimization scheme, the value before the parentheses is the access node of DG, and the value within parentheses is the access capacity (kW) of DG. In the interruptible load scheme, the value before the parentheses is the interruptible load (kW), which is drawn

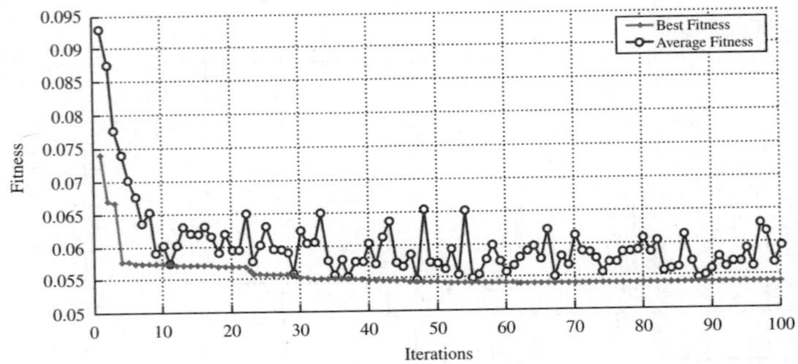

Figure 4.16 SVM parameter optimization process.

Table 4.19 Optimization results.

Algorithm	DG optimization scheme	Interruptible load scheme	Objective value (Yuan)	Planning time (second)
Stochastic scholar	12 (40), 15 (60), 25 (190), 33 (10)	330 (1210),	5112 592,	4978
Hybrid intelligent algorithm	10 (50), 13 (50), 25 (180), 30 (10)	360 (1190)	5119 780	1662

up by the power supply enterprise and the chemical plant, and the value within parentheses is the interruption duration (hour).

The optimization scheme solved with the hybrid intelligent algorithm can achieve valid effect, and the relative error of objective value is only 0.14%. The calculation error caused by the SVM approximation function is still in the tolerance range. At the same time, only few rounds of power flow calculation are needed for the SVM training samples. The planning time reduced from 4978 seconds to 1662 seconds due to avoiding the need for the power flow calculation for each particle of every generation in the PSO.

In order to research the economy and reliability of the integrated planning with DG and DSR, the hybrid intelligent algorithm is used to solve the planning schemes under three cases. The costs of each plan under the three cases are shown in Table 4.20.

Table 4.20 shows that not only does the cost of power loss reduce, but also the problem of overload and blocking is alleviated after integrated planning with DG and DSR. Especially in the case of the global energy shortage, the DG is an environmentally friendly green power, and the interruptible load measures are conducive to

Table 4.20 Costs (Yuan) of different planning schemes.

Cost	Traditional planning without DG or DSR	Planning with DG only	Integrated planning with DG and DSR
C	6863 894	5127 256	5119 780
C_{loss}	6863 894	5330 981	5305 620
C_{DSR}	0	0	257 040
C_{DG}	0	460 835	222 737
C_b	0	534 230	535 080
C_e	0	130 330	130 537

Table 4.21 Reliability of different planning schemes.

Reliability index	Traditional planning without DG or DSR	Planning with DG only	Integrated planning with DG and DSR
SAIFI (interruptions/ customer)	4.3613	4.1286	4.1086
SAIDI (hours/customer)	13.2209	10.5965	10.0965
CAIDI (hours/customer interruption)	3.031	2.567	2.457
ASAI (%)	99.85	99.88	99.88

energy-saving emission reduction work, and bring considerable environmental benefits. In addition, the DG equipment cost will gradually reduce with the development of the technology of DG. The economic advantages of integrated planning with DG and DSR will become more and more obvious.

The ETAP electrical engineering software is used to model the 42-node distribution network, and the results of reliability indices of the planning schemes are shown in Table 4.21.

According to Table 4.21, in the event of substation and/or some line faults, DG can supply power to the important loads to meet the electricity needs for short-term production and life, so that the power supply reliability rate increases. Moreover, on the premise of safe operation of the network and reliable power supply, the access number and capacity of DG will reduce by proper implementation of interruptible load management measures so that the impact of intermittent power output from DG can be avoidable. Then the reliability indices of the distribution network have been further improved.

4.4.4 Summary

The integrated planning model with DG and DSR is established, and the SCCP model based on stochastic simulation for DGs can avoid the conservative optimization results caused by deterministic constraints. In order to improve the computing speed, the hybrid intelligent algorithm combined with SVM and PSO is used in the integrated planning model. The optimization results show that the hybrid intelligent algorithm can save much time for power flow calculation and also achieve valid accuracy. The calculation results show that this algorithm is effective and feasible based on a practical 42-node distribution network. The economy and reliability of three planning schemes are compared to show that the integrated planning with DG and DSR can reduce the cost of network loss and bring about considerable environmental benefits. Then DSR can make up for the defects of DG intermittent output characteristics and improve the reliability of the power supply.

5

Power System Operation

5.1 Unit Commitment Based on Cooperative Coevolutionary Algorithm

Unit commitment (UC) is a consistent research interest of the electric power system because of its potential economic savings. This problem is quite difficult due to its inherent high-dimensional, nonconvex, discrete, and nonlinear nature. Many methods have been developed for solving the UC problem. Besides heuristic methods, the optimization-based methods for unit commitment include dynamic programming (DP) [161, 162], Lagrangian relaxation (LR) [17, 163], integer programming, and Benders' decomposition, etc. Among these methods, dynamic programming and Lagrangian relaxation are the most extensively used. DP searches the solution space consisting of the unit status combinations for an optimal solution. The time periods of the time horizon studied are known as the stages of the DP problem. Typically each stage represents one hour of operation. The combinations of unit status within a time period are known as the states of the DP problem. DP suffers from the "curse of dimensionality" because the problem solution space grows rapidly with the number of generating units to be committed. A framework of the LR method is presented in Guan et al [17]. The basic idea of LR is to relax the systemwide constraints such as the power and spinning reserve requirements by using Lagrange multipliers, then to decompose the problem into individual unit commitment subproblems, which are much easier to solve. LR can overcome the dimensional obstacle and get quite good suboptimal solutions. Another kind of noticeable newly emerging method is genetic algorithms (GAs). GAs are general-purpose optimization algorithms which apply the rules of natural selection and heredity to explore a given search space. GAs can adapt to the nonlinearities and discontinuities commonly found in power systems. Genetic-based optimization methods of UC have been presented [164, 165].

Power System Optimization – Large-scale Complex Systems Approaches, First Edition.
Haoyong Chen, Honwing Ngan and Yongjun Zhang.

A cooperative coevolutionary algorithm (CCA) is presented in this section, which combines the basic ideas of the decomposition–coordination optimization of a large-scale system with evolutionary optimization. The basic idea is to relax the systemwide constraints of the primal problem and form the dual problem, then to decompose the dual problem into a number of subproblems. A subgradient-based stochastic optimization method is used to solve the dual problem. For the low-level subproblems, which are nonconvex, discrete, nonlinear, and with lower dimension, the GA is used. CCA can manage very complicated time-dependent constraints. This method is akin to De Jong's coevolutionary framework [67], and so named after it. Test results show that CCA has good convergence properties and can obtain high-quality solutions. Compared with GA, CCA exhibits significant improvement in computational efficiency.

5.1.1 Problem Formulation

Consider a thermal power system with I units. It is required to determine the start-up, shut-down, and generation levels of all units over a specified time horizon T. The objective is to minimize the total cost subject to system demand and spinning reserve requirements, and other individual unit constraints. The notation to be used in the mathematical model is first introduced:

T	time horizon studied, in hours;
I	number of thermal units;
$p_i(t)$	power generated by unit i at time t, in MW;
$x_i(t)$	state of unit i at time t, denoting number of hours that the unit has been ON (positive) or OFF (negative);
$u_i(t)$	decision variable of unit i at time t, 1 for up, 0 for down;
$C_i(p_i(t))$	fuel cost of unit i for generating power $p_i(t)$ at time t;
$S_i(x_i(t-1), u_i(t))$	start-up cost of unit i at time t;
$p_d(t)$	system demand at time t, in MW;
$r_i(t)$	spinning reserve contribution of unit i at time t, in MW;
$p_r(t)$	system spinning reserve requirement at time t, in MW;
$\overline{p}_i$	maximum generation level of unit i;
$\underline{p}_i$	minimum generation level of unit i;
$\overline{\tau}_i$	minimum up-time of unit i, in hours;
$\underline{\tau}_i$	minimum down-time of unit i, in hours;
Δ_i	maximum allowable change in generation between two consecutive hours;
$\overline{r}_i$	maximum spinning reserve contribution of unit i.

The UC problem can be formulated as the following mixed integer programming problem.

5.1.1.1 Objective Function

$$\min\ F = \sum_{t=1}^{T} \sum_{i=1}^{I} [C_i(p_i(t)) + S_i(x_i(t-1), u_i(t))] \tag{5.1}$$

5.1.1.2 Constraints

System demand

$$\sum_{i=1}^{I} p_i(t) = p_d(t), t = 1, 2, \cdots, T \tag{5.2}$$

Spinning reserve requirement

$$\sum_{i=1}^{I} r_i(t) \geq p_r(t), t = 1, 2, \cdots, T \tag{5.3}$$

Individual unit constraints include:
Unit generation limit

$$\underline{p}_i u_i(t) \leq p_i(t) + r_i(t) \leq \overline{p}_i u_i(t) \tag{5.4}$$

Minimum up-/down-time

$$u_i(t) = \begin{cases} 1 & \text{if} \quad 1 \leq x_i(t) < \overline{\tau}_i \\ -1 & \text{if} \quad -\underline{\tau}_i < x_i(t) \leq -1 \end{cases} \tag{5.5}$$

Ramp rate

$$-\Delta_i \leq p_i(t) - p_i(t-1) \leq \Delta_i \tag{5.6}$$

Minimum generation for the first and last hour

$$\begin{aligned} p_i(t) = \underline{p}_i \quad & \text{if} \quad u_i(t-1) = 0 \quad \text{and} \quad u_i(t) = 1 \\ \text{or} \quad & \text{if} \quad u_i(t) = 1 \quad \text{and} \quad u_i(t+1) = 0 \end{aligned} \tag{5.7}$$

Unit spinning reserve limit

$$0 \leq r_i(t) \leq \overline{r}_i \tag{5.8}$$

5.1.2 Cooperative Coevolutionary Algorithm

A power system is a typical large-scale system with a large number of variables and very complicated constraints. The mathematical models of power system optimal planning and operation are always high-dimensional, nonconvex, discrete, and nonlinear. Thus they are very difficult to solve by conventional methods. The decomposition–coordination technique of large-scale systems decomposes the

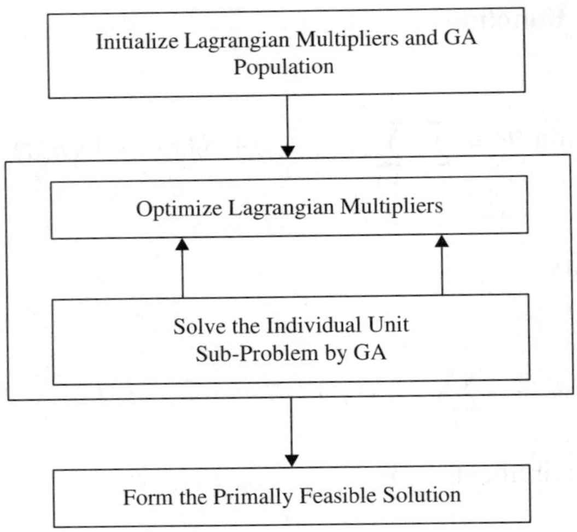

Figure 5.1 Framework of the cooperative coevolutionary algorithm. *Source*: Chen 2002. Reproduced with permission from IEEE.

complex system optimization problem into a set of simpler subsystem optimization problems and then coordinates them from the whole system's view. This method can overcome the dimensional obstacle. The GAs researched and applied extensively at present are general-purpose optimization algorithms. These methods are a highly efficient search strategy and have few requirements on the objective function. This chapter considers combining these two kinds of method: decomposing the large-scale system into a number of subsystems, and taking into account their connections and coordination, and forming a multi-level optimization problem, and then applying evolutionary optimization to the subsystems. This new algorithm is akin to De Jong's coevolutionary framework [67], which consists of a collection of independent subpopulations, each attempting to evolve subcomponents (species) which are useful as modules for achieving more complex structures. So we call this method the cooperative coevolutionary algorithm.

For the UC problem, the system constraints of the primal problem are relaxed by a Lagrangian function. Then a two-level optimization algorithm framework is formed. The low-level problems solve the optimal commitment of each individual unit, and GA is used. The high-level problem optimizes the Lagrange multipliers, and a subgradient-based stochastic optimization method is adopted. The framework of CCA is shown in Figure 5.1.

5.1.2.1 Dual Problem

The mathematical model of (5.1)–(5.8) is a mixed integer programming problem. To decompose the problem, we first relax systemwide constraints (5.2) and (5.3) and formulate the dual problem by the principle of duality:

$$L = \sum_{t=1}^{T} \left\{ \sum_{i=1}^{I} [C_i(p_i(t)) + S_i(x_i(t-1), u_i(t))] + \lambda(t) \left[p_d(t) - \sum_{i=1}^{I} p_i(t) \right] + \mu(t) \left[p_r(t) - \sum_{i=1}^{I} r_i(t) \right] \right\} \tag{5.9}$$

where $\lambda(t)$ and $\mu(t)$ respectively are Lagrange multipliers associated with demand and spinning reserve requirements at time t. For notational convenience, define

$$\lambda = (\lambda(1), \lambda(2), \cdots, \lambda(T))^{\mathrm{T}} \quad \text{and} \quad \mu = (\mu(1), \mu(2), \cdots, \mu(T))^{\mathrm{T}} \tag{5.10}$$

Using the duality theorem and exploiting the decomposable structure of (5.9), we can form a two-level maximum–minimum optimization problem. Given multipliers λ and μ, the low level consists of individual unit subproblems

$$\min L_i = \sum_{t=1}^{T} \{[C_i(p_i(t)) + S_i(x_i(t-1), u_i(t))] - \lambda(t)p_i(t) - \mu(t)r_i(t)\} \tag{5.11}$$

subject to (5.4)–(5.8). The high-level dual problem is

$$\begin{aligned}(P-D) \max_{\lambda,\mu} L(\lambda,\mu) &= \sum_{t=1}^{T} \left\{ \sum_{i=1}^{I} [C_i(p_i(t)) + S_i(x_i(t-1), u_i(t))] \right\} \\ &\quad + \sum_{i=1}^{T} \left\{ \lambda(t) \left[p_d(t) - \sum_{i=1}^{I} p_i(t) \right] + \mu(t) \left[p_r(t) - \sum_{i=1}^{I} r_i(t) \right] \right\} \\ &= \sum_{i=1}^{I} L_i^*(\lambda,\mu) + \sum_{t=1}^{T} [\lambda(t)p_d(t) + \mu(t)p_r(t)]\end{aligned} \tag{5.12}$$

subject to

$$\mu(t) \geq 0, t = 1, 2, \cdots, T \tag{5.13}$$

where $L_i^*(\lambda, \mu)$ is the optimal Lagrangian for problem (P_i) with the given λ and μ.

Although the primal problem is nonconvex, the dual problem can be easily proved concave. Therefore the dual problem has the unique maximum solution. Since the objective function of the dual problem is not differentiable, a subgradient algorithm is used to update λ and μ as follows [63]:

$$\begin{aligned}\lambda^{j+1} &= \lambda^j + \alpha_j g_\lambda \quad \text{for} \quad j = 0, 1, 2, \cdots \\ \mu^{j+1} &= \mu^j + \alpha_j g_\mu\end{aligned} \tag{5.14}$$

Here α_j is a scalar step size, g_λ is the subgradient of $L(\lambda, \mu)$ with respect to λ, and g_μ is the subgradient of $L(\lambda, \mu)$ with respect to μ:

$$g_\lambda = \left(p_d(t) - \sum_{i=1}^{I} p_i(t) \bigg|_{t=1,2,\cdots,T} \right)^{\mathrm{T}} \tag{5.15}$$

with $p_i(t)$ the optimal solution of (P_i), and

$$g_\mu = \left(p_r(t) - \sum_{i=1}^{I} r_i(t) \bigg|_{t=1,2,\cdots,T} \right)^{\mathrm{T}} \tag{5.16}$$

with $r_i(t)$ the optimal solution of (P_i). The subgradients are essentially the ascent directions of the objective function at certain points.

5.1.2.2 Solving Individual Unit Subproblems by GA

When solving the high-level dual problem, for any λ and μ, it is necessary to find the ascent directions g_λ and g_μ by (5.15) and (5.16) respectively, where $p_i(t)$ and $r_i(t)$ can be calculated by solving the subproblems.

The conventional Lagrangian relaxation method solves the subproblems by dynamic programming, which is convenient for handling discrete status constraints. But one obstacle of DP is the difficulty when dealing with time-dependent constraints. In an electrical power system, such constraints are quite common, for example, the ramp rate of a thermal unit. The usual way to solve this problem needs to introduce an additional set of multipliers to relax these constraints [17]. This method will increase the computational burden and duality gap, and even become invalid when the problem's feasible region is narrow.

To overcome these difficulties, GA is used in this chapter to solve the subproblems. The basic principle of GAs is the maintenance of a population of problem solutions that are encoded into "chromosomes" of individuals. These individuals evolve in time. The evolution is based on the laws of natural selection (survival of the fittest) and genetic information recombination within the population. The evolving population samples the search space and accumulates knowledge about good solutions and recombines this knowledge until some satisfactory solutions are found [166].

In a UC problem, we encode the variables of unit status into a binary digit string, with "1" denoting unit start-up and "0" denoting unit shut-down. Such coding may bring about some questions, for example, when two chromosomes of a feasible solution cross over, the new chromosomes may become infeasible because the minimum up-/down-time constraints are violated. Thus some special treatments of genetic operators, viz. crossover and mutation, are required here. For the ramp rate constraints, a direct adjustment approach combined with a penalty function is introduced. To solve the subproblems by dynamic programming, a unique ascent direction is calculated in every iteration—the convergent trajectory of the algorithm is deterministic. On the other hand, to solve the subproblems by GA, the ascent direction in every iteration is random—hence the convergent trajectory of the algorithm is stochastic. So CCA is essentially a stochastic optimization algorithm, and it is less likely to stick to a local optimum.

5.1.2.3 Solving Dual Problem by Subgradient-based Stochastic Optimization Method

In CCA, for any given λ and μ, GA is used to solve the subproblems and find the ascent directions g_λ and g_μ. Because the optimal solutions of GA are stochastic, the new values of λ and μ calculated by (5.14) and these ascent directions may be nonoptimal, namely, reducing the Lagrangian values. To solve this problem, this chapter introduces a subgradient-based stochastic optimization method. This method can be illustrated by Figure 5.2.

Supposing in the jth iteration, we obtain the Lagrange multipliers λ^j and μ^j, and use these multipliers to solve the subproblems, and then we get the Lagrangian value $\underline{L}^j$ and the subgradients $g_{\underline{\lambda}}$ and $g_{\underline{\mu}}$. The optimization process goes forward to the $(j+1)$th iteration, and the new values of Lagrange multipliers $\underline{\lambda}^{j+1}$ and $\underline{\mu}^{j+1}$ are calculated. Then we solve the subproblems again and obtain the Lagrangian value $\underline{L}^{j+1}$. Since $g_{\underline{\lambda}}$ and $g_{\underline{\mu}}$ are random, it may be that $\underline{L}^{j+1} < \underline{L}^j$, namely, the dual solution gets worse. In this case, the optimization process backtracks to the jth iteration, and the subproblems are solved again by using λ^j and μ^j. Then we get the Lagrangian value L^j and the subgradients g_λ and g_μ. The optimization process goes forward to the $(j+1)$th iteration, and the new values of Lagrange multipliers λ^{j+1} and μ^{j+1} are calculated. Then we solve the subproblems and obtain the Lagrangian value L^{j+1}. If $L^{j+1} > L^j$ this time, we can conclude that the optimization process is going in the right direction, and then the iteration proceeds. Otherwise the optimization process backtracks again.

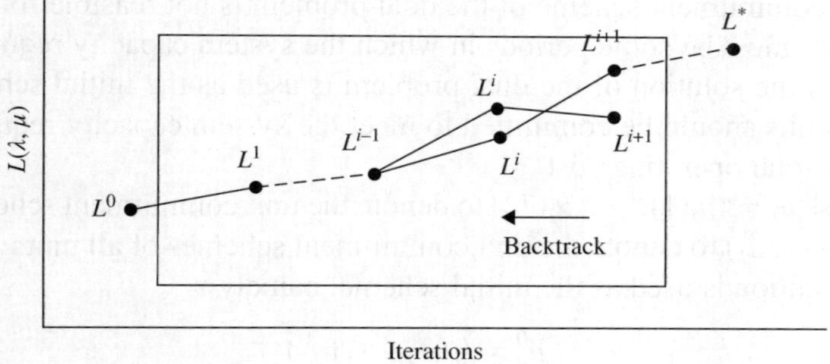

Figure 5.2 Stochastic optimization method for solving dual problem. *Source*: Chen 2002. Reproduced with permission from IEEE.

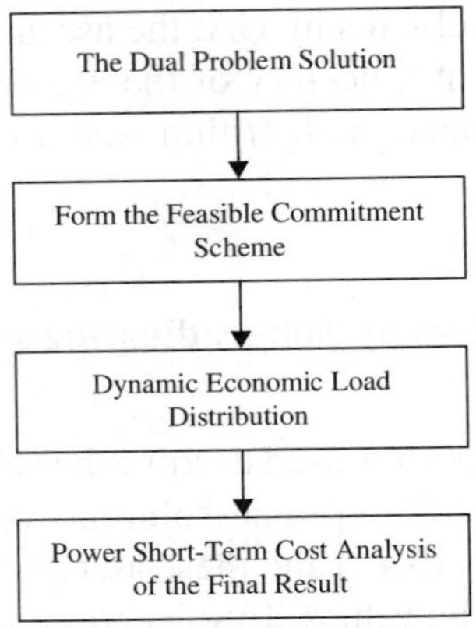

Figure 5.3 Generate the final solution of the original problem from the dual solution. *Source*: Chen 2002. Reproduced with permission from IEEE.

Since the dual solution is generally primally infeasible due to the nonconvexity of the primal problem, a heuristic or optimization method should be used to generate a good primally feasible solution based on the dual results (Figure 5.3).

5.1.3 Form Primal Feasible Solution Based on the Dual Results

First we define the total capacity requirement per hour for the system as the sum of load and spinning reserve in this hour. For each unit, the effective capacity spinning reserve in a given hour is defined as the effective capacity the unit can provide to meet the total system capacity requirement in this hour. The unmet system capacity requirement in a certain hour is equal to the total system capacity required minus the total effective capacity spinning reserve offered by all units in this hour. If the unmet system capacity requirement in each hour during the scheduling horizon is already known, the next unit to be committed is determined by a relative economic indicator. How to define the relative economic indicator is a key issue of the algorithm.

If the unit commitment scheme of the dual problem is not feasible for the original problem, there must be some periods in which the system capacity requirements are not met. Then the solution of the dual problem is used as the initial scheme to consider which units should be committed to meet the system capacity requirement and minimize the total operating cost.

Here we use $u_i = (u_i(1), \cdots, u_i(T))$ to denote the unit commitment scheme of unit i, and $u = (u_1, \cdots, u_I)$ to denote the unit commitment schemes of all units. The scheme of the dual solution is used as the initial scheme, namely

$$u^0 = ({u_1}^0, \cdots, {u_I}^0)$$

The units to be committed for meeting the system capacity requirement are determined through an iterative process. In the dual solution scheme, some units are obviously not able to be selected as the candidate units, such as the units already committed in all hours of the scheduling horizon. The index set of candidate units to be committed is denoted by Θ.

5.1.3.1 Process of Local Optimization

For each unit in Θ, a temporary unit commitment scheme is determined according to the following rules: for the periods in which the system capacity requirements are not met, the unit must be started up as long as the unit constraints can be satisfied, and the unit commitment scheme in the scheduling horizon can be obtained through a special dynamic programming or heuristic method. From the economic model of Lagrangian relaxation, the objective function of the Lagrangian problem can be regarded as the sum of total generation cost of power plants in the system and expenses of power and reserve purchase from outside the system with prices λ^* and μ^* (i.e. the system marginal cost), respectively. When the candidate unit is committed, the total generation cost F increases due to the increase of generation capacity,

$$F = \sum_{t=1}^{T} \sum_{i=1}^{I} [C_i(p_i(t)) + S_i(x_i(t-1), u_i(t))]$$

and decrease of power and reserve purchase expense P from the outside system,

$$P = \sum_{t=1}^{T} \left\{ \lambda(t) \left[p_d(t) - \sum_{i=1}^{I} p_i(t) \right] + \mu(t) \left[p_r(t) - \sum_{i=1}^{I} r_i(t) \right] \right\}$$

Then the relative economic indicator can be defined as

$$E_r = -\frac{\Delta F}{\Delta P}$$

Obviously, the smaller E_r, the more economical the unit commitment scheme will be. The unit commitment scheme of a single unit is still determined by solving sub-problems. The temporary commitment schemes of all candidate units are calculated and the unit with the smallest E_r is chosen. After the commitment of the new unit, the unmet system capacity requirement will be recalculated. This process is repeated until the system capacity requirement can be met in all dispatch periods. Note that after the commitment of the new unit into the system, the marginal costs of power and reserve will change, but λ^* and μ^* are always used as the system marginal costs in the calculation because the duality gap between the primal and dual solutions is often very small.

5.1.3.2 Process of Global Optimization

Although the above optimization process can find a unit commitment scheme that is economical and can meet the system capacity requirement, it is still a local optimization process. At every step of the optimization process, only the unit with the current smallest E_r is considered, which is not necessarily globally optimal. For example, in a certain optimization step, unit i with the smallest E_r can only provide relatively small effective spinning reserve. Assuming its generation cost increment to be ΔF_i, after its start-up, another unit j with generation cost increment ΔF_j should be committed to meet the system capacity requirement. Then the total generation cost increment becomes $\Delta F' = \Delta F_i + \Delta F_j$, which is big.

On the other hand, if another unit k with bigger E_r and also larger effective spinning reserve is used, and the system capacity requirement is able to be met after its commitment, the generation cost increment is only for the incremental cost of this unit set ΔF_k. Because generally $\Delta F_k < \Delta F'$, in a certain step of the optimization process, the unit with the smallest E_r is often not globally optimal. To solve this problem, a more complex global optimization process is used.

The global optimization is developed on the basis of the simpler local optimization process. At each step of the optimization process, the units are not directly chosen by comparing the relative economic indicators. For each unit in Θ, after obtaining its temporary unit commitment scheme, the unmet system capacity requirement is calculated first, and then the above local optimization process is carried out until the system capacity requirement is fulfilled. Assuming the total generation cost increment to be ΔF_s in the optimization process, the unit with the smallest ΔF_s is committed. The solution algorithm is shown in Figure 5.4.

5.1.4 Dynamic Economic Dispatch

After the formation of a feasible unit commitment scheme, we must determine the unit power output in each period of the scheduling horizon and allocate the spinning reserve, and also consider the ramp rate constraints. Therefore, a dynamic economic dispatch algorithm considering spinning reserve is used here.

5.1.4.1 Static Economic Dispatch Considering Spinning Reserve

First, the economic dispatch in a single period (namely static economic dispatch) is carried out. The ramp rate constraints are not considered but the system spinning reserve requirements are taken into consideration. The mathematical model is as follows.

Objective function

$$\min f_p = \sum_{i=1}^{I} C_i(p_i) \tag{5.17}$$

where p_i denotes the power output of unit i.

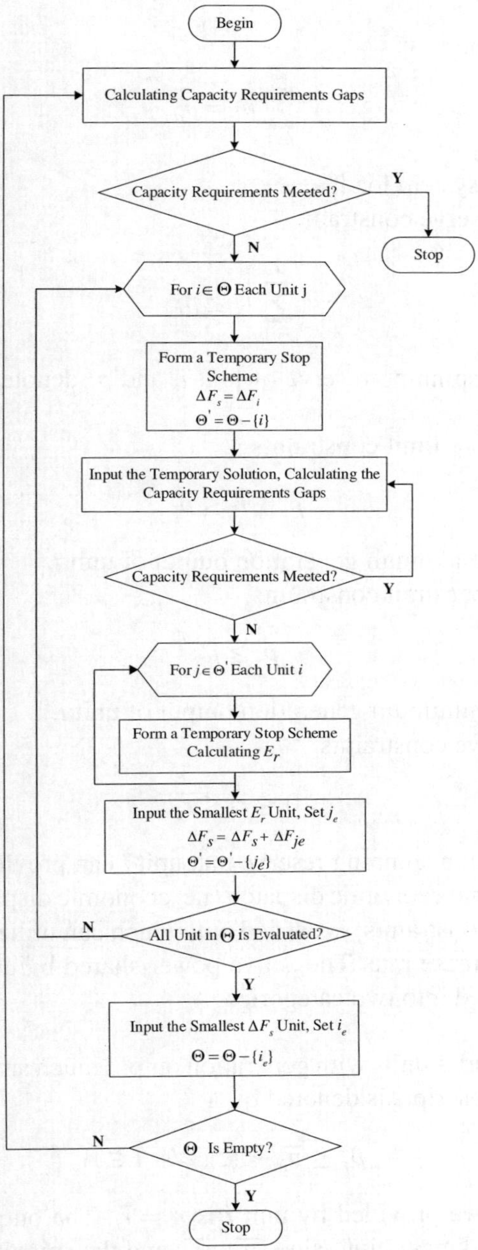

Figure 5.4 Algorithm for feasible unit commitment scheme formation. *Source*: Chen 2002. Reproduced with permission from IEEE.

Constraints:
System power balance

$$\sum_{i=1}^{I} p_i = p_d \tag{5.18}$$

where p_d denotes the system load.
System spinning reserve constraint

$$\sum_{i=1}^{I} r_i \geq p_r \tag{5.19}$$

where r_i denotes the spinning reserve of unit i, and p_r denotes the system spinning reserve requirement.
Unit generation upper limit constraints

$$p_i + r_i \leq \overline{p_i} \tag{5.20}$$

where $\overline{p_i}$ denotes the maximum generation output of unit i.
Unit generation upper limit constraints

$$p_i \geq \underline{p_i} \tag{5.21}$$

where $\underline{p_i}$ denotes the minimum generation output of unit i.
Unit spinning reserve constraints

$$0 \leq r_i \leq \overline{r_i} \tag{5.22}$$

where $\overline{r_i}$ is the maximum spinning reserve that unit i can provide.
First, the conventional economic dispatch (i.e. economic dispatch without considering spinning reserve constraints) is carried out, which can utilize the method of equal coal consumption increase rate. The active power shared by unit i is denoted by p_i^*. All units can be divided into two categories.

1. The first class includes units with generation output much smaller than their capacities. The set of subscripts is denoted by A:

$$p_i^* \leq \overline{p_i} - \overline{r_i}, \quad \forall \quad i \in A \tag{5.23}$$

The spinning reserve provided by unit i is $r_i = \overline{r_i}$. The output of these units can be increased to the knickpoint value $(\overline{p_i} - \overline{r_i})$ and the spinning reserve will not be influenced. Hence, for units in A, the total generation power increment can be

$$p_{up} = \sum_{i \in A} (\overline{p_i} - \overline{r_i} - p_i^*)$$

without changing the spinning reserve.

2. The second class includes units with bigger generation output compared to their capacities. The set of subscripts is denoted by B. The generation output of these units exceeds the knickpoint value, namely,

$$p_i^* > \overline{p_i} - \overline{r_i}, \quad \forall \quad i \in B \tag{5.24}$$

Hence, the spinning reserve is

$$r_i = \overline{p_i} - p_i^* \tag{5.25}$$

The output of these units can be decreased to the knickpoint value $(\overline{p_i} - \overline{r_i})$ and more spinning reserve can be provided. For units in B, the total generation power decrement can be

$$p_{down} = \sum_{i \in B} (\overline{r_i} - (\overline{p_i} - p_i^*))$$

The same quantity of spinning reserve can be increased.

Supposing the total spinning reserve provided by the two classes of units to be r_s, from conventional economic dispatch we have

$$r_s = \sum_{i=1}^{I} r_i \tag{5.26}$$

Supposing the spinning reserve deficiency to be Δr_s, we have

$$\Delta r_s = p_r - r_s \tag{5.27}$$

The calculation results can be divided into three cases as follows:

1. The system spinning reserve can be satisfied by conventional economic dispatch, namely $\Delta r_s \leq 0$. In this case the result of conventional economic dispatch is the optimal solution required.
2. The system spinning reserve cannot be satisfied by conventional economic dispatch, and either the total power increment of units in A is less than spinning reserve deficiency, or the total power decrement of units in B is less than spinning reserve deficiency, namely, $\Delta r_s > 0, p_{up} < \Delta r_s, p_{down} < \Delta r_s$. To meet the system load and provide as much spinning reserve as required, even though we cannot fully meet the spinning reserve requirement, the output power must be transferred from units in B to units in A. The amount of power transfer is the smaller of p_{up} and p_{down}. In this case, we define $\Delta r_s = \min(p_{up}, p_{down})$ and calculate by the method in 3 below.
3. The system spinning reserve can be acquired by transferring power Δr_s from units in B to units in A, namely $\Delta r_s > 0, p_{up} \geq \Delta r_s, p_{down} \geq \Delta r_s$.

Now the problem is how to economically allocate power among the units, which is solved through the following steps.

Step one. Allocate power increment Δr_s among units in A on the basis of conventional economic dispatch results. This step is accomplished by the following procedure:

1. Perform economic dispatch on units in A, not considering units in B.
2. Calculate the total power to be allocated, which is equal to the total power of units in A in conventional economic dispatch plus power increment Δr_s, namely

$$p_{d(A)} = \sum_{i \in A} p_i^* + \Delta r_s$$

3. Set the new generation minimum/maximum power limits for units in A, namely

$$\underline{p}_{i(A)} = p_i^*, \overline{p}_{i(A)} = \overline{p}_i - \overline{r}_i$$

4. Perform conventional economic dispatch with the new constraints.

Step two. Allocate power decrement Δr_s among units in B on the basis of conventional economic dispatch results. This step is accomplished by the following procedure:

1. Perform economic dispatch on units in B, not considering units in A.
2. Calculate the total power to be allocated, which is equal to the total power of units in B in conventional economic dispatch minus power increment Δr_s, namely

$$p_{d(B)} = \sum_{i \in B} p_i^* - \Delta r_s$$

3. Set the new generation minimum/maximum power limits for units in B, namely

$$\underline{p}_{i(B)} = \overline{p}_i - \overline{r}_i, \overline{p}_{i(B)} = p_i^*$$

4. Perform conventional economic dispatch with the new constraints.

It should be noted that

$$p_{d(A)} + p_{d(B)} = \sum_{i \in A} p_i^* + \Delta r_s + \sum_{i \in B} p_i^* - \Delta r_s = \sum_{i=1}^{I} p_i^* = p_d$$

Hence the final results can meet the system load requirement.

5.1.4.2 Dynamic Economic Dispatch Considering Spinning Reserve

The above static economic dispatch is performed on a single dispatch period. To consider economic dispatch for the whole scheduling horizon, the main difficulty is with the ramp rate constraints, which couple power outputs of different periods. The mathematical model of dynamic economic dispatch in the whole scheduling horizon is as follows.

Objective function

$$\min\ F_p = \sum_{t=1}^{T} \sum_{i=1}^{I} C_i(p_i(t)) \tag{5.28}$$

Constraints:

Power balance constraints

$$\sum_{i=1}^{I} p_i(t) = p_d(t), t = 1, \cdots, T \tag{5.29}$$

Spinning reserve constraints

$$\sum_{i=1}^{I} r_i(t) \geq p_r(t), t = 1, \cdots, T \tag{5.30}$$

Maximum unit generation limits

$$p_i(t) + r_i(t) \leq \overline{p_i} \tag{5.31}$$

Minimum unit generation limits

$$p_i(t) \geq \underline{p_i} \tag{5.32}$$

Unit spinning reserve constraints

$$0 \leq r_i(t) \leq \overline{r_i} \tag{5.33}$$

Unit ramp rate constraints

$$-\Delta_i \leq p_i(t) - p_i(t-1) \leq \Delta_i \tag{5.34}$$

The meanings of all the notation is the same as in the mathematical model of the previous section.

Compared with the static economic dispatch model in (5.17)–(5.22), the dynamic economic dispatch model is composed of a static economic dispatch model in all dispatch periods and the unit ramp rate constraints (5.34) coupling different dispatch periods. Hence if we can decouple power connections expressed by (5.34), the problem can be decomposed into T static economic dispatch problems.

A simplified approach is to remove constraints (5.34) and replace constraints (5.31) and (5.32) with the following constraints:

$$p_i(t) + r_i(t) \le \overline{p_i}(t) \tag{5.35}$$

$$p_i(t) \ge \underline{p_i}(t) \tag{5.36}$$

with

$$\overline{p}_i(t) = \min(\overline{p}_i, p_i^*(t-1) + \Delta_i) \tag{5.37}$$

and

$$\underline{p}_i(t) = \max(\underline{p}_i, p_i^*(t-1) - \Delta_i) \tag{5.38}$$

where $p_i^*(t-1)$ is the power output of unit i at dispatch period $t-1$. It is clear that we can only obtain suboptimal solutions this way. More complicated and accurate methods can be used according to practical requirements.

5.1.5 *Example*

The IEEE RTS-96 is used as the test system [167]. The new addition of production cost-related data for the generating units facilitates our use. One notable issue is that, owing to the lack of a standard coal price, the heat consumption (in MBTU) is used in place of the fuel cost as the objective to be optimized. Another problem concerns the cost ratio of per unit heat consumption of a nuclear unit and a coal/steam unit. This section assumes that the ratio is 6, i.e. the cost of 6 units of nuclear heat equals that of 1 unit of coal heat.

5.1.5.1 Statistical Analysis of CCA Solutions

The convergent trajectory of CCA is stochastic. To evaluate the quality of the final solutions, we should compare the statistical result of solutions of repetitive runs. To that end, 50 runs are taken for the one-area system (32 units). The typical daily load data set shown in Table 5.1 is used. Assume the reserve to be 10% of the load.

The distribution of solutions and its normal fit is shown in Figure 5.5. From the statistical result, we can see that the minimal cost is 336 044 MBTU, the maximal cost is 338 723 MBTU, the average cost is 337 261 MBTU, and the standard deviation is 955 MBTU, being 0.3% of the minimal cost. The CCA solutions are highly concentrated, and 52% of them are better than the LR solution, 336 939 MBTU.

5.1.5.2 Influence of Unit Number on CPU Time of CCA

When solving the UC problem by GA, the CPU time will increase remarkably as the unit number increases. Unit commitment of a 100 unit system in a 24 hour time

Table 5.1 System daily load

t (h)	$p_i(t)$ (MW)	t (h)	$p_i(t)$ (MW)
1	1776	13	2518
2	1670	14	2518
3	1590	15	2465
4	1564	16	2491
5	1564	17	2624
6	1590	18	2651
7	1961	19	2651
8	2279	20	2544
9	2518	21	2412
10	2544	22	2200
11	2544	23	1935
12	2518	24	1670

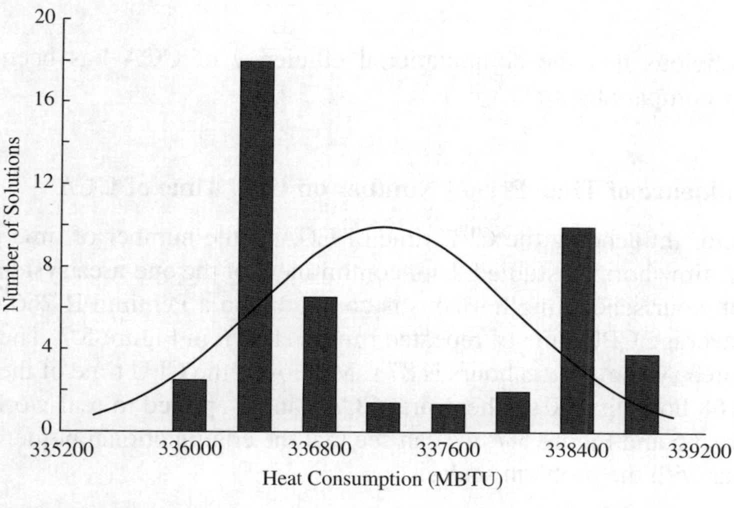

Figure 5.5 Distribution of CCA final solutions. *Source*: Chen 2002. Reproduced with permission from IEEE.

horizon by GA requires 16 000 s CPU time (4 h and 27 min) on an HP Apollo 720 workstation [164]. This drawback hinders this method's application to real-world systems.

To estimate the influence of unit number on computational time of CCA, the one-area (32 unit), two-area (64 unit), and three-area (96 unit) systems of the IEEE RTS-96 are used. The average CPU time of repeated runs on a Pentium II 266 microcomputer is shown in Figure 5.6. The CPU time of the three-area system is

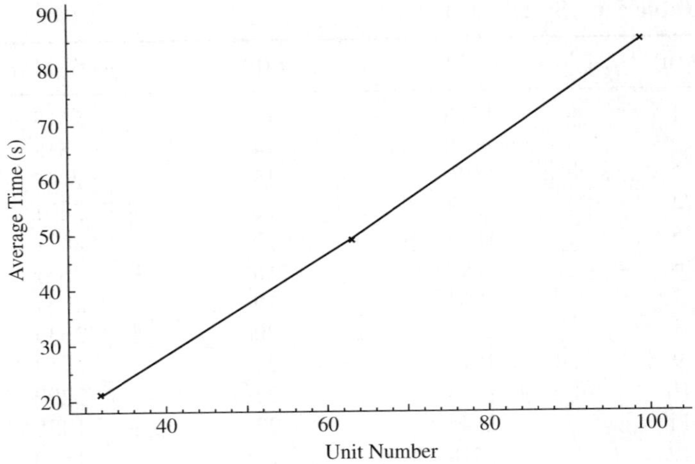

Figure 5.6 Influence of unit number on CPU time of CCA. *Source*: Chen 2002. Reproduced with permission from IEEE.

87 s. It is obvious that the computational efficiency of CCA has been improved significantly compared with GA.

5.1.5.3 Influence of Time Period Number on CPU Time of CCA

Another factor influencing the CPU time of CCA is the number of time periods (in hours) of the time horizon studied. Unit commitment of the one-area system in 24, 72, 120 and 168 hour scheduling horizons is computed on a Pentium II 266 microcomputer. The average CPU time of repeated runs is shown in Figure 5.7. The CPU time of the one-area system in 168 hours is 87 s. Moreover, the CPU time of the three-area system in 168 hours is 370 s. Therefore, CCA can be applied to real-world systems. From Figure 5.6 and Figure 5.7, we can see that the computational burden of CCA is almost linear with the problem scale.

5.1.6 Summary

A new cooperative coevolutionary algorithm is introduced, which appears to have considerable potential for formulating and solving more complex problems by explicitly modeling the coevolution of cooperating species. CCA encourages the parallel evolution of substructures that interact in useful ways to form more complex higher-level structures.

The CCA for the unit commitment problem combines the basic ideas of LR and GA to form a novel two-level approach. The first level uses a subgradient-based stochastic optimization method to optimize Lagrange multipliers. The second level uses a genetic

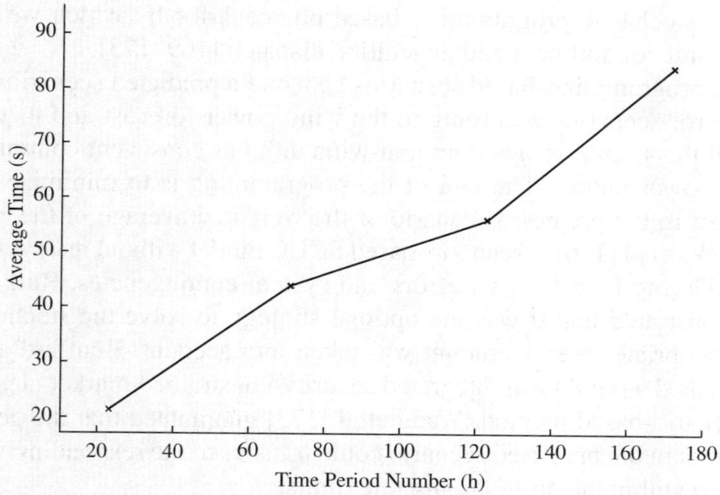

Figure 5.7 Influence of time period number on CPU time of CCA. *Source*: Chen 2002. Reproduced with permission from IEEE.

algorithm to solve the individual unit commitment subproblems. CCA can manage more complicated time-dependent constraints than conventional LR.

Simulation results show that CCA has good convergence properties and a significant speedup over traditional GAs and can obtain high-quality solutions. The "curse of dimensionality" is surmounted, and the computational burden is almost linear with the problem scale.

Based on the methodology presented in this section, different CCAs can be designed according to specific problems, which provides a new approach to large-scale system optimization problems.

5.2 Security-Constrained Unit Commitment with Wind Power Integration Based on Mixed Integer Programming

Wind energy has become increasingly popular across the world in recent years for its properties of cleanness and low cost. However, wind power cannot be forecasted with great accuracy for dispatching purposes [168]. Hence, there is an unavoidable random error between the actual wind power output and its forecasted value. How to deal with the uncertainty is a difficult point in wind farm integrated technology which brings new challenges to security-constrained unit commitment (SCUC) and security-constrained economic dispatch (SCED). For accommodating the random wind power fluctuation by reasonable dispatch of the non-wind generation units, one method is to increase the system reserve. However, it is not easy to get the precise reserve amount which is enough to overcome the wind power uncertainty and guarantee the system security, so the results usually tend to be conservative. A better alternative is stochastic programming based on scenarios by which we can obtain more robust unit commitment and generation dispatch [169–173].

Stochastic programming-based scenarios generate a predicted scenario and a large number of error scenarios according to the wind power forecast and its error distribution. In all the scenarios, a certain non-wind unit has consistent commitment state and diverse power output. The aim of the programming is to minimize the system operating cost in the predicted scenario or the weighted average of the cost in each scenario. In Wu et al [170], a scenario-based SCUC model without reserves was established, considering load forecast errors and system contingencies. Ruiz et al [171] further demonstrated that it was the optimal strategy to solve the stochastic SCUC when an appropriate reserve amount was taken into account. Bouffard and Galiana [172] established a wind farm integrated security-constrained market clearing model with the scenario-based method. Wang et al [173] guaranteed that the generation of non-wind units in the predicted scenario could transfer to the responding value in any error scenario within the unit ramping rate limits.

These papers have two common points. First, they all use a scenario method based on Monte Carlo simulation, and assume that the system contains only one wind farm. When considering multiple wind farms, there will be a sharp increase in the number of scenarios. Another common point is the use of the mixed integer programming method (MIP) to solve the model. In fact, a recent breakthrough in the optimization field (especially the cutting plane) has made the MIP quite effective to solve the optimal operating problems of power systems [174]. Its advantages are as follows:

1. It can ensure that the global optimal solution is found.
2. With precise and tough modeling capabilities, it can express the objective function and constraints directly. It is also suitable for handling security constraints, adding constraints and variables to the original model without complex modification.
3. The user can concentrate on the problem definition rather than the algorithm itself for which we can use commercial software to solve the problem once the model is established.

However, there is hidden trouble when we use MIP to solve the stochastic model which is established by the scenario method based on Monte Carlo simulation. With the increase of model scale, the computing time will increase exponentially [175]. As a result, the massive number of scenarios caused by the consideration of multiple wind farms leads to the consequences of overlarge scale of the model and overweighed computational burden. Usually, we have to use scenario reduction technology [176] for feasibility of MIP while sacrificing accuracy. Bixby et al [175] pointed out that the key to improving the practical effect of MIP is to design the formulation of the model elaborately and control its scale.

The contribution of this section is to introduce generalized generation distribution factors [177] into the multiple wind farm integrated SCUC model, and verify its acceleration to MIP. This section also proves that the set of extreme scenarios is able to completely stand for the value space of wind power output and replace the scenario method based on Monte Carlo simulation for diminishing the loss of computing accuracy. By the improvement of the speed and accuracy of computing, we point out an approach for the use of MIP to solve the multiple wind farm integrated SCUC problem.

5.2.1 Suitable SCUC Model for MIP

5.2.1.1 Source of the Adaptability to MIP

In the DC power flow model, the branch flow constraint is usually expressed as the following inequalities:

$$P = B\theta$$
$$|(\theta' - \theta'')/x_j| \leq L_j, \quad \forall \quad j \in J \tag{5.39}$$

where $P = B\theta$ is the matrix form of the DC power flow equations, and θ' and θ'' represent the voltage phase angles of the nodes at both ends of branch j. By introducing generalized generation distribution factors (GGDFs) proposed by Ng [177], we can express branch flow in the DC flow model as a linear combination of generating unit outputs, so that (5.39) can be transformed into (5.49). section 5.2.2 will prove that this transformation can significantly reduce computing time of the MIP.

On the other hand, unit commitment models usually require the allocation of three binary variables for each unit in each period, i.e. the unit states, the start-up sign, and shut-down sign, respectively. Carrion and Arroyo [178] proposed an alternative method which required only one binary variable for each unit, and proved that the method can alleviate the computational burden of the MIP by reducing the number of nodes and the scale of the search tree.

The multiple wind farm integrated SCUC model is established as follows, combining the previous two references [177, 178]. For the sake of simplicity, it is assumed in this section that there are only thermal units and wind farms in the system.

5.2.1.2 Formulation of SCUC

We start with a list of notation:

a_{gj}	GGDF of unit g in branch j;
a_g, b_g, c_g	coefficients of quadratic production cost function of unit g;
x_j	impedance of branch j;
L_j	flow capacity of branch j;
J	branch set;
g	index of unit;
GT	set of thermal units;

Nt	number of hours within the planning period;
$F(\cdot)$	production cost function;
$S_u(\cdot)$	start-up cost function;
$K(g, m_g)$	cost of interval m_g of piecewise start-up cost function of unit g;
$\overline{p_g}$	upper output limit of unit g;
$\underline{p_g}$	lower output limit of unit g;
$r(g,t)$	spinning reserve provided by unit g in period t;
$R(t)$	spinning reserve requirement of system in time period t;
R_u	ramp-up rate limit of unit;
R_d	ramp-down rate limit of unit;
$T_u(g)$	minimum ON time of unit g;
$T_d(g)$	minimum OFF time of unit g;
GW	set of wind farms;
$D(t)$	system demand in period t;
$I(g,t)$	commitment state of unit in period t;
$p(g,t)$	generation of unit g in period t in predicted scenario;
s	index for wind power generation scenario;
St	set of scenarios in period t;
$p(s,g,t)$	generation of unit g in period t in scenario
t	index for time.

The objective function

$$\min \sum_{g\in GT}\sum_{t\in Nt} F(g,t) + S_u(g,t) \tag{5.40}$$

is the system generating cost in predicted scenario composed of fuel costs and start-up costs of individual units. The wind power generation is assumed to be free.
Set of scenarios St:

$$F(g,t) = a_g p^2(g,t) + b_g p(g,t) + c_g I(g,t), \quad g \in GT, t \in Nt \tag{5.41}$$

Constraints:

$$S_u(g,t) \geq K(g,m_g)\left[I(g,t) - \sum_{h=1}^{m_g} I(g,t-h)\right] S_u(g,t) \geq 0, \quad g \in GT, t \in Nt \tag{5.42}$$

$$\underline{p_g} I(g,t) \leq p(g,t) + r(g,t) \leq \overline{p_g} I(g,t), \quad g \in GT, t \in Nt \tag{5.43}$$

$$\sum_{g\in GT} r(g,t) \geq R(t), \quad t \in Nt \tag{5.44}$$

$$p(g,t) - R_d \leq p(g,t+1) \leq p(g,t) + R_u, g \in GT, t = 1, \cdots, Nt-1 \tag{5.45}$$

$$\sum_{h=0}^{T_u(g)-1} I(g,t+h) \geq [I(g,t) - I(g,t-1)] \times \min[T_u(g), Nt - t + 1] \tag{5.46}$$

$$\sum_{h=0}^{T_d(g)-1} [1 - I(g,t+h)] \geq [I(g,t-1) - I(g,t)] \times \min[T_d(g), Nt - t + 1] \tag{5.47}$$

$$\sum_{g\in GT} p(g,t) + \sum_{g\in GW} p(g,t) = D(t), \quad t \in Nt \tag{5.48}$$

$$\left| \sum_{g\in GT} a_{gj} p(g,t) + \sum_{g\in GW} a_{gj} p(g,t) \right| \leq L_j, \quad j \in J, t \in Nt \tag{5.49}$$

The above constraints, as the first part, refer to the predicted scenario, including the unit generation limits (5.43), system reserve requirements (5.44), unit ramping limits (5.45), unit minimum ON time limits (5.46), unit minimum OFF time limits (5.47), system power balance constraints (5.48), and branch flow limits (5.49).

Then $\forall \quad t \in Nt$ and $\forall \quad s \in St$:

$$\underline{p}_g I(g,t) \leq p(s,g,t) \leq \overline{p_g} I(g,t), \quad g \in GT \tag{5.50}$$

$$\sum_{g\in GT} p(s,g,t) + \sum_{g\in GW} p(s,g,t) = D(t) \tag{5.51}$$

$$\left| \sum_{g\in GT} a_{gj} p(s,g,t) + \sum_{g\in GW} a_{gj} p(s,g,t) \right| \leq L_j, \quad j \in J \tag{5.52}$$

$$-\frac{1}{6} R_d \leq p(s,g,t) - p(g,t) \leq \frac{1}{6} R_u, \quad g \in GT \tag{5.53}$$

The constraints (5.50)–(5.53), as the second part, refer to the error scenarios, representing the unit generation limits (5.50), system power balance constraints (5.51), branch flow limits (5.52), and adjustment time constraint (5.53).

Note that $I(g,t)$ in the two parts of the constraints are consistent and that $p(g,t)$ and $p(s,g,t)$ of wind farms are obtained from the wind power forecast. The difference $p(s,g,t) - p(g,t)$ represents the output adjustment of a unit when the system state transfers from predicted scenario to error scenario s. We can limit the adjustment time within 10 minutes by selecting one-sixth of the ramp rate limit of the unit as the upper and lower bounds of (5.53).

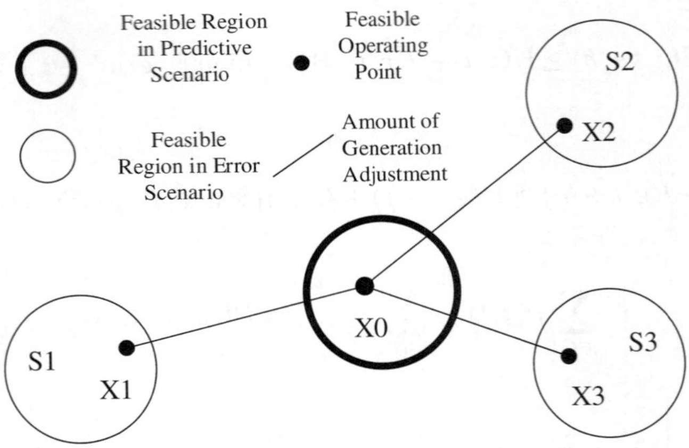

Figure 5.8 Feasible solution under the adjustment time constraint.

5.2.1.3 Wind Power Uncertainty Adaptability

As shown in Figure 5.8, the solution of the model includes a thermal unit generation schedule *X*0 which satisfies all operating constraints and security constraints in the wind power predicted scenario. Suppose that the actual wind power output deviates from its forecasted value in period *t* and represents error scenario *si*. We say that the solution of the model accommodates scenario *si* if the system can transfer to a new feasible operating point *Xi* within the adjustment time limit by redispatching the thermal units. Obviously, it ought to accommodate any scenario *s* within *St* in any period *t*.

Take the case with only one wind farm integrated, for example. The decision-makers can specify a certain confidence level based on the forecast curve and error distribution of wind power, and thus they obtain the confidence interval in Figure 5.9, between the confidence limits. If *St* in each period is selected reasonably, which can stand for all the possible error scenarios within the interval, the solution of the model can accommodate situations when the wind power output represents any value in the specified confidence interval. As a result, the units generation schedule maintains a certain level of adaptability to wind power uncertainty while minimizing the operating costs. The level of adaptability could be controlled by specifying the confidence level.

5.2.2 *Selection of St and the Significance of Extreme Scenarios*

Set *St* should be able to stand for the value space derived from the wind power output confidence interval of a single wind farm or that of multiple wind farms. Suppose that there are two wind farms, for which the forecasted outputs in period *t* are 110 MW and 120 MW. The confidence intervals (40 MW, 180 MW) and (60 MW, 180 MW) at certain confidence levels constitute a rectangular space, as shown in Figure 5.10, which

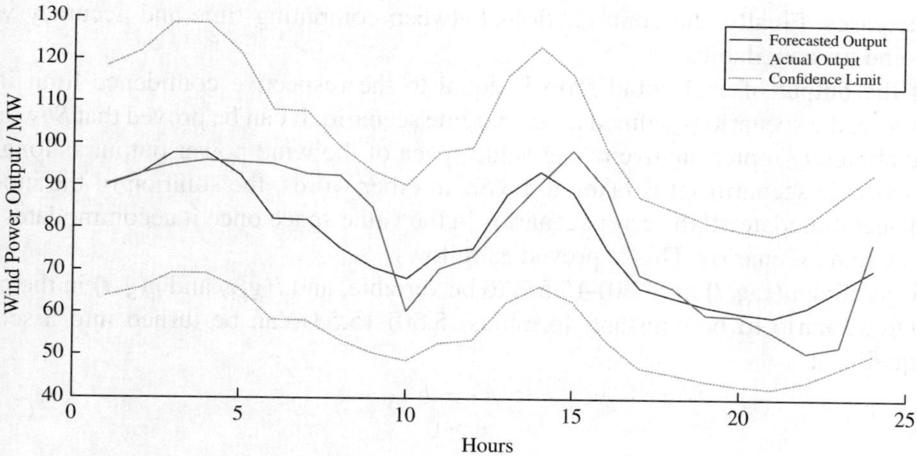

Figure 5.9 The curve of the forecasted wind power output and its confidence interval compared with actual values.

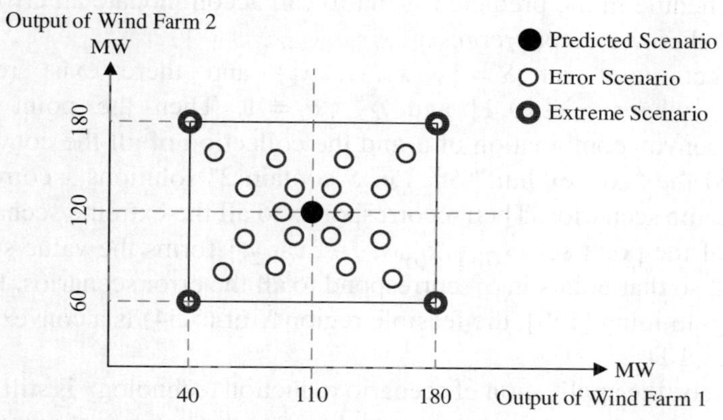

Figure 5.10 The space of wind generation considering two wind farms.

turns out to be the value space of wind power output in period t. The most common method to select St is Monte Carlo simulation, which samples randomly according to the error probability distribution, generates a large number of error scenarios, and then reduces the scenarios for the computing capacity limit by merging similar scenarios using the scenario reduction technology [176], thus forming St. Obviously, this is at the cost of losing the computing accuracy, so the wind power uncertainty adaptability of the solution may be weaker than the specified requirements. Besides, the value space will be an n-dimensional convex polyhedron with 2^n vertices when there are n $(n \geq 3)$ wind power farms. For a certain number of scenarios, the representativeness of St obtained by Monte Carlo simulation will be weakened rapidly when

n increases. Finally, the contradictions between computing time and accuracy will become irreconcilable.

If the output of each wind farm is equal to the respective confidence limit in a scenario, the scenario is defined as an extreme scenario. It can be proved that *St* would be completely representative of the value space of the wind power output as long as the extreme scenario set is taken to be *St*. In other words, the solution of the model will accommodate all the error scenarios in the value space once it accommodates all the extreme scenarios. This is proved as follows.

Regarding $p(s,g,t)$ in (5.50)–(5.53) to be variable, and $I(g,t)$ and $p(g,t)$ in the predicted scenario to be constant, formulas (5.50)–(5.53) can be turned into a set of inequalities:

$$\begin{aligned} Ax &= b \\ x &\geq 0 \end{aligned} \tag{5.54}$$

where A is a coefficient matrix, and b is a coefficient column vector. Each feasible solution $x = (\ldots, x_{GW}, \ldots)^{\mathrm{T}}$ corresponds to an error scenario, where x_{GW} is a vector made up of $p(s,g,t) \quad (g \in GW)$. When x is feasible, it means that the thermal unit generation schedule in the predicted scenario can accommodate an error scenario in which the wind power output represent x_{GW}.

Assume that point set $S = \{x_1, x_2, \ldots, x_k\}$ and there exist real numbers $\alpha_1, \alpha_2, \ldots, \alpha_k$ where $\alpha_i \in [0,1]$ and $\sum_{i=1}^{k} \alpha_i = 1$. Then the point $\sum_{i=1}^{k} \alpha_i x_i$ is defined as a convex combination of S and the collection of all the convex combinations is called the "convex hull" S^{Δ}. Let S contain 2^n solutions x corresponding to different extreme scenarios. Then S corresponds to all the extreme scenarios, and the convex hull of the point set $\{x_{GW1}, x_{GW2}, \ldots, x_{GW2^n}\}$ forms the value space of wind power output, so that points in S^{Δ} correspond to all the error scenarios. By the theory of linear programming [179], the feasible region X of (5.54) is a convex set; if $S \subseteq X$ then $S^{\Delta} \subseteq X$. QED

Unfortunately, the application of scenario reduction technology is still unavoidable, although the extreme scenarios set is used, because the number of extreme scenarios would be raised exponentially with the increase of n. But what is valuable is that the solution based on the extreme scenarios set has greater wind power uncertainty adaptability than that based on Monte Carlo simulation, which will be verified in the next section.

5.2.3 Example

A modified IEEE 118-bus system is used to test the performance of the proposed method. There are 54 thermal units, 186 branches, and 91 demand sides in the original version of the system. The peak load of 3733 MW occurs at 12th, 14th, and 15th hours. Detailed data for the system can be found at http://motor.ece.iit.edu/data/SCUC_118test.xls [176], while data on the wind farms is

Table 5.2 Hourly forecasted output (MW) of wind power

Hours	Nodes				
	26	59	73	77	92
1	207.2	70.0	62.1	100	100.0
2	204.6	72.1	61.2	103	102.5
3	253.3	77.7	66.0	111	104.0
4	257.1	60.9	66.0	113	107.5
5	212.8	81.2	62.4	116	95.0
6	164.1	88.2	55.5	126	90.0
7	215.2	97.3	62.1	139	92.5
8	140.5	109.2	51.0	160	97.5
9	126.6	120.4	48.0	180	85.0
10	115.5	128.1	45.0	183	79.0
11	173.4	134.4	57.0	192	80.0
12	147.1	138.6	51.0	198	90.0
13	170.3	138.6	59.7	198	97.5
14	269.2	130.2	66.0	170	105.0
15	185.2	105.0	60.0	150	95.0
16	181.5	116.2	51.0	166	87.5
17	131.3	138.6	45.0	198	80.0
18	204.3	138.6	43.5	198	75.0
19	122.4	138.6	42.0	195	70.0
20	133.5	138.6	45.0	192	75.0
21	105.6	138.6	40.5	196	77.5
22	100.8	138.6	42.0	198	80.0
23	126.7	130.2	45.0	186	85.0
24	146.1	113.4	48.0	162	80.0

shown in Table 5.2. A day-ahead SCUC model is established and solved by the MIP software CPLEX 10.1 on a personal computer (3.0 GHz, 2.0 GB RAM).

Table 5.2 lists the forecasted wind power output in each period. The confidence limits of wind farms at nodes 26, 59, and 92 are set as the forecasted value ±30%, while that of wind farms at nodes 73 and 77 are set as the forecasted value ±50%. Accordingly, the value space of wind power output is formed. Requiring that the thermal unit generation schedule in the predicted scenario can accommodate as many error scenarios in the value space as possible, we take the following three methods into consideration.

Method 1. Use (5.39) to express the branch flow constraint, choose the extreme scenarios set as *St* in each period, and use the scenario reduction technology to reduce the number of scenarios to 10 when the number of wind farms is more than three (namely, the number of extreme scenarios is over 10).

Method 2. Use GGDF to express the branch flow constraint (5.49), sample the wind power output by Monte Carlo simulation, setting the number of samples to be 10 000, then use the scenario reduction technology to reduce the number to 10 so as to form *St*.

Method 3. Use GGDF to express the branch flow constraint (5.49), choose the extreme scenarios set as *St* in each period, and use the scenario reduction technology to reduce the number of scenarios to 10 when the number of wind farms is more than three (namely, the number of extreme scenarios is over 10).

Replace thermal units on nodes 59, 26, 92, 77, and 73 with the wind farms in Table 5.2 in succession. Replace one each time from left to right. Thus we obtain the performance curves of the three methods while the number of wind farms increases, which is shown in Figure 5.11.

The model scale could be characterized by the number of model variables in Figure 5.11(a) to a certain extent. With the same number of scenarios, method 1 produces a significantly larger model than method 3, and the gap widens with the increase of the number of wind farms and scenarios. The reason is that a lot of variables, such as the node voltage phase and so on, have to be added to the model by method 1. When there are more than four wind power farms, the number of scenarios will remain at 10 by using scenario reduction technology in methods 3 and 4, resulting in the number of variables beginning to decline due to the reduction of the thermal power units. In method 2, the number of scenarios remains at 10, so that the scale of the model continues to decline when the number of thermal power units reduces. When there are more than four wind farms, the model scale of method 2 becomes similar to that of method 3. Figure 5.11(b) shows the computing time of the three methods, which display the corresponding rule with Figure 5.11(a).

Discretize the value space of wind power output to create 10 000 scenarios (240 000 scenarios in total in 24 hours), and verify the thermal units generation schedule. The proportion of accommodated scenarios in Figure 5.11(c) means the proportion of scenarios which satisfy (5.53) in all the 240 000 scenarios, by which the wind power uncertainty adaptability of a certain method could be characterized. Because of the complete representativeness of the extreme scenarios set for the value space, the proportion of accommodated scenarios reaches 100% in methods 1 and 3 when scenario reduction technology is not used. The adaptability begins to decline when there are five wind farms for the number of extreme scenarios reducing from 32 to 10.

The scenarios set in method 2 based on the Monte Carlo simulation does not have as clear and efficient representativeness as that in methods 1 and 3, and always has a lower adaptability, which grows even lower when the number of wind power farms exceeds four. As shown in Figure 5.11(d), the generation costs get close, because the three methods all obtain a feasible generation schedule in a predicted scenario. However, the cost of method 2 occurs relatively significantly decreased when the number of wind power farms exceeds four, which is at the cost of lower adaptability shown in Figure 5.11(c).

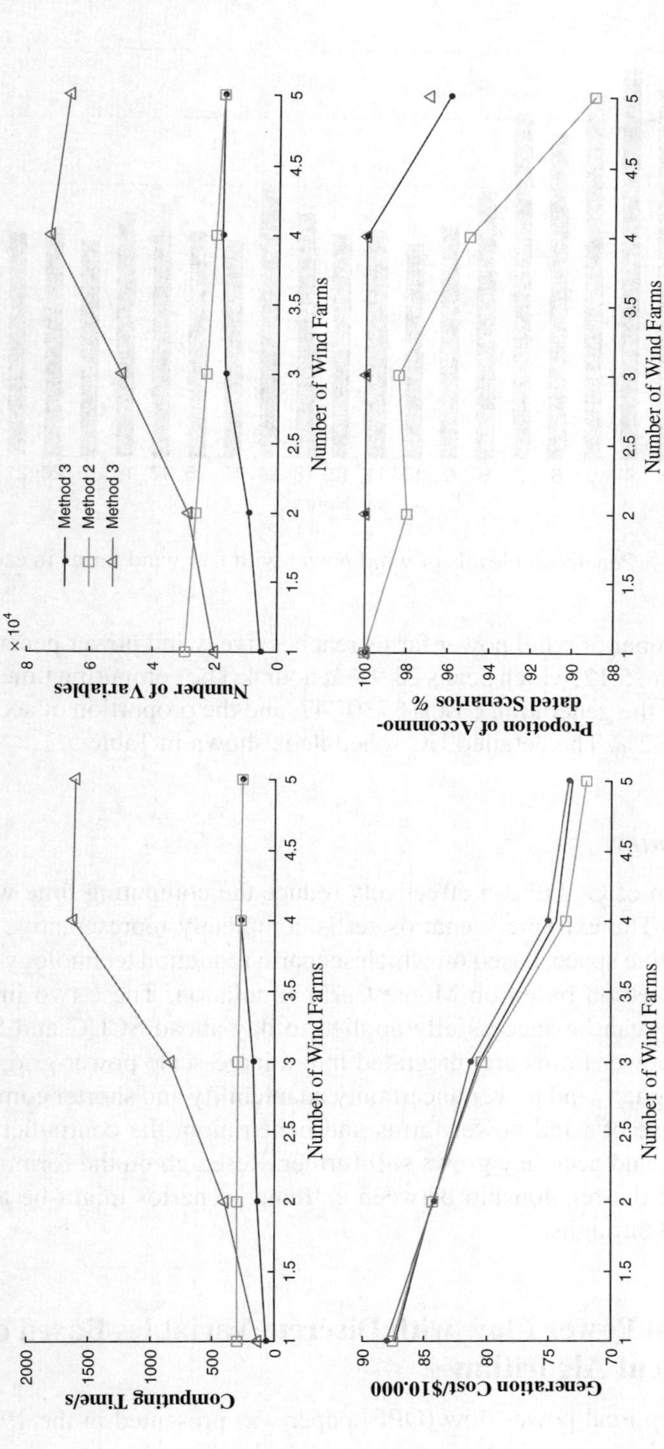

Figure 5.11 Performance curves of the three methods.

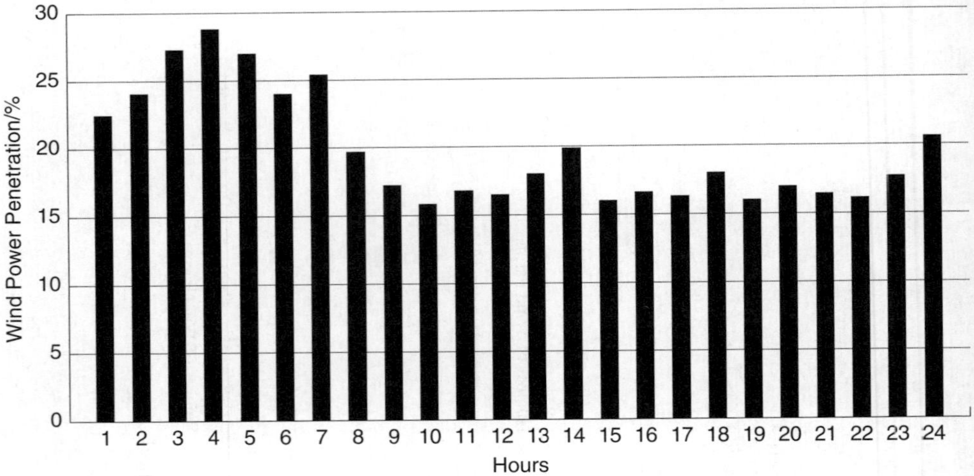

Figure 5.12 Penetration levels of wind power with five wind farms in each hour.

When the number of wind power farms reaches five, wind power penetration [180] is shown in Figure 5.12, which peaks 28.9% at hour 4. The computing time of method 3 is 253.8 s, while the generation cost is \$730 747, and the proportion of accommodated scenarios is 96.32%. The detailed UC schedule is shown in Table 5.3.

5.2.4 Summary

The introduction of GGDF can effectively reduce the computing time when solving SCUC by MIP. The extreme scenarios set is completely representative of the wind power output value space, based on which scenario reduction technology is more efficient than the method based on Monte Carlo simulation. These two improvements ensure that MIP can be successfully applied to day-ahead SCUC and SCED when multiple wind power farms are integrated into a large-scale power grid, and obtains solution with higher wind power uncertainty adaptability and shorter computing time. With the increase of wind power farms and penetration, the contradiction between computing time and accuracy grows still further. Research on the formulation of the MIP model and the relationship between extreme scenarios might be able to bring about new breakthroughs.

5.3 Optimal Power Flow with Discrete Variables Based on Hybrid Intelligent Algorithm

Since the first optimal power flow (OPF) paper was presented in the 1960s, various optimization techniques have been proposed to solve this complex optimal problem,

Table 5.3 SCUC schedule with five wind farms solved by method 3

Nodes	Hours (0–24)[a]
6	0 0 0 0 0 0 0 0 0 0 0 0 1 0 0 0 0 0 0 0 0 0 0 0 0
10, 89	1 1 1 1 1 1 1 1 1 1 1 1 1 1 1 1 1 1 1 1 1 1 1 1 0
12, 87	1 0 0 0 0 0 0 0 0 0 0 0 0 0 0 0 0 0 0 0 0 0 0 0 1
18	0 0 0 0 0 1 1 1 1 1 1 1 1 1 1 1 1 1 1 1 1 1 1 1 1
19	0 0 0 0 0 0 0 0 0 0 0 0 0 0 0 0 0 0 1 0 0 0 0 0 0
25	1 1 0 0 0 0 0 0 0 0 1 1 1 1 1 1 1 1 0 0 0 0 0 0 0
27	0 0 0 0 0 0 0 0 0 1 0 0 0 0 0 0 0 0 0 0 0 1 0 0 0
32	0 0 0 0 0 0 0 0 0 1 1 1 1 0 0 0 0 0 0 0 0 0 0 0 1
34	0 0 0 0 0 0 0 0 0 1 0 0 0 0 0 0 0 0 0 0 0 0 0 0 0
36	0 0 0 0 0 0 0 0 0 0 1 1 1 1 1 1 1 1 1 1 1 1 1 0 0
42	0 0 0 0 0 0 0 0 1 0 0 0 0 0 0 0 0 0 0 0 0 0 0 0 0
46	0 0 0 0 0 0 0 0 0 1 1 1 1 1 0 0 0 0 0 0 0 0 0 0 0
49	1 0 0 0 0 0 0 0 0 0 0 1 1 1 1 1 1 1 1 1 1 0 0 0 0
54	0 0 0 0 0 0 0 0 1 1 1 1 1 1 1 1 1 1 1 1 1 1 0 0 0
55	0 0 0 0 0 0 0 0 0 0 1 1 1 1 1 0 0 0 0 0 0 0 0 0 0
56	0 0 0 0 0 0 1 1 1 1 1 0 0 0 0 0 0 0 0 1 1 1 1 1 0
62	0 0 0 0 0 0 0 1 1 1 1 1 1 0 0 0 0 0 0 0 0 0 0 0 0
65	1 1 1 1 1 0 0 0 0 0 0 0 0 0 0 0 0 0 0 0 1 1 1 1 1
66	1 0 0 0 0 0 0 0 0 0 0 0 0 0 1 1 1 1 1 1 1 1 1 1 0
69, 80	1 1 1 1 1 1 1 1 1 1 1 1 1 1 1 1 1 1 1 1 1 1 1 1 1
70	0 0 1 1 1 1 0 0 0 0 0 1 1 1 1 1 1 1 1 0 0 0 0 0 0
82	0 0 0 0 0 0 1 1 1 1 1 0 0 0 0 0 0 0 0 0 0 0 0 0 0
90	0 0 0 0 0 0 0 1 0 0 0 1 1 0 0 0 0 0 0 0 1 0 0 0 0
91	0 0 0 0 0 0 0 0 1 0 0 1 0 0 1 1 0 0 0 0 0 0 1 0 0
99	0 0 0 0 0 0 0 0 0 0 0 0 0 0 0 0 1 1 1 1 1 1 1 1 0
100	1 0 0 0 0 0 0 0 0 0 0 1 1 1 1 1 1 1 1 0 0 0 0 0 0
103	0 0 0 0 0 0 0 0 0 0 0 0 0 0 0 0 1 0 0 0 1 0 1 0 0
104	0 0 0 0 0 0 0 0 0 0 0 0 0 0 0 0 0 1 1 1 1 1 0 0 0
105	0 0 0 0 0 0 0 0 0 0 0 0 0 0 0 0 0 0 0 0 0 0 0 0 1
107	0 0 0 0 0 0 0 0 1 1 0 0 0 0 0 0 0 1 0 0 0 0 1 1 0
110	0 0 1 1 0 0 0 0 1 1 1 0 0 0 0 0 1 1 1 1 0 0 1 1 0
111	0 0 0 0 0 1 1 1 1 1 1 0 0 0 0 0 0 0 0 0 0 0 0 1 1
112	0 0 0 0 0 1 1 1 1 1 1 0 0 0 0 0 0 0 1 1 1 1 1 0 0
113	0 0 0 0 0 0 0 0 0 0 0 1 1 1 1 1 0 0 0 0 0 0 1 1 1
116	0 0 0 0 0 1 1 1 1 1 0 0 0 1 1 1 1 1 1 1 1 0 0 0 1

[a]Data of hour 0 refer to the initial state of the unit.

such as simplified gradient technique [22], nonlinear programming technique [20], Newton's method [181], linear programming technique [182]. However traditional optimization techniques require the optimal problem to be differentiable, and they cannot accurately deal with discrete variables and rely on convexity to obtain the global optimum solution. Recently, the application of the evolutionary programming technique in OPF solves such problems. Because it searches for an optimal solution by a stochastic technique in an evolving population, it inevitably consumes a great deal of engineering time to obtain the global optimum. Moreover the execution time will rapidly increases with the size of the problem.

This section presents a new approach, incorporating the annealing genetic algorithm (AGA) as well as the modern interior point (MIP) algorithm for OPF. The advantages of MIP include: (1) it can treat equality and inequality constraints conveniently; (2) it can start from either a feasible point (interior point) or an infeasible point (noninterior point); (3) it searches along the Newton direction and can solve optimization problems more quickly and more reliably; and (4) because of its polynomial-time characteristic, it can solve large-scale optimization problems more effectively than the simplex handles a flock of continuous variables. At the same time, we deal with discrete variables through modified AGA, which is superior to the SGA in convergence and escaping local optima, and AGA can obtain the global optimum with great probability, especially in large-scale problems. Numerical simulations on IEEE 14- to 118-bus standard test systems have shown that the proposed method is very promising for OPF problems of large-scale power systems.

5.3.1 *Formulation of OPF Problem*

The optimal power flow procedure consists of determining the optimal steady-state operation of an electricity power generation–transmission system, which simultaneously minimizes the value of a chosen objective function and satisfies certain physical and operating constraints. The following formulation is used in this section.

1. Object function (operation cost minimization):

$$\min \sum_i (a_{2i} P_{Gi}^2 + a_{1i} P_{Gi} + a_{0i}) \tag{5.55}$$

$$(i \in \text{set of all active power sources})$$

   A different format can be used according to the purpose of the calculation, such as loss minimization, shut load minimization, voltage level maximum, minimization magnitude of operation change, etc.

2. Equality constraints (power flow equations):

$$P_{Gi} = P_{Di} + \sum_{j=1}^{n} [e_i(e_j G_{ij} - f_j Bij) + f_i(f_j G_{ij} + e_j B_{ij})] \tag{5.56}$$

$$Q_{Gi} = Q_{Di} + \sum_{j=1}^{n} [f_i(e_j G_{ij} - f_j Bij) + e_i(f_j G_{ij} + e_j B_{ij})] \tag{5.57}$$

$$(i, J \in \text{set of all buses})$$

3. Inequality constraints (including all operating constraints of power system in normal conditions):
   - upper and lower bounds on the active power

$$\underline{P}_{Gi} \leq P_{Gi} \leq \overline{P}_{Gi}$$

   - upper and lower bounds on the reactive power

$$\underline{Q}_{Ri} \leq Q_{Ri} \leq \overline{Q}_{Ri}$$

   - upper and lower bounds on the tap position

$$\underline{T}_i \leq T_i \leq \overline{T}_i$$

   - upper and lower bounds on the magnitude of voltage in bus i

$$\underline{V}_i^2 \leq (e_i^2 + f_i^2) \leq \overline{V}_i^2, \quad (i \in \text{set of all buses})$$

   - upper and lower bounds on the flow of transmission line

$$\underline{P}_{ij} \leq P_{ij} \leq \overline{P}_{ij}, (i < j, i, j \in \text{set of all buses})$$

$$P_{ij} = -(e_i^2 + f_i^2 - e_i e_j - f_i f_j)G_{ij} - (e_i f_j - e_j f_i)B_{ij}$$

The following notation is used:

P_{Gi}, Q_{Ri}	active and reactive power generation at bus i;
T_i	tap position of transformer i with tap-changer;
a_{0i}, a_{1i}, a_{2i}	fuel cost coefficients of generator i;
e_i, f_i	real and imaginary parts of voltage in bus i;
G_{ij}, B_{ij}	real and imaginary parts of transfer admittance between nodes i and j;
P_{ij}	power flow on transmission line connecting with nodes i and j.

The types and numbers of constraints 2 and 3 will increase or decrease based on the application situation of OPF; for example, some constraints which express the contract rule and relationship of participants will be reckoned in under the power market competition circumstance [183].

5.3.2 Modern Interior Point Algorithm (MIP)

In view of the aforementioned advantages of MIP, recently it has been introduced to solve optimal problems in electric power systems, for instance reservoir optimal dispatch, security-constrained economic dispatch, steady-state voltage stability analysis, spot price, VAR optimization and OPF etc.

5.3.2.1 MIP Based on Perturbed KKT Conditions

MIP based on perturbed Karush–Kuhn–Tucker (KKT) conditions includes four steps. In order to explain the problem explicitly, we use the following generalized nonlinear optimal model:

$$\begin{aligned} &\min\ f(x) \\ \text{s.t.}\quad &h(x) = 0 \\ &\underline{g} \le g(x) \le \overline{g} \end{aligned} \tag{5.58}$$

where

$$h(x) = [h_1(x),\ \ldots\ , h_m(x)]^{\mathrm{T}} \tag{5.59}$$

$$g(x) = [g_1(x),\ \ldots\ , g_r(x)]^{\mathrm{T}} \tag{5.60}$$

The steps are as follows:

1. Introduce slack variables to transform the functional inequality constraints to equality constraints

$$g(x) + u = \overline{g}$$

$$g(x) - l = \underline{g}$$

where

$$l = [l_1,\ \ldots\ , l_r]^{\mathrm{T}}$$

$$u = [u_1,\ \ldots\ , u_r]^{\mathrm{T}}$$

2. Append constrained conditions to the objective function to form the Lagrange function

$$L = f(x) - y^{\mathrm{T}}h(x) - z^{\mathrm{T}}[g(x) - l - \underline{g}] - w^{\mathrm{T}}[g(x) + u - \overline{g}] - \tilde{z}l - \tilde{w}u \tag{5.61}$$

3. Educe perturbed KKT first-order optimality conditions

$$\begin{aligned} L_{x0} &\equiv \nabla_x f(x) - \nabla_x h(x)y - \nabla_x g(x)(z + w) = 0 \\ L_{y0} &\equiv h(x) = 0 \\ L_{w0} &\equiv g(x) + u - \overline{g} = 0 \\ L_{z0} &= g(x) - l - \underline{g} = 0 \end{aligned} \tag{5.62}$$

and perturbed complementary conditions

$$L_l^{\mu} = LZe - \mu e = 0, L_u^{\mu} = UWe - \mu e = 0 \tag{5.63}$$

where

$$(l, u, z) \geq 0, w \leq 0, y \neq 0$$

$$L_{x0} = \frac{\partial L}{\partial x} \tag{5.64}$$

Other similar forms have similar meanings: $L = \text{diag}(l_1, \ldots, l_r)$, Z, W and U have the same expression; $y \in R^m$, $(z, w) \in R^r$, $e = [1, \ldots, 1]^T \in R^r$.

4. Solve the perturbed KKT equations by the Newton method:

$$\begin{bmatrix} H & \nabla_x h(x) & \nabla_x g(x) & \nabla_x g(x) & 0 & 0 \\ \nabla_x^T h(x) & 0 & 0 & 0 & 0 & 0 \\ \nabla_x^T g(x) & 0 & 0 & 0 & -I & 0 \\ \nabla_x^T h(x) & 0 & 0 & 0 & 0 & I \\ 0 & 0 & L & 0 & Z & 0 \\ 0 & 0 & 0 & U & 0 & W \end{bmatrix} \begin{bmatrix} \Delta x \\ \Delta y \\ \Delta z \\ \Delta w \\ \Delta l \\ \Delta u \end{bmatrix} = \begin{bmatrix} L_x \\ -L_y \\ -L_z \\ -L_w \\ -L_l^\mu \\ -L_u^\mu \end{bmatrix} \tag{5.65}$$

Exchange some lines and rows to obtain the following equivalent matrix:

$$\begin{bmatrix} L & Z & 0 & 0 & 0 & 0 \\ 0 & -I & 0 & 0 & \nabla_x^T g(x) & 0 \\ 0 & 0 & U & W & 0 & 0 \\ 0 & 0 & 0 & I & \nabla_x^T g(x) & 0 \\ \nabla_x g(x) & 0 & \nabla_x^T g(x) & 0 & H & \nabla_x h(x) \\ 0 & 0 & 0 & 0 & \nabla_x^T h(x) & 0 \end{bmatrix} \begin{bmatrix} \Delta z \\ \Delta l \\ \Delta w \\ \Delta u \\ \Delta x \\ \Delta y \end{bmatrix} = \begin{bmatrix} -L_l^\mu \\ -L_z \\ -L_u^\mu \\ -L_w \\ L_x \\ -L_y \end{bmatrix} \tag{5.66}$$

where

$$H = -[\nabla_x^2 f(x) - \nabla_x^2 h(x) y - \nabla_x^2 g(x)(z + w)] \tag{5.67}$$

The reduced matrix is obtained by use of the transform technique:

$$\begin{bmatrix} I & L^{-1}Z & 0 & 0 & 0 & 0 \\ 0 & I & 0 & 0 & -\nabla_x^T g(x) & 0 \\ 0 & 0 & I & U^{-1}W & 0 & 0 \\ 0 & 0 & 0 & I & \nabla_x^T g(x) & 0 \\ 0 & 0 & 0 & 0 & H' & \nabla_x h(x) \\ 0 & 0 & 0 & 0 & \nabla_x^T h(x) & 0 \end{bmatrix} \begin{bmatrix} \Delta z \\ \Delta l \\ \Delta w \\ \Delta u \\ \Delta x \\ \Delta y \end{bmatrix} = \begin{bmatrix} -L^{-1}L_l^\mu \\ L_z \\ -U^{-1}L_u^\mu \\ -L_w \\ L_x' \\ -L_y \end{bmatrix} \tag{5.68}$$

where

$$\begin{aligned} H' &= H - \nabla_x g(x)[L^{-1}Z - U^{-1}W]\nabla_x^T g(x) \\ L_x' &= L_x + \nabla_x g(x)[L^{-1}(L_l^\mu + ZL_z) + U^{-1}(L_u^\mu - WL_w)] \end{aligned} \tag{5.69}$$

Here, we can speed up the calculation and simplify the program by factorizing the undersized symmetrical matrix that lies at the bottom right corner of (5.68). Finally, we obtain a new approximate solution by solving (5.68) for offsets of all variables in step k:

$$x^{(k+1)} = x^{(k)} + \alpha_p \Delta x^{(k)}, l^{(k+1)} = l^{(k)} + \alpha_p \Delta l^{(k)}$$
$$u^{(k+1)} = u^{(k)} + \alpha_p \Delta u^{(k)}, y^{(k+1)} = y^{(k)} + \alpha_d \Delta y^{(k)}$$
$$z^{(k+1)} = z^{(k)} + \alpha_d \Delta z^{(k)}, w^{(k+1)} = w^{(k)} + \alpha_d \Delta w^{(k)}$$

where

$$\alpha_p = 0.9995\min\left\{\min_i\left(\frac{-l_i}{\Delta l_i}, \Delta l_i < 0; \quad \frac{-u_i}{\Delta u_i}, \Delta u_i < 0\right), 1\right\}$$
$$\alpha_d = 0.9995\min\left\{\min_i\left(\frac{-z_i}{\Delta z_i}, \Delta z_i < 0; \quad \frac{-w_i}{\Delta w_i}, \Delta w_i > 0\right), 1\right\}$$
$$i = 1, 2, \ldots, r$$

At every iteration step, barrier parameter $\mu = \sigma\, Gap/(2r)$, where $Gap = l^{\mathrm{T}}z - u^{\mathrm{T}}w$, is regarded as the dual-gap, and $\sigma \in (0, 1)$ is called the "central parameter". It is proved that $\sigma = 0.1$ will bring on better convergence for MIP in most situations.

It is obvious that the major computation efforts of the algorithm lie in the formation and factorization of the modifying equation coefficient matrix. Therefore, excellent data structure and sparse matrix techniques made such an algorithm more attractive and successful: (1) voltage expression in rectangular coordinates is used to make the Hessian matrix constant; (2) a reduced correction equation is derived by rearranging and blocking of the problem variables to speed up the computation; and (3) a 44 block submatrix has been realized by rearranging the modifying equation, which corresponds to the elements of the nodal admittance matrix so that the algorithm can be programmed conveniently. By the way, the number of fill-ins is reduced extremely, which can decrease the calculation efforts, and speed up the calculation.

5.3.2.2 Flow of OPF with MIP

STEP 0. Initialization:
(1) Let initial slack variables l, u *satisfy* $[l, u]^{\mathrm{T}} > 0$;
(2) Let initial dual variables z, w, y *satisfy* $z > 0, w < 0, y = 0$;
(3) Set $\sigma \in (0, 1)$, *convergence precision* $\varepsilon = 10^{-8}, k = 0$;
(4) Set initial continuous variables:

$$x = [P\ Q\ e\ f] = (x_{min} + x_{max})/2 \tag{5.70}$$

Do While $k < K_{max}$
Step 1. Compute dual-gap. If dual-gap $< \varepsilon$*, then stop, else go to STEP 2;*
Step 2. Compute the barrier parameter μ*;*
Step 3. Solve (5.68) *for* $\Delta x, \Delta y, \Delta l, \Delta u, \Delta z, \Delta w$*;*
Step 4. Search the minimal step length of primal and dual α_p, α_d*;*
Step 5: Update variables;
$k = k + 1$;
End Do

5.3.3 Genetic Algorithm with Annealing Selection (AGA)

5.3.3.1 Genetic Algorithm and Simulated Annealing Algorithm

The simulated annealing algorithm (SA) and the simple genetic algorithm (SGA) have some resemblances: (1) they only need the objective function value (or the fitness of each individual) to search for the optimum; and (2) both optimization processes use a stochastic operator to search for the global optimum of the optimal problem. From another point of view, SA is different from AGA in the following respects: (1) SA is globally convergent because it adopts a Boltzmann policy which manages the optimization process by a parameter. On the contrary, GA is pre-convergent due to not having such a strategy. (2) During each iteration, GA synchronously operates on each individual in the population. As a result, GA can keep down all the useful information which relates to the former population. However, SA just saves one feasible solution in the vicinity of the former solution. (3) SA keeps down some faintish solutions with certain probability in every iteration step. GA only reserves the individuals that are better than the former population. Consequently, SA can escape from local optima and tend toward the global optimum. GA likely fails to find the global optimum because of gene loss. Thus it can be seen that incorporating GA and SA can be more effective.

5.3.3.2 Genetic Algorithm with Annealing Selection

In order to incorporate GA and SA, one needs to introduce an annealing factor into the selection scheme of SGA to form an entirety selection scheme (ESS). ESS constructs a father population F_k by randomly selecting some individuals in the population P_k, where the probability of $b_i \in P_k$ being selected is

$$P = e^{f(b_i)/T_k} \bigg/ \sum_{b_j \in P_k} e^{f(b_j)/T_k}$$

where $f(x)$ is a merit function, and T_k is the annealing temperature, which tends to zero during the iteration. It can be proved that a necessary and sufficient condition that genetic algorithm with ESS is convergent is that the father population joins the competition. Therefore this kind of genetic algorithm is called a genetic algorithm with annealing selection (AGA).

In numerical tests, AGA takes on unstable convergence speed and ability to obtain the global optimum, which consequently harm the performance of AGA. For the sake of improving the algorithm, the following techniques are used.

(1) Semi-determination ESS In the selection scheme of SGA, the probability of an individual which is selected into the father population relies on the value of its merit function. In this regard, ESS is uniform except that it modifies the fitness by use of the annealing temperature. By reason of a randomizer, individuals in the father population likely do not accord with our imagination. If the expected amount of an individual in the new father population is

$$N_F = \left(e^{f(b_i)/T_k} \Big/ \sum_{b_j \in P_k} e^{f(b_j)/T_k}\right)$$

then, in order to reduce the impact of the randomizer, we exactly select the father individual that depends on the amount which is calculated by the above formula. We call this selection strategy "semi-determination ESS".

(2) Using multifarious operators in turn In AGA, multifarious operators are used in turn, depending on their function. For example, the action of a random mutation operator is to remodel every bit of chromosome within its bound. A self-adjusted mutation operator can modify every bit of chromosome depending on the difference of fitness between this individual and the best individual—if the difference is less, the mutation region is narrower. So the random mutation operator is used at the beginning of the iteration to ensure the multiplicity of population, and the self-adjusted mutation operator is used at the last stage of the iteration to improve the solution. On balance, using multifarious operators in turns helps to improve the performance of the algorithm.

(3) Father population unchangeably appended to new population The new population completely consists of the mutated individuals whose tie to the father population will induce the loss of a better individual in the original population, and more iterations and longer calculation time. Otherwise, sort the individuals of the father population and the mutated population by their fitness, and then select the better one of them to form the new population. This process will cut down the multiplicity of the population and result in pre-constringency and stagnation phenomenon. To count the cost, half of the new population consists of the unchanged father population, and the other part of the new population consists of the mutated individuals who tie to the father population. So the multiplicity of the population ensures that the algorithm obtains the global optimum with fairly large probability. At the same time, in every iteration step, the

best individual is held, along with the generations that contain more and more better individuals, so the efficiency of the algorithm is ensured.

5.3.4 Flow of Presented Algorithm

The main idea of the proposed algorithm is to combine the advantages of AGA and MIP to respectively deal with continuous and discrete variables of the OPF problem, which not only guarantees the calculation speed of the new algorithm, but also improves the ability to seek the global optimum. The flow of the new algorithm is as follows.

STEP 0. Initialization: Form the original chromosome group by taking every discrete variable of the optimal problem within their bounds, let $k = 0$;
STEP 1. Do While $k <$ *size of group;*
    STEP 2. Determine discrete variables of optimal problem according to kth chromosome, then modify nodal admittance matrix; thus there are only continuous variables in optimal problem;
    STEP 3. Turn to run MIP program of OPF for the objective function value, let the fitness of kth chromosome be this value; $k = k + 1$;
End Do
    STEP 4. If convergence condition is satisfied, then stop; else go to STEP 5;
    STEP 5. Modify fitness of every individual in group by annealing temperature, compute expectation amount N_F *of individual which will append into the father group,* $k = 0$;
Do While $k <$ *size of group;*
    STEP 6. Select two father individuals depend on N_F*, and record their numbers;*
    STEP 7. According to P_c*, crossbreed two father individuals to obtain two new individuals;*
    STEP 8. According to P_m*, mutate two new individuals by use of multifarious operators in turn;* $k = k + 2$;
End Do
STEP 9. Form the new population by special strategy, $k = 0$*, go to STEP 1.*

5.3.5 Example

The proposed algorithm was implemented using the Visual C++6.0 language and the software program was executed on an 800 MHz Pentium Pro computer. Numerical simulations on IEEE 14- to 118-bus standard test systems have been done to test the performance of this new algorithm.

Table 5.4 Different settings of ratio percentage

Percentage	2.5% ($)	2% ($)	1.5% ($)	0.5% ($)
AGA	11 841.63	11 841.56	11 841.27	11 842.16
Before rounding	11 753.83	11 747.28	11 746.95	11 746.25
After rounding	11 869.73	11 850.76	11 850.56	11 850.69

5.3.5.1 Necessity of Accurately Handling Discrete Variables

There are many methods to handle discrete variables, among which are: (1) the continued method and (2) the branch-and-bound method. The continued method treats discrete variables as continuous variables during the solution process. Once the continuous solution is found, each discrete variable moves to its nearest discrete setting; this is called the "rounding process". Owing to its complexity and tremendous execution time, the application of and research on the branch-and-bound method in power systems are limited. From Table 5.4 we find that the difference between the results of the continued method before and after the rounding process is considerable, so its optimal characteristic is lost. On the other hand, the rounding process will introduce an unequal error on account of the different settings of discrete variables. However, AGA is insensitive to change in the discrete variable setting, and it can steadily find the optimum and secondary optimum in various instances. Moreover, during the iteration, the solution of AGA is always superior to that of the continued method after the rounding process. All in all, it is absolutely essential to accurately handle discrete variables.

5.3.5.2 Average Convergence Characteristic

In order to evaluate the convergence characteristic of an algorithm, we have the following definitions. For running the algorithm once, we denote the convergence characteristic of the algorithm by use of the best solution change curve during iteration, and that average convergence characteristic is an average among the results of running the algorithm many times. The average convergence characteristics of AGA and SGA for running the program 50 times are summarized in Figure 5.13. In Figure 5.13 the two curves initially slope steeply, which illustrates that the solution improves rapidly before the 20th iterative step. Thereafter the slopes of curves tend toward zero, which illustrates that the solution hardly changes. Because the slope of SGA is less than that of AGA in the first part of the curve, SGA always obtains local optima. On the other hand, the average convergence curve of AGA is below that of SGA during iterations, which is to say that the ultimate solution of AGA is always superior to that of SGA.

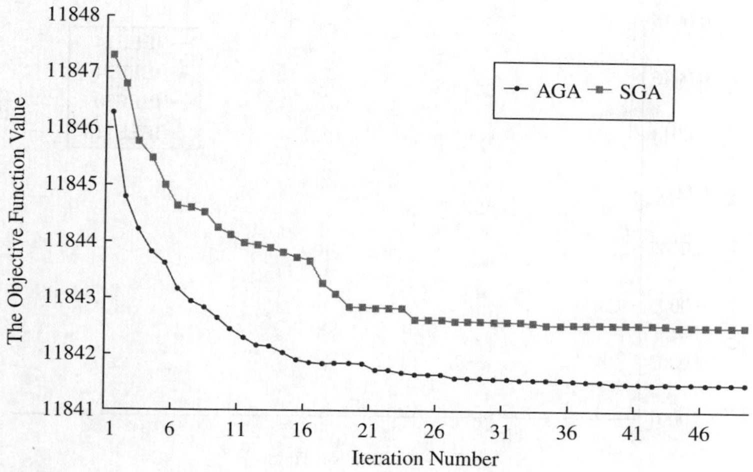

Figure 5.13 The average convergence characteristics of AGA and SGA.

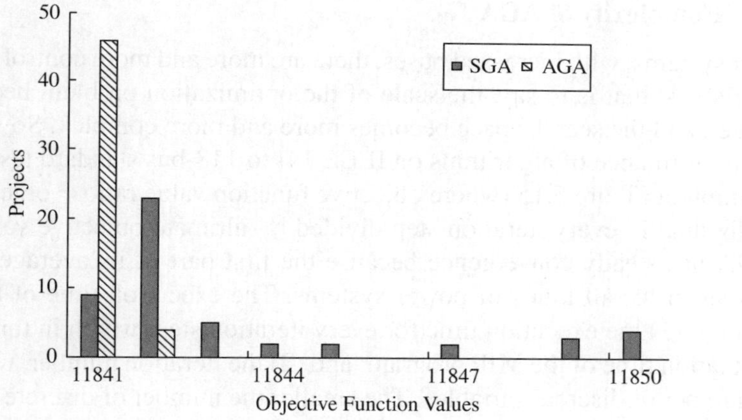

Figure 5.14 The statistics of ultimate solution for AGA and SGA.

5.3.5.3 Statistic of Ultimate Solution

In order to review the impact of different original chromosome groups, the distributions of the ultimate solution for running the program 50 times are summarized in Figure 5.14. The statistics show that: the average ultimate solution of SGA is 11 843.9277, variance is 8.3555; the average ultimate solution of AGA is 11 841.5213, variance is 0.07431. Comparing AGA and SGA, the solution of AGA is better and the distribution is more concentrated; thus it is proved that AGA has robust ability to search for the global optimum.

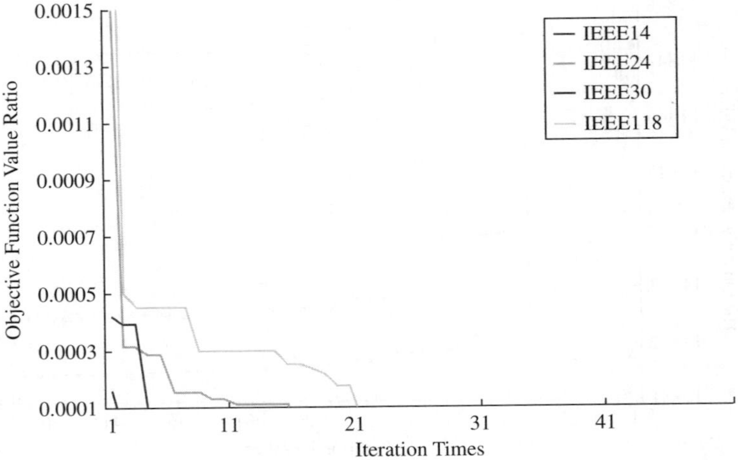

Figure 5.15 The average convergence characteristic of the presented algorithm.

5.3.5.4 Complexity of AGA

In power systems, with increased buses, there are more and more control variables and state variables, that is to say, the scale of the optimization problem becomes bigger and bigger, and the search space becomes more and more complex. So we separately test the performance of algorithms on IEEE 14- to 118-bus standard test systems.

The graph in Figure 5.15 (where objective function value ratio = objective value of best individual in every iteration step divided by ultimate objective value) indicates that AGA has steady convergence because the first part of its average convergence curve is steep for all kinds of power systems. The execution time of the algorithm depends on: (1) the execution time for every iteration step, which in turn depends on the calculation time of the MIP program; and (2) the iteration number, which depends on the number of discrete variables. The smaller the number of discrete variables, the more quickly the optimum will be found, and accordingly the iteration numbers are lower. The steady convergence characteristic and fast calculation speed of MIP ensure the performance of the proposed algorithm. Furthermore, the graph in Figure 5.16 shows that the number of discrete variables has less impact on the iteration numbers of the algorithm. Altogether, the execution time of the proposed algorithm does not obviously change as the scale of the system becomes larger and larger.

5.3.6 Summary

The deregulated power industry requires OPF calculations that are more multifunctional, more real-time, and more reliable, so it is pressing to find a fast and efficient algorithm for OPF. Because the OPF problem includes a lot of continuous and discrete

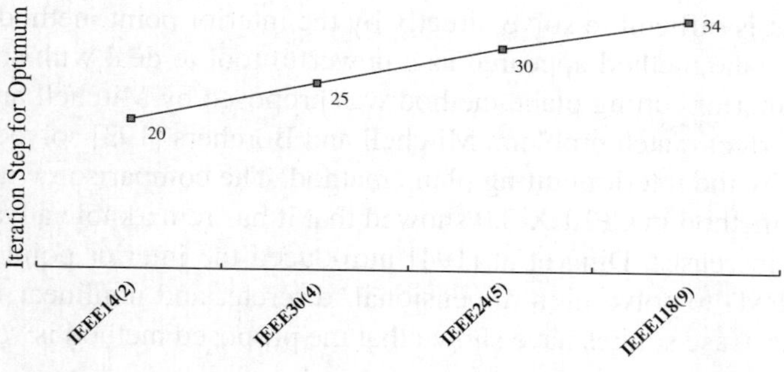

Figure 5.16 The complexity of the proposed algorithm.

variables, it is impossible to accurately and effectively handle all of them by using a single method. Using AGA to deal with discrete variables, we proposed a new algorithm to solve the OPF problem through MIP based on perturbed KKT conditions. The advantages of this new algorithm are as follows.

1. Dealing with continuous variables by MIP to ensure the whole calculation speed and steady convergence characteristic, even in large-scale power systems.
2. Accurately dealing with discrete variables by AGA to obtain better solution.
3. Modified AGA is superior to SGA in convergence and escaping local optima, especially for large-scale problems.

Numerical simulations on IEEE 14- to 118-bus standard test systems have shown that the proposed method is very promising in solving OPF problems for large-scale power systems.

5.4 Optimal Power Flow with Discrete Variables Based on Interior Point Cutting Plane Method

Optimal power flow (OPF) has become a noticeable research area since it was presented by Carpentier in the 1960s [184]. As an extension of economic dispatch, it combines economy with security and deals well with various types of constraints. There are two types of algorithms for solving OPF. The fist is an analytic algorithm, including a simplified gradient method, Newton's method, linear programming method, interior point method, etc. The second [185, 186] is a heuristic algorithm, including genetic algorithm, tabu search, etc.

In recent years, the interior point method 187–189 has been widely used to solve OPF as it has fast convergence characteristics and can deal with inequalities conveniently. Generally, the interior point method can deal with problems in which all variables are continuous. If adjustable transformer taps, parallel capacitors, and reactors are introduced, OPF becomes a mixed integer nonlinear programming problem

[190, 191] that is difficult to solve directly by the interior point method. The interior point cutting plane method appeared as a powerful tool to deal with such problems. In 1992, the interior cutting plane method was proposed by Mitchell and Todd [192] to solve the perfect match problem. Mitchell and Borchers [193] solved linear ordering problems by the interior cutting plane method. The comparison with the simplex cutting plane method in CPLEX 3.0 showed that it has remarkable advantages as the problem size increases. Ding et al [194] introduced the interior point cutting plane method (IPCPM) to solve high-dimensional, discrete, and nonlinear mixed integer OPF problems. Case studies have shown that the proposed method is very efficient in solving OPF problems for large-scale power systems.

The cutting plane method is used in IPCPM to solve the linearized mixed integer programming of OPF. There are several kinds of cutting plane methods, including analytic center cutting plane method (ACCPM) [195, 196], Kelly's cutting plane [197], extended cutting plane [198], bundle methods [199], and Gomory cutting plane [200]. The Gomory cutting plane is the very kind of cutting plane used in IPCPM. Two optimization problems are solved alternately. The main procedures are as follows:

1. Obtain a feasible solution by solving OPF by ignoring integer restrictions.
2. The OPF problem is linearized at the feasible point obtained from the first optimization problem.
3. The mixed integer linear programming problem is solved by IPCPM to determine the values of the integer variables. First, the linear programming relaxation is solved by ignoring the integrality restrictions. If the solution to the relaxation is feasible, stop with optimality. Otherwise, find one or more cutting planes that separate the optimal solution from the convex hull of feasible integral points, and add a subset of these constraints to the relaxation. Then resolve the relaxation. The procedure is repeated until integer variables are obtained.

The three steps are repeated alternately until the optimal solutions are obtained. The detailed procedures of IPCPM are shown in Figure 5.17.

However, if the optimal solution of the relaxation problem solved by interior point method is a degenerate solution or a convex combination solution, the cutting planes fail to be generated, which leads to the failure of IPCPM. So in this section, IPCPM is improved. A new base identification method is proposed. In this method, the optimal solution is moved from the interior of the optimal face to the vertex point of the convex hull through transformation of coordinates, and then the optimal base is obtained during this procedure. The improved algorithm can find the optimal base for various types of optimal solutions. Also, the efficiency of the base identification procedure is improved. A long computation time may be consumed in matrix rank calculation and row–column transformation. A perturbation method and some linear algebra skills are introduced to IPCPM, which can significantly improve computational efficiency. The

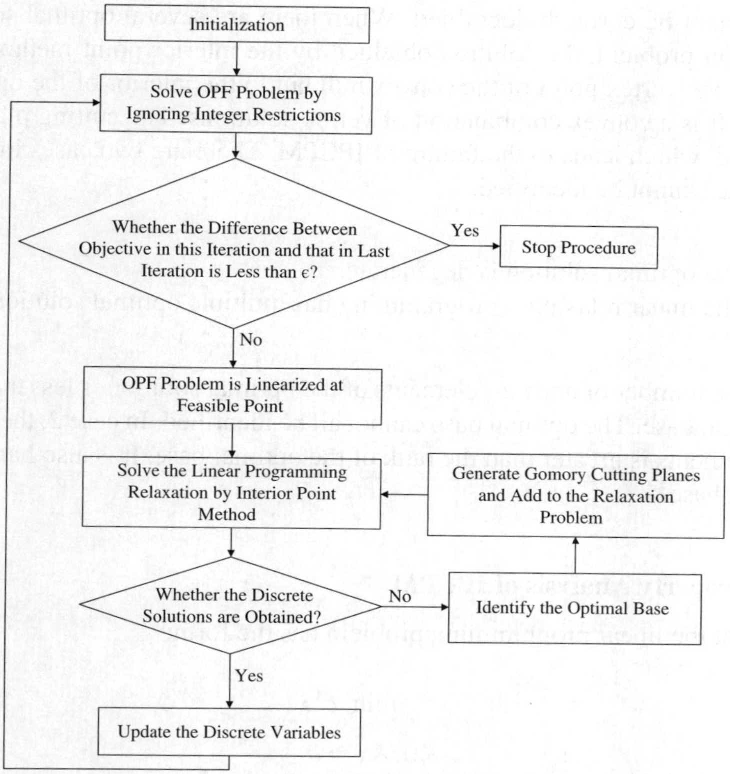

Figure 5.17 Flowchart of solving OPF by IPCPM. *Source*: Liu 2009. Reproduced with permission from IEEE.

tests of IEEE systems show the validity of the improved IPCPM. The improved algorithm can deal with various types of optimal solutions efficiently and is much more useful in application.

5.4.1 IPCPM and Its Analysis

There are two key steps for IPCPM:

1. Identify optimal base matrix from the solutions.
2. Generate cutting planes based on optimal base.

The Gomory cutting plane can be generated once the optimal base is obtained. For the traditional simplex cutting plane method (SCPM), the optimal solution converges to the vertex point of the convex hull. The optimal base can be obtained based on the simplex tableau and then cutting planes are generated. However, for IPCPM, if the relaxation linear program is a degenerate problem or problem with multiple solutions, the base cannot be correctly identified. When there are several optimal solutions to the relaxation problem, the solution obtained by the interior point method may not converge to the vertex point of the convex hull but to the interior of the optimal face [193, 201]. It is a convex combination of vertex solutions. The cutting planes fail to be generated, which leads to the failure of IPCPM. There are two cases in which the optimal base cannot be identified.

- Case 1: the optimal solution is degenerate.
- Case 2: the linear relaxation programming has multiple optimal solutions.

In case 1, the number of nonzero elements of the optimal solution is less than the rank of the optimal base. The optimal base cannot all be identified. In case 2, the number of nonzero elements is greater than the rank of the optimal base. It is also hard to obtain the optimal base.

5.4.1.1 Property Analysis of IPCPM

Assume that the linear programming problem has the form:

$$\begin{aligned} \min\ & c^{\mathrm{T}}x \\ \text{s.t.}\ & Ax = b \\ & x \geq 0 \end{aligned} \tag{5.71}$$

Dual problem:

$$\begin{aligned} \min\ & b^{\mathrm{T}}y \\ \text{s.t.}\ & Ay + s = c \\ & s \geq 0 \end{aligned} \tag{5.72}$$

where $c, x \in R^n$, $b \in R^m$, and $A \in R^{m\times n}$. Assume that A has full row rank.

An example is given to illustrate the difficulty caused by the above two cases:

$$\max\ 2x_1 + 4x_2 \tag{5.73}$$

$$\text{s.t.}\ x_1 + 2x_2 + x_3 = 8 \tag{5.74}$$

$$x_1 + x_4 = 8 \tag{5.75}$$

$$x_2 + x_5 = 3 \tag{5.76}$$

$$x_1, x_2, x_3, x_4, x_5 \geq 0$$

Equations (5.73)–(5.76) are equivalent to

$$\max \ 2x_1 + 4x_2 \tag{5.77}$$

$$\text{s.t.} \ \ x_1 + 2x_2 \leq 8 \tag{5.78}$$

$$x_1 \leq 8 \tag{5.79}$$

$$x_2 \leq 3 \tag{5.80}$$

$$x_1 \geq 0, x_2 \geq 0$$

This is a multiple-solution linear programming problem. There are three types of solutions.

1. Nondegenerate solution: $x' = (2, 3, 0, 6, 0)$, the number of nonzero elements equals the number of constraints, three.
2. Degenerate solution: $x' = (8, 0, 0, 0, 3)$, the number of nonzero elements is less than the number of constraints, three.
3. Convex combination solution:

$$x''' = \alpha x' + (1 - \alpha)x''' \quad \text{and} \quad \alpha \in (0, 1)$$

For example $x''' = (4.5, 1.75, 0, 3, 5, 1.25)$ when $\alpha = 7/12$. The number of nonzero elements is greater than the number of constraints, three.

The geometric meanings of the above optimal solution types are shown in Figure 5.18. Line AC represents (5.80), line CD represents (5.79), line BD represents (5.78), and the area enclosed by ABDO is the feasible region constructed by (5.77)–(5.80). Because the line BD is parallel to the line represented by the objective function, any point on line BD is an optimal solution. Points B, D, and P correspond to x', x'', and x''', respectively.

As the simplex method searches for optimal solutions through vertices, the solution is definitely a vertex point of the optimal face. The vertex solution corresponds to a non-degenerate solution (point B) or a degenerate solution (point D) as shown in Figure 5.19. The Gomory cutting planes can be generated from the final tableau. Obviously, a simplex tableau is not available when using an interior point method. The interior point method searches for the optimal solution through the interior of the feasible region, and all three types of solutions may be obtained. It can be seen in Figure 5.18 that the optimal solution is more likely to converge on line BD. In other words, the solution is probably a convex combination solution. In this situation, IPCPM would fail, as the optimal base cannot be identified.

5.4.1.2 Multiple-Solution Characteristic of OPF

The mixed integer optimal power flow (MIOPF) is a multiple-solution problem. The above difficulties stated in section 5.4.1.1 will be encountered as IPCPM is applied.

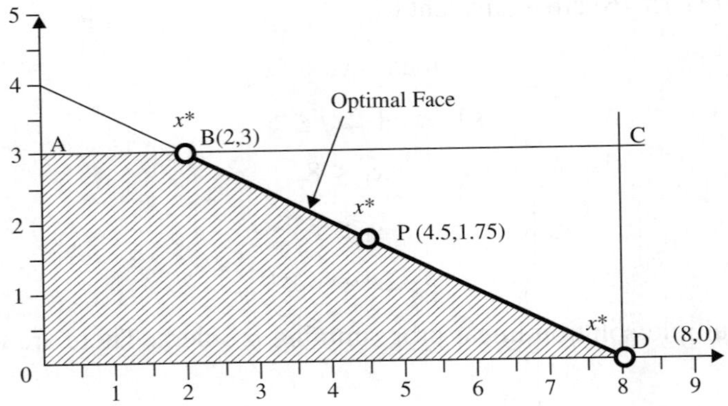

Figure 5.18 Schematic diagram of optimal solutions. *Source*: Liu 2009. Reproduced with permission from IEEE.

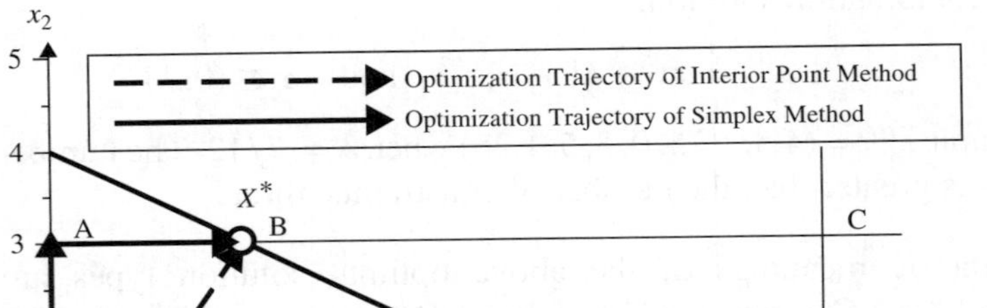

Figure 5.19 The optimizing trajectory comparison of the simplex method to the interior point method. *Source*: Liu 2009. Reproduced with permission from IEEE.

Assume that the MIOPF has the form:

$$\min \; F(P_G) = \sum_{i \in S_G} (a_{2i} P_{Gi}^2 + a_{1i} P_{Gi} + a_{0i}) \tag{5.81}$$

$$P_{Gi} - P_{Di} - \sum (e_i(e_j G_{ij} - f_j B_{ij}) + f_i(f_j G_i j + e_j B_i j)) = 0 \tag{5.82}$$

$$Q_{Ri} - Q_{Di} - \sum (f_i(e_j G_{ij} - f_j B_{ij}) - e_i(f_j G_i j + e_j B_i j)) = 0 \tag{5.83}$$

$$\underline{P_{Gi}} \leq P_{Gi} \leq \overline{P_{Gi}}, i \in S_G \tag{5.84}$$

$$\underline{Q_{Ri}} \leq Q_{Ri} \leq \overline{Q_{Ri}}, \quad i \in Q_R \tag{5.85}$$

$$\underline{V_i^2} \leq (e_i^2 + f_i^2) \leq \overline{V_i^2} \tag{5.86}$$

$$\underline{t_i} \leq t_i \leq \overline{t_i}, \quad i \in T \tag{5.87}$$

$$\underline{c_i} \leq (c_i) \leq \overline{c_i}, \quad i \in C \tag{5.88}$$

$$\underline{l_i} \leq l_i \leq \overline{l_i}, \quad i \in L \tag{5.89}$$

$$\underline{P_{ij}} \leq (P_{ij}) \leq \overline{P_{ij}} \tag{5.90}$$

$$P_{ij} = (-e_i^2 - f_i^2 + e_i e_j + f_i f_j)G_{ij} + (e_j f_i - e_i f_j)B_i j$$

where the notation is as follows:

$F(P_G)$	total generation cost;
a_{0i}, a_{1i}, a_{2i}	fuel cost coefficients of generator i;
S_G	set of all active power sources;
Q_R	set of all reactive power sources;
P_{Gi}, Q_{Ri}	active and reactive power generation at bus i;
P_{Di}, Q_{Di}	active and reactive demand at bus i;
t_i	tap position of transformer i;
c_i	bank number of capacitor at bus i;
l_i	bank number of reactor at bus i;
T, C, L	sets of adjustable transformers, capacitors, and reactors;
e_i, f_i	real and imaginary parts of voltage vector at bus i;
G_{ij}, B_{ij}	real and imaginary parts of admittance between nodes i and j;
P_{ij}	power flow on transmission line between nodes i and j.

Equations (5.82) and (5.83) relate to the system active and reactive power balance; (5.84) shows the output limitations of active power; (5.85) shows the output limitations of reactive power; (5.86) relates to limitations of bus voltage; (5.87) relates to the tap limitations of adjustable transformers; (5.88) and (5.89) are the capacitor and reactor bank number limitations; and (5.90) shows the limitations of power flows in transmission lines.

The tap of adjustable transformers and the numbers of capacitor banks and reactor banks mainly affect the system reactive power flow. However, the objective function is to minimize the active power cost. It is probable that the objective function can maintain optimality when these discrete variables change in certain ranges. That is the reason why there are multiple solutions for MIOPF. The five-bus system is taken as an example. The relationship between the objective function and transformer tap position is shown in Figure 5.20.

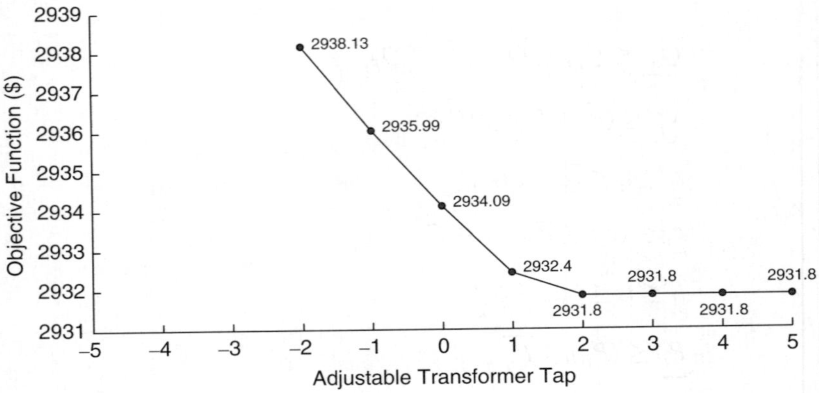

Figure 5.20 Relationship between objective function and transformer tap in five-bus system. *Source*: Liu 2009. Reproduced with permission from IEEE.

Table 5.5 The type of optimal solutions

Systems	Dimension of coefficient matrix A	Nonzero element number of solution	Optimal solution type
5	38×46	40	Convex combination solution
14	99×114	101	Convex combination solution
24	169×194	172	Convex combination solution
30	197×214	199	Convex combination solution
57	366×390	368	Convex combination solution
118	794×880	795	Convex combination solution
300	1917×2034	1923	Convex combination solution

When the transformer tap positions are 2, 3, 4, and 5, the objective values are kept constant, indicating that this is a multiple-solution problem.

This is also the case for large-scale systems. The test results of IEEE 14- to 30-bus systems are listed in Table 5.5.

The rank of the coefficient matrix is less than the number of nonzero elements in the optimal solutions. In other words, these are all convex combination solutions. The results indicate that MIOPF is a multiple-solution problem. There is an urgent need to make improvements to IPCPM to deal with such problems.

5.4.2 *Improvement of IPCPM*

According to the above analysis in section 5.4.1, the key is to identify the optimal base when MIOPF with multiple solutions is solved by IPCPM. Currently there are several approaches to identify the optimal base [202].

1. The first approach is the standard purification procedure to find a feasible vertex where the objective value is at least as good as the current point. This approach can be done in strongly polynomial time [203].
2. The second approach is a theoretical effort to identify the optimal base [204]. If the LP problem is nondegenerate, the unique optimal bases can be identified before the worst-case time bound. The procedure seems to work for some LP problem; however, it has difficulty for degenerate LP problems. The difficulty arises simply because any degenerate LP problem has multiple optimal bases.
3. The third approach is to slightly randomize the data. The new LP problem is nondegenerate. Its optimal base remains one of the optimal bases of the original LP problem [205].

The approach used in Ding et al [194] is the second method above. The improved IPCPM applies the approach in Megiddo [203].

5.4.2.1 Purification Method

Based on the first approach, the optimal base identification procedure is as follows. If the optimal solution obtained by the interior point method is in the interior of the optimal face, the optimal solution is moved through pivot steps to a vertex point and the objective value is at least as good as the current one; meanwhile the optimal base is identified.

Theorem 5.1 *There exists a strongly polynomial-time algorithm that finds an optimal base, given optimal solutions for both the primal and the dual [203].*

The procedures are shown in Figure 5.21 and described in detail as follows.

The columns of matrix A and elements of c are classified into three parts; and the index set $\{1, \ldots, n\}$ is partitioned into subsets J_1, J_2, and J_3:

$$A_1 = \{A_{\cdot j} \mid j \in J_1\}, \quad c_1 = \{c_j \mid j \in J_1\}$$

$$A_2 = \{A_{\cdot j} \mid j \in J_2\}, \quad c_2 = \{c_j \mid j \in J_2\}$$

$$A_3 = \{A_{\cdot j} \mid j \in J_3\}, \quad c_3 = \{c_j \mid j \in J_3\}$$

$$J_1 = \{j \mid x_j > 0, s_j = 0\}, \quad J_2 = \{j \mid x_j = 0, s_j = 0\}, \quad J_3 = \{j \mid x_j = 0, s_j > 0\}$$

The steps in the algorithm are as follows:

1. Classify the type of solution: nondegenerate solution, degenerate solution, or convex combination solution. The index sets J_1, J_2, and J_3 are also obtained.
   (a) If it is a nondegenerate solution, the index set of nonzero elements is the index set of the optimal base, let $B = A_1$, go to Step 10.
   (b) If it is a degenerate solution, let $B = A_1$, go to Step 6.

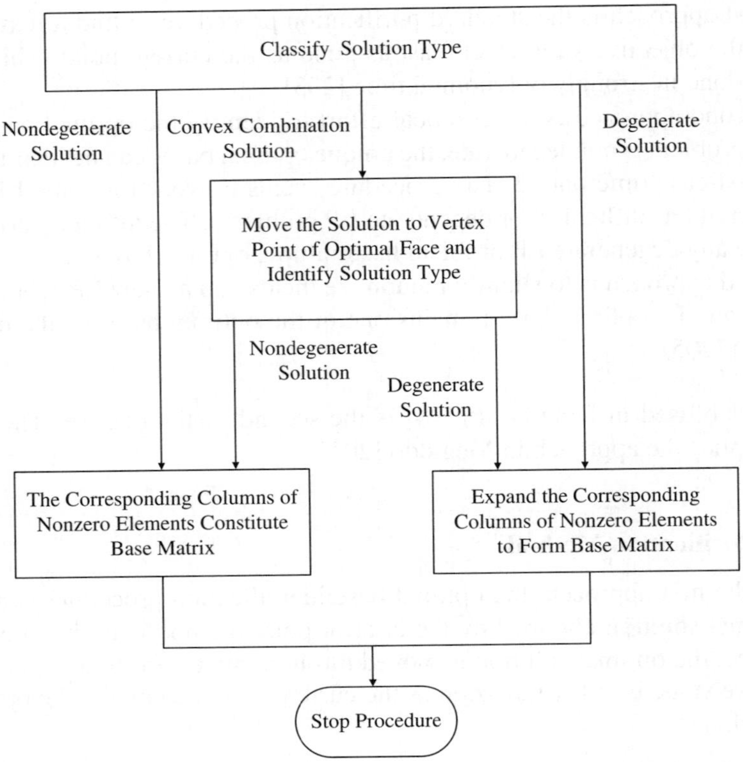

Figure 5.21 Flowchart of optimal base identification. *Source*: Liu 2009. Reproduced with permission from IEEE.

   (c) If it is a convex combination solution, go to Step 2.
2. The movement of primal optimal solution:
   (a) Solve equation $A_1 z = 0$ to get z. As A_1 has linearly dependent columns, z may have multiple solutions. Choose one of nonzero vector z.
   (b) Solve $x_1' = x_1 + tz \geq 0$, calculate the range of scalar t: $t_{min} \leq t \leq t_{max}$.
3. Choose any one from t_{min} and t_{max}, let $x_1' = x_1 + tz$. For certain $x_{1j}' = 0$ (x_{1j}' represents the jth element of x_1'), x_{1j}' is removed from x_1' and added to x_2, the column corresponding to x_{1j}' is removed from A_1 and added to A_2, let $x_1 = x_1'$, $x = (x_1', x_2, x_3)$.
4. If A_1 is column-dependent, go to Step 2, else let $B = A_1$, go to next step.
5. If $\text{rank}(B) < \text{rank}([A_1, A_2])$ go to Step 8, else go to next step.
6. Add the columns of A_2 which are linearly independent of A to B.
7. If $\text{rank}(B) = m$, go to Step 10, else go to next step.
8. The movement of dual optimal solution:
   (a) Solve equation $B^{\mathrm{T}} z = 0$ to get z, choose any one of nonzero vector z.
   (b) According to $A_3^{\mathrm{T}} y' \leq c_3$, calculate the range of t: $t_{min} \leq t \leq t_{max}$.

9. Choose any one of t_{min} and t_{max}, let $y' = y - tz$, certain j satisfies $a_{3j}^{\mathrm{T}} y' = c_{3j}$ (where a_{3j} represents the jth column of A_3, c_{3j} is the jth element of c_3), a_{3j} is removed from A_3 and added to A_2 and B, meanwhile B should be kept column linearly independent, go to Step 7.
10. Stop the procedure, B is the optimal base matrix.

5.4.2.2 Improve the Efficiency of Proposed Approach

The above optimal base identification procedure can deal with various solution types. However, for a large-scale system, the above optimal base identification procedure may encounter the following problems.

1. If the optimal solution is a convex combination solution, the primal solution is moved to the vertex point of the optimal face through pivot steps. During this process, primal optimality should be kept. Only one linearly dependent column in A_1 can be removed each time. If the nonzero elements of optimal solutions are much greater than the rank of the coefficient matrix, large computation times are consumed in pivot steps. Taking the IEEE 300-bus system as an example, the dimension of the coefficient matrix is 1917×2034, and the nonzero element number is 1923. This indicates that six coordinate conversion steps are needed.
2. In Steps 5–9 of section 5.4.2.1, matrix B is expanded until the optimal base is obtained. In each step, only one column can be added meanwhile maintaining the linear independence of B, and the dual optimality is kept. If the solution is highly degenerate, many coordinate conversion steps are needed to obtain the optimal base.

These two problems may be solved in the following ways.

Problem 1. Introduction of perturbation method. The first problem is solved by a perturbation method [205]. A controlled random perturbation is made to the cost vector to generate a vertex solution. The approach makes a controlled random perturbation to the cost vector. It improves the likelihood that the perturbed linear program has unique solution (at a vertex). Then the computational time used in removing optimal solutions from the interior of the optimal face to the vertex point can be saved.

Assume the linear programming problem with perturbation vector has the form

$$
\begin{aligned}
\min \quad & z = (c + r)^{\mathrm{T}} x \\
\text{s.t.} \quad & Ax = b \\
& x \geq 0
\end{aligned}
\tag{5.91}
$$

$$
\begin{aligned}
\min \quad & b^{\mathrm{T}} y \\
\text{s.t.} \quad & A^{\mathrm{T}} y + s = c + r \\
& s \geq 0
\end{aligned}
\tag{5.92}
$$

Equations (5.71) and (5.72) are first solved to a desired accuracy (say 0.1ε); then the cost vector of (5.71) is perturbed by r to obtain (5.91) and (5.92). The key steps of the perturbation method are to decide how and when the perturbation is introduced.

(a) Choice of perturbation vector. Let z^* denote the optimal objective value of (5.71). The vertex solution $\tilde{v}^*$ of (5.71) should satisfy

$$c^{\mathrm{T}}\tilde{v}^* - z^* \leq \varepsilon \tag{5.93}$$

For a prespecified tolerance ε, if the choice of ε is large, then a vertex solution satisfying (5.93) is not necessarily optimal. On the other hand, for an appropriate (small) choice of ε, a vertex solution satisfying (5.93) is optimal.

The size of the perturbation has to be controlled so that the objective value at the optimal solution of the perturbed problem is within the specified tolerance of z^*.

Proposition 5.1 *Assume that an interior point algorithm generates* $\{x^k\}$. *Let* $r_i = \varepsilon\rho_i/(4nx_i^k)$, $1 \leq \rho_i \leq 2$, *and* $\varepsilon > 0$. *Then, for some large k*

$$r^{\mathrm{T}}x^* \leq \varepsilon \tag{5.94}$$

A perturbation vector r is generated as

$$r_i = \frac{\varepsilon}{40n}\frac{RAND(1,2)}{x_i^k} \tag{5.95}$$

Here $RAND(1,2)$ is a function generating a random variable which is uniformly distributed in $(1,2)$. For a randomly generated r as in (5.95) it is expected that (5.91) will almost always have a unique vertex solution.

(b) When the perturbation vector is introduced. The perturbation vector is introduced at the iteration when $\beta^k \leq 0.1\varepsilon$ is met:

$$\beta^k = y^{k\ \mathrm{T}}\xi_x^k + s^k - \xi_s^k + M\sum_{i=1}^{n}\xi_s^k \tag{5.96}$$

where $\xi_x^k = Ax^k - b$, $\xi_s^k = A^{\mathrm{T}}y^k + s^k - c \geq 0$. Thus x^* is the optimal solution of P and satisfies $x_i^* \leq M$, $i = 1, \ldots, n$.

Problem 2. Extension step of optimal base. The second problem is solved as follows. In order to generate Gomory cutting planes, only the primal optimal base is needed. The dual optimality needs not be guaranteed. Then the optimal base can be obtained by expanding B through one row simplest form transformation. First, the matrix is transformed to row simplest form through elementary row transformation; the columns corresponding to the echelon elements form the base matrix.

5.4.2.3 Procedure of IPCPM with Improved Optimal Base Identification Approach

The modified procedures of IPCPM are described as follows.

1. Solve the linearization of the optimal power flow equations with primal–dual interior point. At each iteration, a test is made to find whether the perturbation should be introduced. Once the perturbation vector is introduced, the problem is solved to the optimal solution.
2. Classify the type of solution: nondegenerate solution or degenerate solution (because of the introduction of the perturbation vector, the solution cannot be a convex combination solution). The index sets J_1, J_2, and J_3 are also obtained.
   (a) If it is a nondegenerate solution, the index set of nonzero elements of the optimal solution is the index set of the optimal base, let $B = A_1$, stop the procedure.
   (b) If it is a degenerate solution, go to Step 3.
3. Construct matrix $A' = [A_1, A - A_1]$, where $A - A_1$ is a submatrix of A except for the columns in A_1. In other words, the matrix A' is obtained by reordering the columns of A. As A_1 is already column linearly independent, the columns of A_1 are arranged in the front of A', and the remaining columns of A are arranged subsequently.
4. Through elementary row transformations, A' is transformed to row simplest form. The maximum linearly independent subset B can be obtained. Apparently $\text{rank}(B) = m$, B is the base matrix.
5. If the solutions of integer variables are not integer, Gomory mixed integer cutting planes are generated based on base matrix B. New constraints are added into the relaxation problem, go to Step 11. If the solutions of integer variables are integer, the solutions are the optimal solutions to the mixed integer program, and then stop the procedure.

5.4.3 *Example*

The five-bus system and IEEE 14- to 300-bus systems are taken for computation tests. The results are listed in the following tables. The settings of discrete variables of test systems are listed in Table 5.6.

Table 5.7 shows the results of a comparison between the former algorithm and the improved one. Owing to the convex combination solution, the base matrix cannot be obtained to generate the cutting plane, which leads to the failure of IPCPM. The improved algorithm can deal with various solution types well and get accurate results. The work in this section is a very important supplement for IPCPM.

The convergence times with perturbation vector introduced can be seen in Table 5.8. The convergence times of the five- and 24-bus systems are the same as without the perturbation vector; the convergence times of the 14-, 57- and 58-bus systems

Table 5.6 Setting of discrete variables of test systems

Systems	Number of discrete variables	Number of adjustable transformers	Number of capacitor banks	Number of reactor banks
5	3	1	1	1
14	7	2	3	2
24	9	3	1	5
30	8	3	2	3
57	13	5	4	4
118	16	6	7	3
300	26	10	9	7

Table 5.7 Comparison of results between former algorithm and improved algorithm in one iteration

Systems	Algorithm	Gomory cutting plane number	Adjustable transformer tap	Capacitor bank	Reactor bank
5	Former	Failed	Failed	Failed	Failed
	Improved	2	−1	3	3
14	Former	Failed	Failed	Failed	Failed
	Improved	1	−5, 5	0, 0, 0	3, 3
24	Former	Failed	Failed	Failed	Failed
	Improved	2	−2, −5, 0	0	3, 3, 3, 3, 3
30	Former	Failed	Failed	Failed	Failed
	Improved	0	−10, −10, −10	0, 0, 0	3, 3, 0
57	Former	Failed	Failed	Failed	Failed
	Improved	0	−10, −1, −10,−10, −10	3, 1, 3, 3	3, 0, 0, 3
118	Former	Failed	Failed	Failed	Failed
	Improved	2	−10, −7, 3,−10, −10, −9	3, 3, 3, 3, 3, 0, 3	3, 3, 3
300	Former	Failed	Failed	Failed	Failed
	Improved	2	−10, −10, −10, −9, −5, −6, −1, −10, −10, −4	0, 3, 3, 0, 0, 0, 3, 3, 3	3, 0, 3, 3, 0, 0, 0

increase by one; and those of the 30- and 300-bus systems increase by two. With the increase of system size, the convergence times do not change significantly. Also the differences between objectives with and without the perturbation vector are less than 1×10^{-5}.

Table 5.8 Comparison between results with and without perturbation

Systems	Iteration time in which perturbation is introduced	Convergence times with perturbation vector	Convergence times without perturbation	Difference between objective functions with and without perturbation
5	10	18	18	2.98775×10^{-6}
14	13	21	20	3.22109×10^{-6}
24	11	18	18	3.21519×10^{-6}
30	14	22	20	3.33946×10^{-6}
57	19	26	25	3.41566×10^{-6}
118	22	29	28	3.27533×10^{-6}
300	24	33	31	3.46589×10^{-6}

Table 5.9 Optimal solution type with perturbation vector

Systems	Number of discrete variables	Perturbation time	Number of nonzero elements with perturbation vector	Optimal solution type with perturbation vector
5	3	1	38	Nondegenerate solution
14	7	1	99	Nondegenerate solution
24	9	1	169	Nondegenerate solution
30	8	1	197	Nondegenerate solution
57	13	1	366	Nondegenerate solution
118	16	1	794	Nondegenerate solution
300	26	1	1917	Nondegenerate solution

The difference between the number of nonzero elements in the optimal solutions and the rank of the coefficient determines the efficiency of the optimal base identification procedure. The greater the difference, the more calculations are needed. The computational results in Table 5.9 show that only one perturbation is needed to force the optimal solutions to converge to the vertex point of the optimal face. So in the improved algorithm, only one perturbation is introduced and one row simplest form is calculated through elementary row transformation—the optimal base can thus be identified. The degenerate degree and the difference between number of nonzero elements and the rank of the matrix do not affect computational efficiency.

5.4.4 Summary

In this chapter, the mixed integer optimal power flow based on IPCPM has been analyzed. A new base identification method is proposed to solve the problems of degenerate solutions and convex combination solutions. The improved algorithm can solve the difficulty brought by multiple solutions.

Theoretical analysis shows that the efficiency of the optimal base procedure depends on the difference between the number of nonzero elements of the optimal solutions and the rank of the coefficient matrix. The following changes are made to improve base identification efficiency:

1. A perturbation vector is introduced, which saves the computations that are consumed to convert the convex combination solutions to vertex point solution.
2. Dual optimality need not be guaranteed. Row simplest form is calculated once through elementary row transformation, and then the optimal base matrix can be obtained. The computational efficiency is independent of degree of degeneracy.

The computational results show that the mixed integer optimal power flow is a multiple-solution problem, and the optimal solutions obtained by interior point are convex combination solutions in test systems. The improved algorithm not only solves the difficulty successfully, but also has good computational efficiency. Only one perturbation and one row simplest form transformation are needed to obtain the base matrix. The computational steps do not increase significantly with increase in the system size. The improved IPCPM is very suitable for large-scale discrete optimization problems of power system.

6

Power System Reactive Power Optimization

6.1 Space Decoupling for Reactive Power Optimization

Recently, reactive power dispatch has received ever-increasing interest from electric utilities because of its significant influence on the secure and economic operation of power systems. In order to improve the voltage profile and to decrease the active power losses along the transmission lines under various operating conditions, power system operators can select a number of control tools such as switching reactive power compensators, changing generator voltages, and adjusting transformer tap settings. The objective of optimal reactive power dispatch (ORPD) is to allocate reactive power generation so that the active power transmission loss is minimized and the voltage stability margin is maximized, while satisfying a number of constraints. The problem is a combinatorial optimization problem with various characteristics: multi-objective, multi-uncertainty, multi-restriction, multi-extremum, and nonlinear property. Under real-time circumstances, ORPD algorithms are faced with the challenges of solution quality, response speed, robustness of the start point, infeasibility detection and handling, and discreteness of control variables.

Extensive computational experience indicates that the nonlinear programming technique is a very demanding task, especially for large-scale power systems [206]. An artificial neural network-based approach has been used in a 62-bus network for reactive power optimization [206]; a hybrid stochastic search technique has been tested on an IEEE 30-bus [207]; and a hybrid approach combining heuristic and numerical techniques has been applied to the Algerian 220/60 kV system with 68 buses [208].

It is not easy to find the global optimum within a limited time not only by using the deterministic optimization methods but also by using heuristic search approaches, so

Power System Optimization – Large-scale Complex Systems Approaches, First Edition.
Haoyong Chen, Honwing Ngan and Yongjun Zhang.

that up to now the technique of real-time ORPD is applied popularly to nothing more than small-scale local electric networks.

In view of the fact that the power dispatch systems are distributed geographically, distributed solving is more suitable for reactive power optimization because it can greatly enhance the solving efficiency and reliability. Distributed artificial intelligence (DAI) is concerned with situations in which several computer systems interact in order to solve a common problem. Multi-agent system (MAS) is an advanced branch of DAI. A new technique based on MAS for the large-scale real-time ORPD problem is presented in this chapter, and a coordinative optimization method with a built-in cataclysmic genetic algorithm is adopted to solve the problem. The numerical simulation results demonstrate that the proposed approach is effective and efficient for solving the ORPD problem of large-scale power systems in real-time mode.

6.1.1 Multi-agent System-based Volt/VAR Control

6.1.1.1 Problem Formulation

The main objective of optimal reactive power dispatch is to improve the voltage profile and to minimize network power losses while satisfying numerous operating constraints. This objective can be achieved by properly adjusting the reactive power distribution of the network. Thus, the multi-objective function can be formulated as follows:

$$\min f_Q = \sum_{k\in N_e} \Delta P_k \tag{6.1}$$

$$\min f_V = \sum_{i=1}^{N_b} \left(\frac{V_i - V_i^s}{V_{i.\max} - V_{i.min}} \right)^2 \tag{6.2}$$

$$\min f_O = \boldsymbol{c_u \Delta u} = C_T \sum_{k\in N_T} \Delta u_{TK} + C_S \sum_{i\in N_C} \Delta u_{Ci} \tag{6.3}$$

The following notation is used:

- k number of branches in the network;
- N_e set of numbers of network branches;
- ΔP_K power losses of kth branch;
- N_b set of total numbers of buses;
- V_i nodal voltage magnitude at bus i;
- V_i^s expectation of V_i;
- $\boldsymbol{\Delta u}$ increment vector of control variables, including those of tap position of transformer k, Δu_{TK} ($K \in N_T$), and size of compensator i, ΔC_i ($i \in N_C$);
- $\boldsymbol{c_u}$ readjustment cost vector of control variables per unit;
- C_T readjustment cost of a transformer;
- C_S readjustment cost of a break for a compensator.

So, f_Q, f_V, and f_Q represent the total network power losses, the total deviations of bus voltages, and the total readjustment cost, respectively.

Clearly, introducing the readjustment cost to the objective function can avoid immoderate frequent change of control variables under real-time dynamic conditions. The above functions are subjected to power flow equations and the following operating constraints:

$$Q_{Ci.min} \leq Q_{Ci} \leq Q_{Ci.max}, \quad i \in N_C \tag{6.4}$$

$$Q_{Gi.min} \leq Q_{Gi} \leq Q_{Gi.max}, \quad i \in N_G \tag{6.5}$$

$$T_{k.min} \leq T_k \leq T_{k.max}, \quad k \in N_T \tag{6.6}$$

$$V_{i.min} \leq V_i \leq V_{i.max}, \quad k \in N_b \tag{6.7}$$

where Q_{Ci} is the reactive power compensation at bus i; Q_{Gi} is the reactive power generation at bus i; T_k is the tap position of transformer k; and N_C, N_G, and N_T are the set of numbers of reactive power source installation buses, the set of numbers of generator buses, and the set of numbers of transformer branches, respectively.

In most nonlinear optimization problems, the constraints are considered by generalizing the objective function using penalty terms. The control variables, including the generator bus voltages, $\boldsymbol{V_{PV}}$ and $\boldsymbol{V_{slack}}$, the tap position of transformer, $\boldsymbol{T}$, and the number of reactive power source installations $\boldsymbol{Q_C}$, are self-constrained. Voltages of PQ buses, $\boldsymbol{V_{PQ}}$, and reactive power generation, $\boldsymbol{Q_G}$, are constrained by adding them as penalty terms to the integrative objective function. Furthermore, the voltage quality is good enough when a solution of ORPD satisfies the constraint (6.7), so the objective (6.2) may not be involved in the integrative objective function as follows:

$$\begin{aligned} \min F_Q = {} & \alpha \sum_{k \in N_e} \Delta P_k + C_T \sum_{k \in N_T} \Delta u_{Tk} + C_s \sum_{i \in N_C} \Delta u_{Ci} \\ & + \sum_{i \in N_{PQ}} \lambda_{Vi} |V_i - Sat(V_i)| + \sum_{i \in N_G} \lambda_{Gi} |Q_{Gi} - Sat(Q_{Gi})| \end{aligned} \tag{6.8}$$

where λ_{Vi} and λ_{Gi} are the penalty factors, α is the cost factor of power loss, and $Sat(x)$ is defined as

$$Sat(x) = \begin{cases} x_{min} & \text{if } x < x_{min} \\ x & \text{if } x_{min} < x < x_{max} \\ x_{max} & \text{if } x_{max} < x \end{cases} \tag{6.9}$$

Nonlinear programming is used to solve the ORPD problem. The application of nonlinear techniques, however, may become cumbersome and require comprehensive computation as the size of the network and the number of available VAR control variables increases. GA is also an excellent algorithm for these global optimization problems but simulation results based on the IEEE 30-bus system [209] cannot evaluate its performance for large-scale power systems.

The methods presented in the previous references [206–209] have a common drawback for real-time global optimal reactive power dispatch, which is severe conflict between the solution efficiency and the optimization quality. As a result, the technique of real-time ORPD is difficult to apply to large-scale systems.

The interior point method (IPM) is also a fine one for solving various optimization problems because of its excellent convergence speed and its insensitivity to the scale of the problem. A modified IPM [210] for optimal reactive power control has been tested on the IEEE 14-bus system and validated for its better performance than linear programming. In order to solve the problem of reactive power optimization of large-scale power systems, an extended nonlinear primal–dual interior point method has been presented [189]. Despite its good convergence, centralized dispatch of large-scale multi-voltage-level power systems is not practicable.

6.1.1.2 Multi-agent Systems

Multi-agent systems provide a novel approach to address complex problems, as ORPD, where decisions should be based on processing of information from various sources of diverse nature. Each of these sources can be viewed as an agent and the whole system as a community of agents.

An agent is anything that is viewed as perceiving its environment through sensors and acting upon that environment through effectors, and is believed to be a system component capable of: (a) perceiving and acting at a certain level; (b) communicating in some fashion with other agents; (c) attempting to achieve particular goals or perform particular tasks; and (d) maintaining an implicit or explicit model of its own state and the state of its world [211]. For example, ORPD software for local power systems can be seen as an agent, the goal of which is the optimization of the local power systems but not that of the global networks.

A multi-agent system could be defined as "a loosely coupled network of [asynchronous] problem solvers that work together to solve problems that are beyond their individual capabilities" [212]. A MAS consists of a set of non-centralized mutually cooperating elements, namely agents, each of which acts autonomously.

The problem solvers operate in parallel. Both control and data are logically and often geographically scattered among different units. In order to solve problems coherently, the agents must communicate, coordinate, and negotiate with each other once they find themselves in conflict. MAS provides a series of advantages, in that [213]:

(a) MAS constitutes a natural way to handle distributed problems logically and physically;
(b) they need neither direct global control nor global data storage;
(c) they enhance the overall reliability and robustness;
(d) the modularity of such systems allows their extensibility;
(e) even complicated problems can be solved rapidly; and
(f) their architecture is simple and the way they work is transparent.

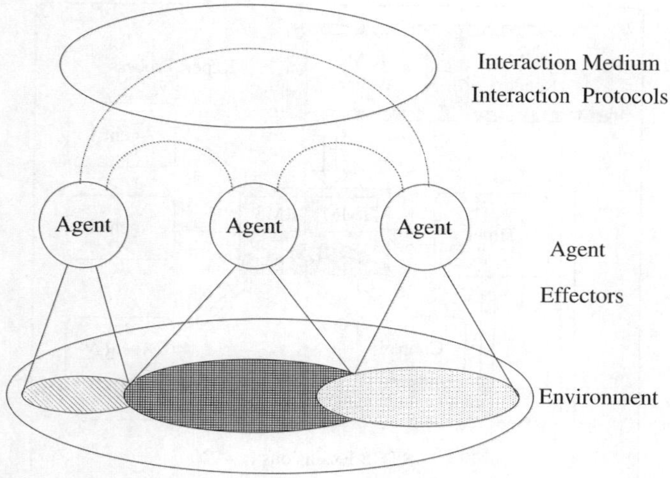

Figure 6.1 Characterization of MAS. *Source*: Zhang 2004. Reproduced with permission from Elsevier.

The organizational perspective leads to a general characterization of MAS as depicted in Figure 6.1 [214]. To accomplish their role in the organization, agents typically need to exchange knowledge with each other and to coordinate their activities.

The system architecture of MAS based on real-time knowledge (named "knowledge data manager") is shown in Figure 6.2 [215]. The main components in the architecture are:

- intelligent communication manager (ICM);
- knowledge data manager (KDM);
- time map manager (TMM);
- reason maintenance system (RMS);
- agents;
- control;
- timer; and
- real-time operating system extensions (RTOSE).

6.1.2 Coordination Optimization Method

In view of the fact that the dispatch of power systems is decentralized in regions rather than controlled as a whole, the ORPD system of the whole network is decomposed into some interacting subsystems according to process distributed problem solving, as shown in Figure 6.3. A network control structure (NCS) is adopted for the global dispatch system and a layer control structure (LCS) for each subsystem. The NCS is an

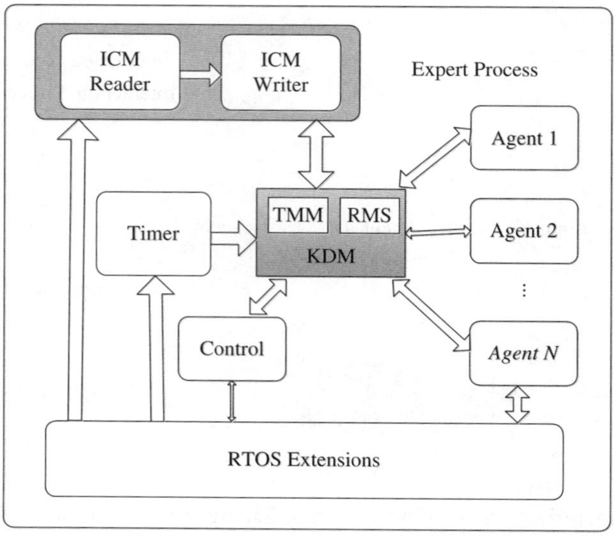

Figure 6.2 MAS architecture. *Source*: Zhang 2004. Reproduced with permission from Elsevier.

absolute distributed system with communication media, in which each agent exists as a crunode, as shown in Figure 6.4. Through communication among the agents MAS_i $(i = 0, 1, 2, \ldots)$, i.e. the local ORPD subsystems, the global optimization is achieved in phase.

6.1.2.1 Layer Control Structure

Local subsystem MAS_i $(i = 0, 1, 2, \ldots)$ employs layer control structure as shown in Figure 6.5. There are three layers of agents: the reaction layer, the coordination layer, and the organization layer.

The reaction layer at the bottom, responsible for the processing of local data and for the adjustment of the local control variables, is represented as

$$L_R = \{A_S, A_D, A_E, A_M\} \tag{6.10}$$

where A_S is the sampling agent to collect the operating data; A_D is the distinguishing agent to estimate the bus voltages and to determine whether a control instruction should be given; A_E is the executing agent to perform control or dispatch instructions; and A_M is the mobile agent to transmit data and to share resource with other agents.

The coordination layer in the middle, which is to fulfill the ORPD computation of the local power network, is represented as

$$L_C = \{A_{RM}, A_{GA}, A_M\} \tag{6.11}$$

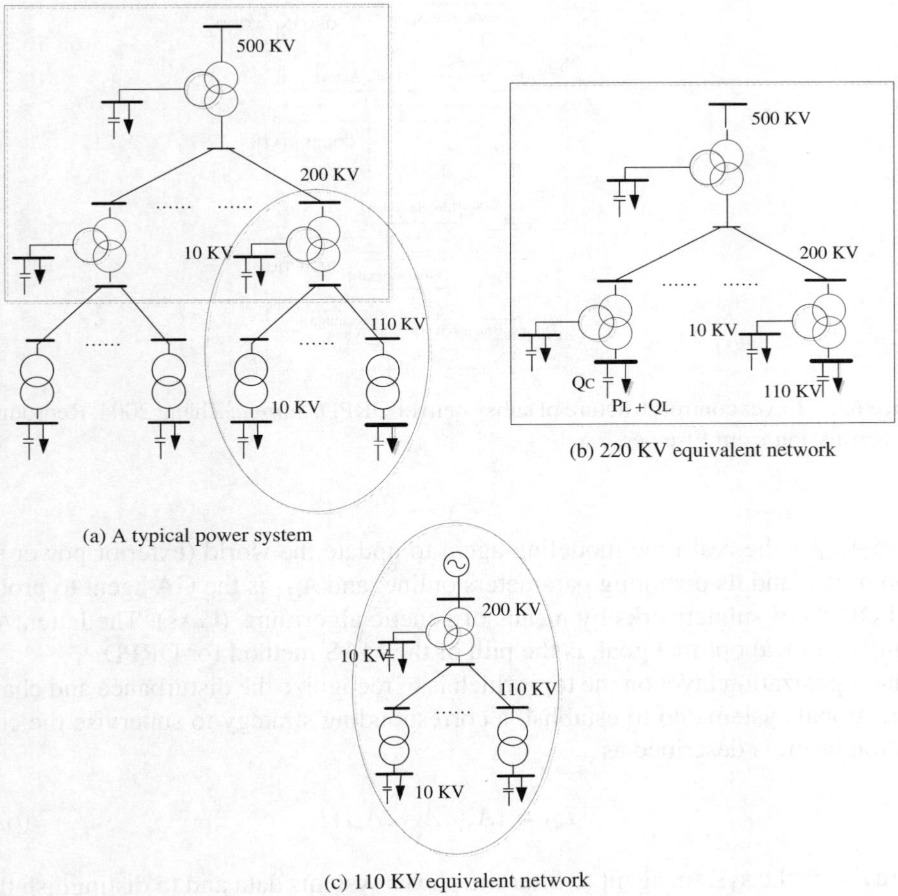

Figure 6.3 Hierarchical model of ORPD. *Source*: Zhang 2004. Reproduced with permission from Elsevier.

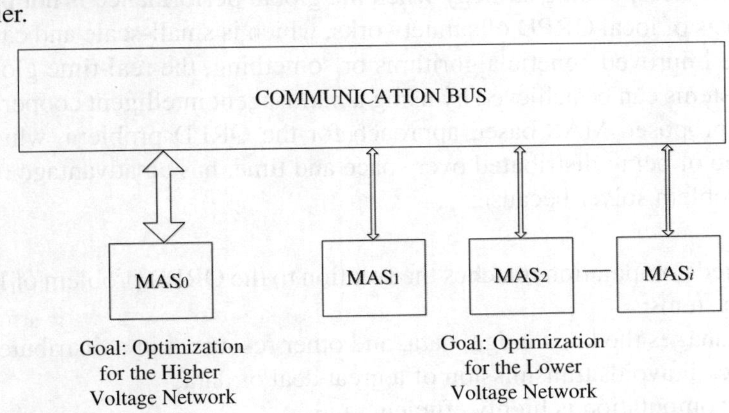

Figure 6.4 Network control structure of global ORPD. *Source*: Zhang 2004. Reproduced with permission from Elsevier.

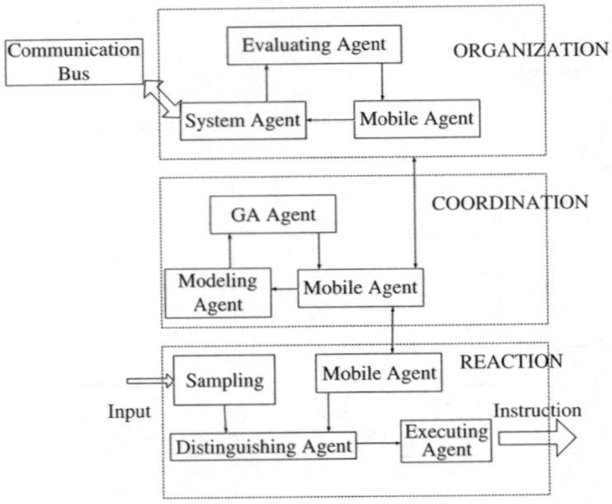

Figure 6.5 Layer control structure of subsystem of ORPD. *Source*: Zhang 2004. Reproduced with permission from Elsevier.

where A_{RM} is the real-time modeling agent to update the world (exterior power systems) model and its operating parameters online, and A_{GA} is the GA agent to process local ORPD of subnetworks by means of genetic algorithms (GAs). The latter, A_{GA}, aiming at a local optimal goal, is the pith of the MAS method for ORPD.

The organization layer on the top, which is to recognize the disturbance and change of the global system and to establish a corresponding strategy to supervise the coordination layer, is described as

$$L_O = \{A_{SY}, A_{EV}, A_M\} \tag{6.12}$$

where A_{SY} is the system agent to filter the power systems data and to distinguish their disturbances, and A_{EV} is the evaluating agent to evaluate the global performance and to modify its corresponding strategy when the global performance is not proper.

On the basis of local ORPD of subnetworks, which is small-scale and easy to solve by using the improved genetic algorithms or something, the real-time global ORPD of power systems can be achieved by using a multi-agent intelligent cooperation technique. The proposed MAS-based approach for the ORPD problem, which has the characteristic of being distributed over space and time, has an advantage over a concentrated problem solver because:

(a) distributed computation matches the solution to the ORPD problem of large-scale power systems;
(b) MAS manages the knowledge, data, and other resources in a distributed manner, and hence it avoids transmission of a great deal of data;
(c) parallel computation is highly efficient; and
(d) MAS has satisfactory dynamic adaptation and reliability.

6.1.2.2 Coordination Optimization

The key for MAS is the coordination mechanism for the sake of sound cooperation among the agents. Thus the key for the reactive power dispatch of large-scale power systems is coordination optimization among subsystems. Concretely, there are two kinds of subsystems of ORPD: one for the higher-voltage network, as shown in Figure 6.3(b), named MAS_0 here; and one for the lower-voltage network, as shown in Figure 6.3(c), named MAS_i here ($i = 1, 2, \ldots$).

For MAS_0, the loads and reactive power generations of lower-voltage networks, both in operation and out of operation, should be transformed into equivalent ones of lower-voltage buses of the transformer.

As shown in Figure 6.3(c), supposing there are l 110 kV lines connecting m 110 kV substations and g power plants at the 110 kV side of a 220 kV substation, the equivalent loads $P_{Le} + jQ_{Le}$ at the 110 kV bus of a 220 kV substation are formulated as follows:

$$\begin{cases} P_{Le} = \sum_{i=1}^{m} P_{Li} + P_{loss} - \sum_{i=1}^{g} P_{Gi} = \sum_{i=1}^{l} P_{Bi} \\ Q_{Le} = \sum_{i=1}^{m} Q_{Li} + Q_{loss} - \sum_{i=1}^{g} Q_{Gi} \approx \sum_{i=1}^{l} Q_{Bi} + \sum_{i=1}^{k_1} Q_i \end{cases} \tag{6.13}$$

where $P_{Bi} + jQ_{Bi}$ ($i = 1, 2, \ldots, l$) are the power outputs of the ith 110 kV feeder; $P_{Li} + jQ_{Li}$ ($i = 1, 2, \ldots, m$) are the loads of the ith 110 kV substation; $P_{Gi} + jQ_{Gi}$ ($i = 1, 2, \ldots, g$) are the power outputs of the ith power plant; $P_{loss} + jQ_{loss}$ are the total power losses of 110 kV networks; and Q_i ($i = 1, 2, \ldots, k_1$) and Q_j ($j = k_1 + 1, k_1 + 2, \ldots, k_1 + k_2$) are the reactive power generation of the ith compensator in operation and that of the jth compensator out of operation, respectively, in all 110 kV substations.

On the other hand, all reactive power compensator banks are "shifted" to the 110 kV bus of the 220 kV substation. The equivalent size Q_C, therefore, includes Q_i ($i = 1, 2, \ldots, k_1$) and Q_j ($j = k_1 + 1, k_1 + 2, \ldots, k_1 + k_2$).

As a result, the rough equivalent higher-voltage network includes several 220 kV substations, several lines, and possibly several power plants. The 220 kV bus of the 500 kV substations is usually regarded as a slack bus. The network's size does not exceed 40 buses.

For MAS_i, the higher-voltage buses (220 kV) of the transformer should be regarded as slack buses, the voltages of which are determined by the higher-voltage network. The size of any lower-voltage network does not exceed 40 buses likewise.

Thus, the ORPD computation of every MAS can be done through a modified genetic algorithm or other approach autonomously and a satisfactory optimization result can be obtained within 2 s.

The process of distributed real-time global ORPD computation can be described as follows:

1. The GA agent of MAS_0 processes ORPD computation (within 2 s). And then the voltage magnitudes of the 220 kV buses V_i $(i = 1, 2, \ldots)$ are to be provided to MAS_i via communication.
2. The GA agent of MAS_i $(i = 1, 2, \ldots)$ process ORPD computation simultaneously (within 2 s). It should be noted that their slack bus voltages equal V_i $(i = 1, 2, \ldots)$ from the first step. And then the tap settings and the inject powers of the 110 and 10 kV buses of the main transformer are to be provided to MAS_0 via communication.
3. The GA agent of MAS_0 processes load flow computation. And then the new voltage magnitudes of 220 kV buses V_i $(i = 1, 2, \ldots)$ are to be provided to MAS_i via communication.
4. The GA agent of MAS_i $(i = 1, 2, \ldots)$ processes ORPD computation (within 2 s). It should be noted that their slack bus voltages equal V_i' $(i = 1, 2, \ldots)$ from the third step.

Obviously, those 220 kV substations presented in both the higher- and lower-voltage equivalent networks concurrently are helpful to the coordination computation of each other. By the above iteration, the ORPD results of each MAS will lead to the global optimization.

One second is set apart for communication and data updating between the anterior step and the posterior one. The total time of global ORPD computation is less than 10 s even if the size of the power system is hundreds of buses. Hence, the proposed approach has a good performance for real-time control.

6.1.2.3 Simulation Results

The proposed approach is tested on two power systems: one 125-bus system with 13 generation plants, 83 VAR compensation buses, and 51 transformers; and another 199-bus system with 14 generation plants, 97 VAR compensation buses, and 111 transformers. Test results under certain load levels have been compared with those found by the cataclysmic genetic algorithm (CGA) presented in [216], and shown in Table 6.1. It is obvious that optimization by the MAS approach has less loss saving and consumes less time than that by CGA.

The nodal voltages of the two systems are shown in Figure 6.6 and Figure 6.7, respectively. It is clear that the voltage profiles have been improved after optimization calculations by CGA but improved further by the MAS approach.

6.2 Time Decoupling for Reactive Power Optimization

Optimal reactive power dispatch (ORPD) has attracted worldwide attention for its significant influence on secure and economic operation of power systems. Reactive power dispatch has been researched extensively as a static snapshot problem, and the

Table 6.1 ORPD simulation results: comparison between CGA and MAS approaches for two bus systems.

Variable	125-bus system		199-bus system	
	CGA	MAS	CGA	MAS
P_{loss} (MW)	14.561	14.037	18.017	16.724
ΔP_{loss} (%)	7.72	11.04	10.63	17.04
V_{avg}	1.0293	1.0371	1.0147	1.0389
MSD[a] of voltage	0.01357	0.01254	0.01797	0.01655
Time (s)	55	10	124	10

[a]Mean square deviation.

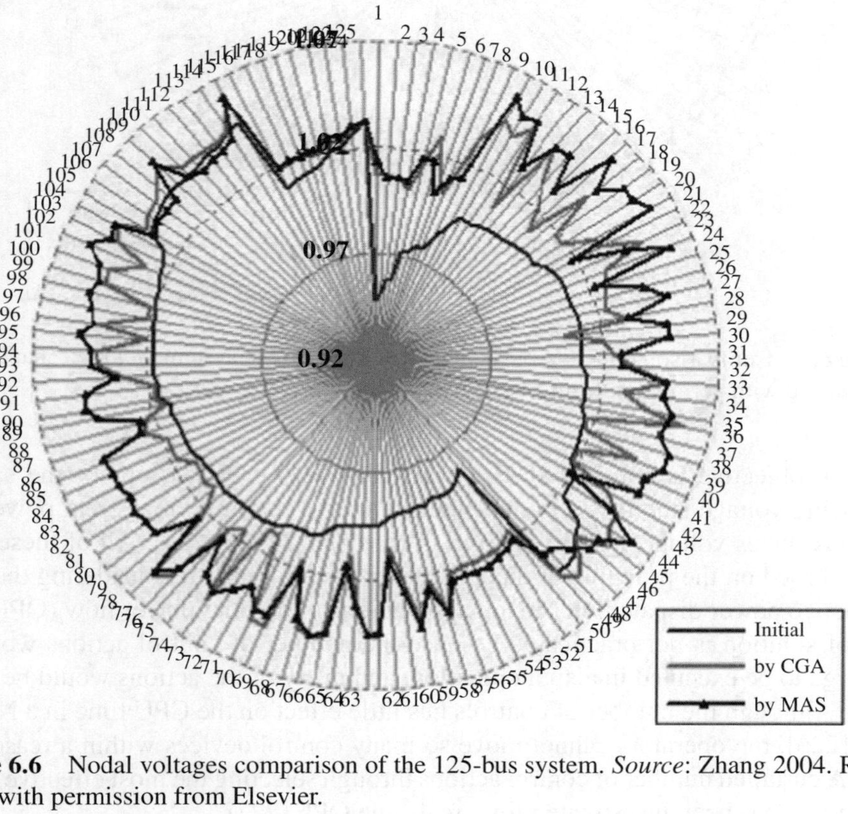

Figure 6.6 Nodal voltages comparison of the 125-bus system. *Source*: Zhang 2004. Reproduced with permission from Elsevier.

objective of ORPD is to minimize the active power transmission loss by means of dispatching reactive power sources while satisfying a lot of constraints, such as the reactive power generation limits of generators, the voltage limits of load buses, tap ratio limits, reactive power compensation limits, and power flow balance [189, 217–222].

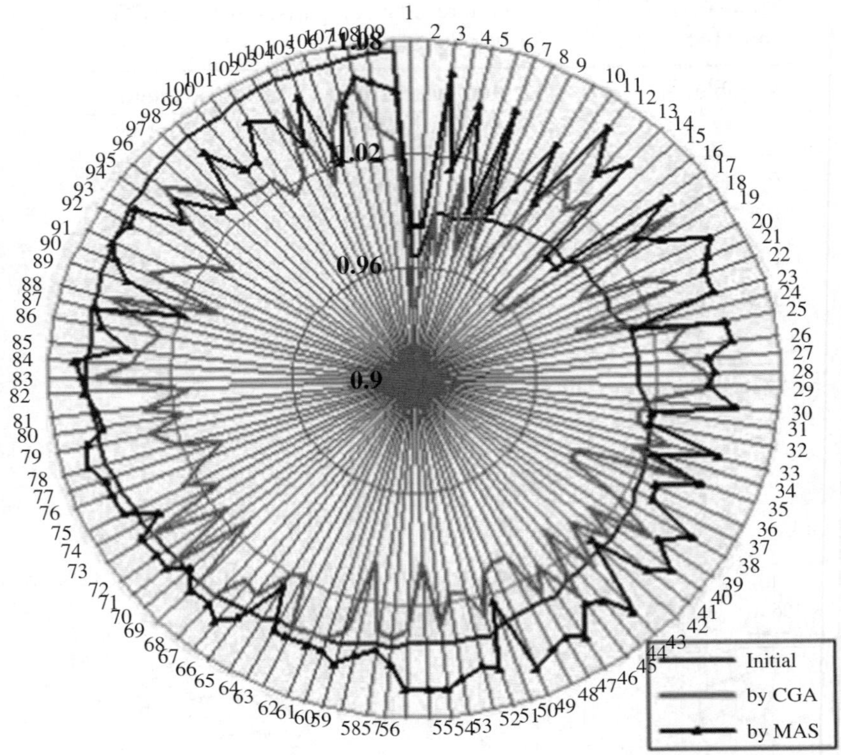

Figure 6.7 Comparison of 10 kV bus voltages of the 199-bus system. *Source*: Zhang 2004. Reproduced with permission from Elsevier.

Such an objective is considered as a classic model of ORPD, or, for the sake of enhancing voltage stability, a multi-objective model that minimizes real power loss and maximizes voltage stability margin is considered [223, 224]. All of these models are based on the principle of income maximization without considering the costs of reactive power dispatch. For many applications in optimal power flow (OPF), this kind of solution is not practicable because the number of control actions would be too large to be executed in actual operation, and many of the actions would be trivial [225]. Although the number of controls has little effect on the CPU time in a Newton OPF [226], the operators cannot move so many control devices within a reasonable time. A curtailed number of control actions through selecting the most effective subset of controls has been investigated for a real-time OPF [227].

Taylor *et al.* [228] suggest that ORPD should be seen as a time-based scheduling problem with the intention of avoiding unnecessary changes in status and output of a reactive control plant. They consider some transition constraints such as the number of control actions allowable within a time domain and the time interval required between actions performed. Hu *et al.* [229] introduce the constraints of maximum allowable switching operations for an on-load tap changer (OLTC) and capacitor of distribution systems. Its strategy is to minimize the power loss and improve the voltage profile for a whole day across the whole system and at the same time ensure that the number of operations is less than the maximum daily allowance.

A model to minimize the energy loss over time intervals in which the transition of discrete variables is governed by the selection of time intervals has been described [230]. In this manner, a near-optimal transition of discrete variables can be achieved by ensuring that they remain constant over a time interval. In Deng *et al.* [231], reactive power optimization with time-varying load demand in distribution systems is investigated. The objective is to minimize the system energy loss of 24 hours subject to the constraint of the maximum allowable daily operating times.

In fact, the curtailed number of control actions and the maximum allowable operations for discrete control devices are presented to avoid their excessive operations, which lead to shorter life expectancy of devices and more maintenance. Hong and Liao [232] also consider that the purpose of minimizing control actions is to reduce the depreciation cost of taps and compensators. From the viewpoint of economics, the better voltage profile and the power-loss saving are the income from ORPD, whereas the investment of control devices and the increased operations and maintenances are costs. It is not felicitous consequently just to consider the income without taking the costs into account in the objective function of ORPD. In other words, reactive power dispatch is to improve the voltage profile and decrease the active power loss at the cost of switching control devices. Therefore we can minimize their costs of adjusting control devices (CACDs) to realize the goal of avoiding their excessive operations while minimizing active power loss.

Up to now, many numerical optimization techniques have been proposed, such as nonlinear programming, linear programming, and interior point methods [189, 217, 218, 220]. However, when these existing continuous optimization algorithms are adopted to solve the ORPD problem, discrete variables are treated as continuous ones until they are almost optimized and then rounded off to their nearest discrete values. This not only introduces some numerical approximations but also causes violations of the inequality constraints and/or an unnecessary increase in the objective function. This serious deficiency limits their practical application [189]. Artificial intelligence methods, such as the simulated annealing algorithm, genetic algorithm (GA), evolutionary programming, and evolutionary strategy [219, 221, 233], have been developed for the discrete ORPD problem. GA is the focus of researchers' attention in this area. GA uses a multiple-point search instead of a single-point search, and the only information required is the objective function, which makes the search simple and strong for the ORPD problem. Moreover, with no special demand for a starting point, GA is good at detecting and handling the infeasible solutions.

6.2.1 *Cost Model of Adjusting the Control Devices of Volt/VAR Control*

6.2.1.1 Problem Formulation

In order to improve the voltage profile and decrease the active power losses along transmission lines under various operating conditions, power system operators should select and change control measures, such as switching reactive power compensators, changing generator voltages, and adjusting transformer tap settings, as little as possible. If the cost of each discrete control device for an ORPD operation, namely, unit adjustment cost (UAC), is specified, the costs of adjusting the discrete control devices for reactive power dispatch can be attained when multiplying the devices' UACs by their operating times. Here, the operating times of an OLTC is defined as the difference of tap positions when implementing reactive power dispatch, and that of a shunt reactive compensator is defined as the difference of the bank number switched in. For example, if the current tap position of an OLTC transformer is 4 and the optimal value is 6, it is necessary to adjust the OLTC twice from 4 to 6. Then, we name the operating times of this transformer as 2.

Since continuous control variables such as generator voltages can be adjusted more frequently by the excitation circuit, the costs of adjusting them can be neglected. From the viewpoint of Su and Lin [223], "Many of the controls whose actions need to be limited in number are adjustable only in discrete steps."

For an N-bus power system with n_Q compensation buses, n_T transformers, and n_G generators, the proposed objective of optimal reactive power dispatch is to minimize the total cost including the cost of energy loss of the network and the costs of adjusting the discrete control devices, which can be formulated as follows:

$$\min f_D = \beta\tau\ \Delta P + \boldsymbol{C}_A \boldsymbol{\Delta u} \tag{6.14}$$

where

- ΔP active power loss of network at current time interval (kW);
- β electricity price (\$/kWh);
- τ time interval, usually 1 h;
- $\boldsymbol{C}_A$ row vector that consists of unit adjustment costs of discrete control variables $\boldsymbol{C}_A \in \boldsymbol{R}^{n_T+n_Q}$;
- $\boldsymbol{\Delta u}$ column vector that consists of deviations of discrete control variables, namely operating times $\boldsymbol{\Delta u} \in \boldsymbol{I}^{n_T+n_Q}$.

Vectors $\boldsymbol{C}_A$ and $\boldsymbol{\Delta u}$ are defined as follows:

$$\boldsymbol{C}_A = [C_{T1}, C_{T2}, \ldots, C_{T_{nT}}, C_{S1}, C_{S2}, \ldots, C_{Sn_Q}] \tag{6.15}$$

$$\boldsymbol{\Delta u} = [\boldsymbol{\Delta u}_T^{\mathrm{T}}, \boldsymbol{\Delta u}_B^{\mathrm{T}}]^{\mathrm{T}} \tag{6.16}$$

$$\Delta \boldsymbol{u}_{\mathrm{T}} = |\boldsymbol{T}_C - \boldsymbol{T}_O| \tag{6.17}$$

$$\Delta \boldsymbol{u}_B = |\boldsymbol{B}_C - \boldsymbol{B}_O| \tag{6.18}$$

where

C_{Ti} unit adjustment cost of adjusting the OLTC tap i, $i = 1, 2, \ldots, n_T$ ($/times);
C_{Sj} unit adjustment cost of switching the compensator j, $j = 1, 2, \ldots, n_Q$ ($/times);
$\Delta \boldsymbol{u}_T$ column vector that consists of operating times of transformer taps, $\Delta \boldsymbol{u}_T \in \boldsymbol{I}^{(n_T)}$ (times);
$\boldsymbol{T}_C$ column vector that consists of current tap positions before optimization, $\boldsymbol{T}_C \in \boldsymbol{I}^{(n_T)}$;
$\boldsymbol{T}_O$ column vector that consists of optimal (or regarded as optimal) tap positions, $\boldsymbol{T}_O \in \boldsymbol{I}^{(n_T)}$;
$\boldsymbol{B}_C$ column vector that consists of current numbers of compensator banks switched in before optimization, $\boldsymbol{B}_C \in \boldsymbol{I}^{(n_Q)}$;
$\boldsymbol{B}_O$ column vector that consists of optimal (or regarded as optimal) numbers of compensator banks switched in, $\boldsymbol{B}_O \in \boldsymbol{I}^{(n_Q)}$.

It is assumed that the system loads and the status of discrete control variables are constant during a time interval (usually an hour). The assumption is reasonable if the time interval is short enough so that there is no great change in operation modes and load flows.

The above functions are subjected to power flow equality constraints and the physical and operating limits as follows:

$$\begin{cases} \boldsymbol{g}(\boldsymbol{u},\boldsymbol{x}) = 0 \\ \underline{\boldsymbol{Q}}_{\mathrm{G}} \le \boldsymbol{Q}_G \le \overline{\boldsymbol{Q}}_{\mathrm{G}} \\ \underline{\boldsymbol{V}} \le \boldsymbol{V} \le \overline{\boldsymbol{V}} \\ \underline{\boldsymbol{T}} \le \boldsymbol{T} \le \overline{\boldsymbol{T}} \\ \underline{\boldsymbol{B}} \le \boldsymbol{B} \le \overline{\boldsymbol{B}} \end{cases} \tag{6.19}$$

where

$\boldsymbol{u}$ control variable, $\boldsymbol{u} = [\boldsymbol{V}_G^{\mathrm{T}}, \boldsymbol{T}^{\mathrm{T}}, \boldsymbol{B}^{\mathrm{T}}]$;
$\boldsymbol{x}$ state variable, $\boldsymbol{x} = [\boldsymbol{V}_{PQ}^{\mathrm{T}}, \boldsymbol{Q}_G^{\mathrm{T}}, \boldsymbol{\Theta}^{\mathrm{T}}]$;
$\boldsymbol{g}(\boldsymbol{u},\boldsymbol{x})$ power flow equations;
$\boldsymbol{\Theta}$ column vector that consists of angles of all buses except slack bus, $\boldsymbol{\Theta} \in \boldsymbol{R}^{(N-1)}$;

$\boldsymbol{Q}_G$ column vector that consists of reactive power outputs of generators, $\boldsymbol{Q}_G \in \boldsymbol{R}^{(N_G)}$;

$\boldsymbol{V}$ column vector that consists of nodal voltage magnitudes, $\boldsymbol{V} \in \boldsymbol{R}^{(N)}$;

$\boldsymbol{V}_G$ column vector that consists of generator voltage magnitudes of slack bus and PV buses, $\boldsymbol{V}_G \in \boldsymbol{R}^{(n_G)}$;

$\boldsymbol{V}_{PQ}$ column vector that consists of nodal voltage magnitudes of PQ buses, $\boldsymbol{V}_{PQ} \in \boldsymbol{R}^{(N-n_G)}$;

$\boldsymbol{T}$ column vector that consists of transformer tap positions, $\boldsymbol{T} \in \boldsymbol{I}^{(n_T)}$;

$\boldsymbol{B}$ column vector that consists of numbers of reactive compensator banks switched in, $\boldsymbol{B} \in \boldsymbol{I}^{(n_Q)}$.

The corresponding reactive power injection vector $Q_C = q_c B$, where q_c is the row vector that consists of standard sizes of a single compensator, $\boldsymbol{q}_c \in \boldsymbol{R}^{(n_Q)}$.

In the objective function (6.14), minimizations of the cost of energy loss and the CACDs represent the income of ORPD maximization and the expenditure minimization, respectively, and minimization of f_D indicates maximization of the energy saving after the CACDs are recouped. Therefore we consider the proposed mathematical model is based on the principle of profit maximization as compared with the traditional model that only minimizes power loss.

6.2.1.2 Unit Adjustment Cost

When implementing reactive power dispatch, reactive power control devices are operated under high voltage, and impact currents will be generated into the control devices. If they are switched too frequently, their insulating intensity must be reduced, and, consequently, some hidden damage to power systems will emerge. On the other hand, too frequent operations overburden operators and are likely to lead to some mistakes. From these viewpoints, each operation of reactive power dispatch has its cost. These above factors and the investment of control devices should be introduced into the ORPD model as the costs of adjusting the discrete control devices, which involve the five aspects as follows.

1. The investment of discrete control devices, such as those of the on-load tap changers and the switched reactive compensators with breakers.
2. The extra investment in transformers due to reactive power control causing their service duration to be shorter. The investments in transformers are not involved entirely because they serve for transmission but not only for reactive power dispatch. However, a transformer's life expectancy must be reduced by the regular adjustment of OLTCs.
3. The overhaul costs related to reactive power control only, including those of OLTCs and compensators.

4. The system loss due to faults of reactive power control devices. If the relay protection installations can work and the damage of transformers or capacitors is avoided, this part is mainly the loss due to cuts in power. This aspect can be ignored if the power system can satisfy the "$N - 1$" operation mode.
5. The pay for operators related to reactive power control.

To divide the sum of the above costs during the service durations of the control devices by their total allowable adjustment times, the unit adjustment cost of discrete control devices can, therefore, be formulated as follows:

$$C_{Ti} = \frac{1}{T_{Ti}}\left(F_{pi} + \frac{a_{Ti} - a'_{Ti}}{a_{Ti}}F_{Ti} + \frac{F_{OTi}a'_{Ti}}{t_{OTi}} + F_{LTi}\right) + F_M \tag{6.20}$$

$$C_{Sj} = \frac{1}{T_{Sj}}\left(F_{Bj} + F_{Cj} + \frac{F_{OSj}a'_{Sj}}{t_{OSj}} + F_{LSj}\right) + F_M \tag{6.21}$$

Here we have used the following notation:

- T_{Tj} design value of total allowable adjustment times of OLTC of transformer i (times), $i = 1, 2, \ldots, n_T$;
- T_{Sj} design value of total allowable adjustment times of compensation bus j (times), $j = 1, 2, \ldots, n_Q$;
- a_{Ti} life expectancy of transformer i when the tap is never adjusted (year);
- a'_{Ti} life expectancy of transformer i after the tap is changed T_{Ti} times (year), $a'_{Ti} < a_{Ti}$;
- T_{Sj} life expectancy of compensation bus j after being switched T_{Sj} times (year);
- t_{OTi} overhaul periods of OLTC of transformer i (year);
- t_{OSj} overhaul periods of compensator j (year);
- F_{Pi} investment in OLTC of transformer i (\$);
- F_{Ti} investment in transformer i (\$);
- F_{OTi} unit overhaul cost of OLTC of transformer i (\$/times);
- F_{LTi} system loss due to fault of the OLTC of transformer i during a'_{Ti} years (\$);
- F_{Bj} investment of breakers and operating equipment of compensation bus j (\$);
- F_{Cj} investment of shunt capacitors/reactors of compensation bus j (\$);
- F_{OSj} unit overhaul cost of compensator j (\$/times);
- F_{LSj} system loss due to the fault of compensator j during a'_{Sj} years (\$);
- F_M unit cost for operators' salaries related to reactive power control (\$/times).

If m operators are added for reactive power control and each operator's annual salary is S (\$/year), then F_M is formulated as

$$F_M = \frac{mS}{\sum_{i=1}^{n_T}(T_{Ti}/a'_{Ti}) + \sum_{j=1}^{n_Q}(T_{Sj}/a'_{Sj})} \tag{6.22}$$

It is assumed that each control device has adjusted its total allowable adjustment times during its whole service term. The UAC indicates the investment, maintenance cost, and latent system loss for each control of reactive power dispatch. If a switching action cannot reduce the amount of power loss more than its UAC or is no help to eliminate the deviation of state variables, we would rather stop this movement.

Although different devices have different UAC, from the above-mentioned formulas, it can be concluded that the UAC of each control device is a constant represented by its average cost of adjustment at any time. It is valuable for ORPD application because the constant UAC means there is no coupling between time intervals mentioned in the models in Deng *et al.* [231] and Hong and Liao [232]. Then, the proposed decoupling ORPD model is simpler and easier to solve than the models of prior works, and it is more suitable for real-time application, which emphasizes the solving speed rather more than the actual optimal solution [227].

6.2.1.3 Normalization of Unit Adjustment Cost

If we define

$$\boldsymbol{C}'_A = \frac{\boldsymbol{C}_A}{\beta\tau} \tag{6.23}$$

as the unit adjustment cost vector based on power (kW/times), then function (6.14) can be transformed to

$$\min f'_D = \frac{f_D}{\beta\tau} = \Delta P + \boldsymbol{C}'_A \boldsymbol{\Delta u} \tag{6.24}$$

whose objective is to minimize the sum of the active power loss of the whole network and the costs of adjusting control devices represented in power. Such an objective function is consistent with (6.14) in substance. Although the UAC of discrete control devices can be calculated theoretically by (6.20) and (6.21), it is difficult and impracticable to determine an exact value of the UAC for each control device since the life of the device is a probability expectation, and the system loss due to the fault of reactive control devices is an estimate, too. Usually, if one ignores the loss of control device fault that is difficult to count due to lack of statistics, the estimated value of UAC of an OLTC varies from 3 to 10 kW/times, and that of a compensation bus from 2 to 6 kW/times.

In terms of the definition of UAC, the transformers, especially those at the same voltage level, have similar values of UAC in a power system. The UAC of shunt compensators has such a characteristic, too. It is, therefore, practicable to normalize the

UAC of the same kind of control devices by their average value in the real-time engineering application. Then, the average unit adjustment cost of all OLTCs C_T and that of all compensators C_s can be defined as

$$C_T = \frac{1}{n_T} \sum_{i=1}^{n_T} C'_{Ai} \tag{6.25}$$

$$C_S = \frac{1}{n_Q} \sum_{j=n_T+1}^{n_T+n_Q} C'_{Aj} \tag{6.26}$$

and one can replace the elements of vector C'_A in (6.24). So the objective function can be written as follows:

$$\min f'_D = \Delta P + C_T \sum_{i=1}^{n_T} \Delta u_{Ti} + C_S \sum_{j=1}^{n_Q} \Delta u_{Cj} \tag{6.27}$$

Although such a model is not so exact as compared with (6.14) or (6.24), its feasibility and rationality are testified by the application in a real network in China, Hainan Power Systems.

6.2.2 *Time-Decoupling Model for Reactive Power Optimization Based upon Cost of Adjusting the Control Devices*

6.2.2.1 Cataclysmic Genetic Algorithm (CGA)

CGA [233] is adopted to solve ORPD problems. This algorithm is one of the modified genetic algorithms with the following characteristics.

1. The cataclysm operator is adopted to improve the diversity of small-size populations and to avoid premature convergence. Mimicking the cataclysm of the evolution course of ecosystems, in which most species are extinct and very few survive, the cataclysm operator updates all individuals randomly, except for the current optimum after tens of generations.
2. An adjacent mutation operator integrating with grouping integer–encoding strategy is used to elude the void solutions and the Hamming cliffs. These techniques redound to improve the search efficiency and local search capability of genetic algorithms.
3. Without special demand for a starting point, it can detect and handle the infeasible solutions and is fit for solving the nonlinear ORPD problem involving multi-objective, multi-restriction, multi-extremum, and discreteness.
4. It has fine performance of global optimization and high search efficiency.

6.2.2.2 Handling Constraints

When adopting genetic algorithms to solve nonlinear optimization problems, the constraints are considered into the objective function using penalty terms. The control variables, including generator voltages, tap positions of transformers, and numbers of compensator banks switched in (directly proportional to the reactive power outputs of compensators), are self-constrained when encoding them as chromosomes. In other words, the chromosomes are generated just within the limits of all control variables so that the constraints of the control variables are always satisfied. The state variables, including voltages of PQ buses, V_i ($i \in N_{PQ}$, where N_{PQ} is the set of number of PQ buses), and reactive power generation, Q_{Gj} ($j \in N_G$, where N_G is the set of number of generator buses), however, are constrained as penalty terms into the integrative objective function as follows:

$$\min F'_D = \Delta P + C_T \sum_{i=1}^{N_T} \Delta u_{Ti} + C_S \sum_{j=1}^{n_Q} \Delta u_{Cj} + \lambda p \tag{6.28}$$

Here λ is a penalty factor proportional to generation (iteration) number when using GA to solve ORPD problems, and p is the maximum agent of constraints, i.e.

$$p = max\{D(V_1), D(V_2), \ldots, D(V_i), \ldots, D(Q_{G1}), D(Q_{G2}), \ldots, D(Q_{Gj}), \cdots\} \tag{6.29}$$

where $i \in N_{PQ}, j \in N_G$, function max is to find the maximum of its elements, and D(x) is defined as

$$D(x) = \begin{cases} x_{min} - x, & x < x_{min} \\ 0, & x_{min} \le x \le x_{max} \\ x - x_{max}, & x_{max} < x \end{cases} \tag{6.30}$$

If $V_i (i \in N_{PQ})$ and $Q_{Gj} (j \in N_G)$ are all within their limits, each element of (6.29) equals zero and $p = 0$. So, function (6.28) just consists of power loss and the CACD. However, any deviation of state variable makes $D(x) > 0$, and consequently $p > 0$. Also, then, the objective function of the infeasible solution is enlarged with such a positive penalty term.

Moreover, CGA uses the objective function (6.28) as the fitness function directly and adopts a tournament selection operator [233]. Individuals with a small value of F'_D have high fitness.

As we know, the optimization result of such an objective function usually makes some state variables close to the boundaries of restrictions. Under real-time circumstances, these variables are likely to exceed their restrictions when the system's operation mode shifts remarkably at the time interval between two optimization computations that run hourly in a general way. Thus, a threat to the security of power systems comes into being. Moreover, the difference between ORPD computational results and the actual values of power flow also cause some state variables to be over their limits.

Therefore, it is necessary to reduce the restriction bandwidths of state variables, which involve nodal voltage magnitudes of PQ buses and reactive power outputs of generators, as follows:

$$x_{\max} = \overline{x} - \eta(\overline{x} - \underline{x})$$
$$x_{min} = \underline{x} + \eta(\overline{x} - \underline{x}) \tag{6.31}$$

Here $\overline{x}$ and $\underline{x}$ are initial upper/lower limits of state variable x, which are set from the viewpoint of the power system; $x_{\max}$ and x_{min} are adjusted upper/lower limits of state variable x in terms of the requirement of real-time ORPD; and η is an adjustment coefficient varying from 3% to 8%. Because $x_{\max} < \overline{x}$ and $x_{min} > \underline{x}$, the state variables optimized get more margins from the initial limit boundaries, and the security margin can be augmented with the varying loads under real-time conditions.

In addition, there are two measures to avoid the deviations of nodal voltage magnitudes from their limits when the system loads varying remarkably, i.e. to regulate voltages of generators, and to shorten the optimization time interval.

6.2.2.3 Simulation Study

A real power system is used to evaluate the proposed model, which is shown in Figure 6.8. The hourly loads of the tested network are shown in Figure 6.9. The tap ratio of transformer T1 is 220 (1.0381.5%), and those of other transformers are all 110 (1.0081.5%). There are six load buses, at which some shunt capacitor banks are installed. From bus 1 to bus 6, the compensation sizes are 44.008, 23, 23, 14.8, 13.3, and 22.016 MVAR, respectively. Three schemes with different UACs and the restriction ranges of nodal voltage magnitudes are listed in Table 6.2. The UACs of scheme #1 are all zero, so its objective function is only the active power loss as a

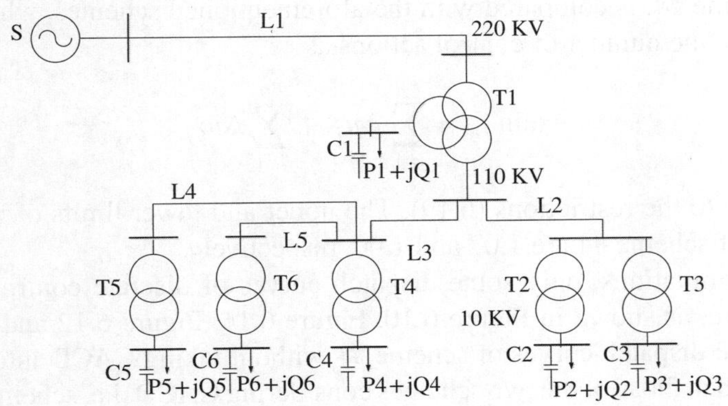

Figure 6.8 Tested network. *Source*: Zhang 2005. Reproduced with permission from Elsevier.

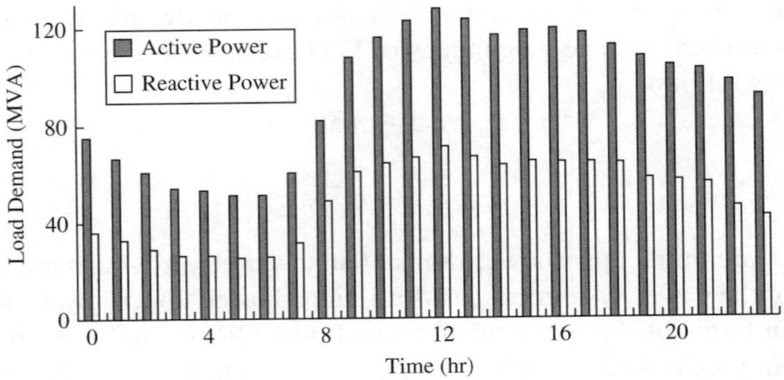

Figure 6.9 Hourly total load curves. *Source*: Zhang 2005. Reproduced with permission from Elsevier.

Table 6.2 Parameters of three tested schemes

Scheme	#1	#2	#3
C_T	0	6	10
C_S	0	3	6
$V_{\max}$	1.07	1.07	1.0679
V_{min}	1.00	1.00	1.0021

traditional model, i.e.

$$\min f_1 = \Delta P \tag{6.32}$$

In Table 6.2, $V_{\max}$ and V_{min} are the upper and lower limits of nodal voltages of PQ buses, respectively. The initial upper and lower limits of nodal voltages are adopted in schemes #1 and #2 directly and, in contrast, the reduced limit bandwidths in scheme #3, in which $\eta = 3\%$. Furthermore, an ordinary voltage and reactive power control, denoted scheme #4, is compared with the aforementioned schemes, whose objective is to minimize the number of control actions as

$$\min f_4 = \sum_{i=1}^{n_T} \Delta u_{Ti} + \sum_{j=1}^{n_Q} \Delta u_{Cj} \tag{6.33}$$

and subjected to the restrictions (6.19). The upper and lower limits of nodal voltage magnitudes of scheme #4 are 1.07 and 1.00, respectively.

Owing to limitations, only some dispatch curves of discrete control devices are illustrated here, as shown in Figure 6.10, Figure 6.11, Figure 6.12 and Figure 6.13. Obviously, the dispatch curves of scheme #1 without taking CACD into account are more zigzag than those of the two schemes considering CACD, i.e. schemes #2 and #3.

The total operating times of all four schemes within 24 hours of all transformer taps and capacitor banks are shown in Table 6.3 and Table 6.4, respectively. In scheme #1,

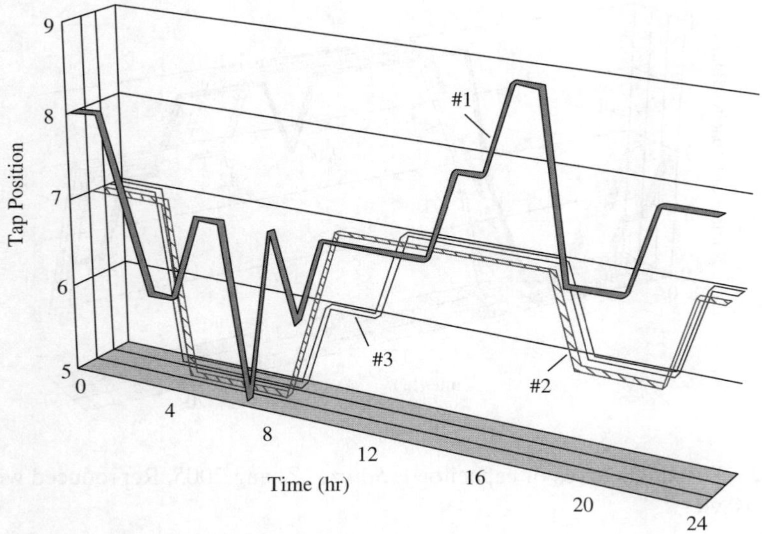

Figure 6.10 Dispatch curves of tap 1. *Source*: Zhang 2005. Reproduced with permission from Elsevier.

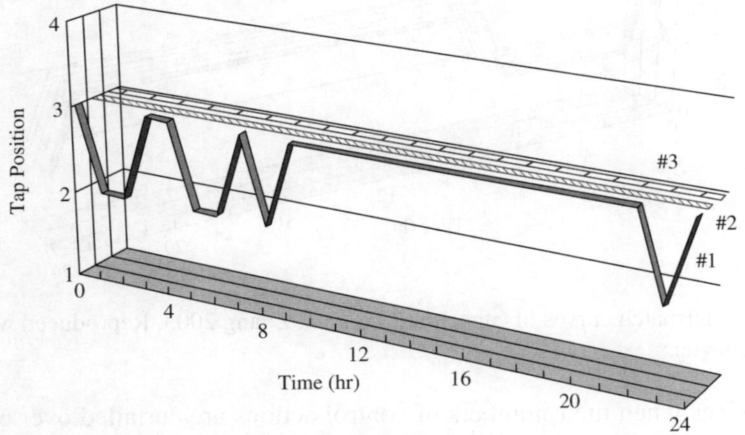

Figure 6.11 Dispatch curves of tap 2. *Source*: Zhang 2005. Reproduced with permission from Elsevier.

the intra-day operating times of tap 1 and tap 6 are all 14, which exceed 10, the maximum allowable switching operation. However, with consideration of CACD, schemes #2 and #3 can curtail the number of control actions to a different extent and eliminate all violations of allowable switching times. In the four schemes, the total operating times within 24 hours of all discrete control variables are 72, 24, 23, and 32, respectively. It is apparent that the introduction of CACD can avoid excessive operations of

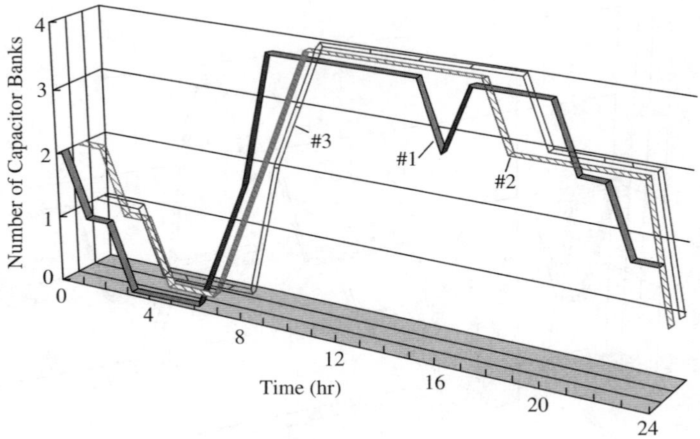

Figure 6.12 Dispatch curves of capacitor 1. *Source*: Zhang 2005. Reproduced with permission from Elsevier.

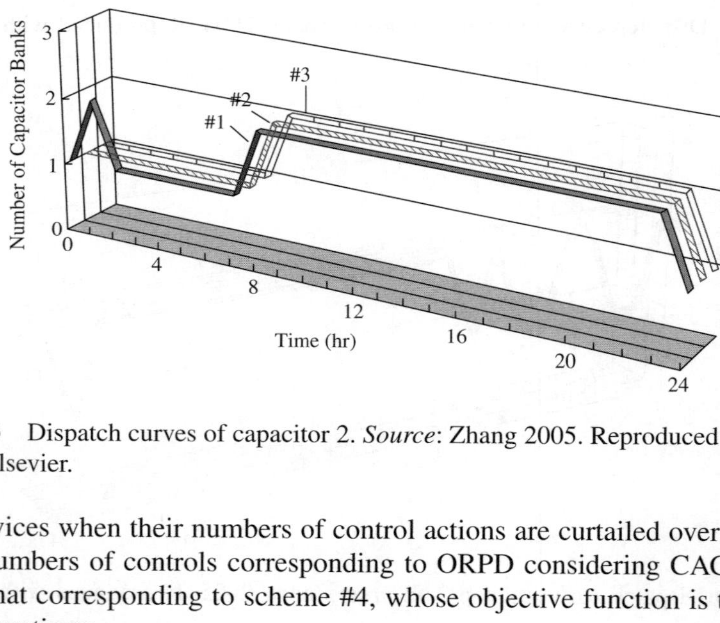

Figure 6.13 Dispatch curves of capacitor 2. *Source*: Zhang 2005. Reproduced with permission from Elsevier.

control devices when their numbers of control actions are curtailed over 60%. Also, the total numbers of controls corresponding to ORPD considering CACD are even less than that corresponding to scheme #4, whose objective function is to minimize the operating times.

On account of different UACs, the dispatch curves of schemes #2 and #3 are not entirely in step; nevertheless, almost all of intra-day operating times of the control devices are equal except that of capacitor banks 1: the number in scheme #2 is 9 whereas it is 8 in scheme #3. On the whole, the larger the UACs, the less operations can be anticipated. From Table 6.3 and Table 6.4, another conclusion is that the different UACs would not change the intra-day operating times remarkably if they are near their average like in schemes #2 and #3.

Table 6.3 Total operating times of transformer taps.

	T1	T2	T3	T4	T5	T6	Σ
#1	14	8	6	6	4	14	52
#2	6	0	0	1	0	0	7
#3	6	0	0	1	0	0	7
#4	4	0	0	0	0	10	14

Table 6.4 Total operating times of capacitor banks.

	C1	C2	C3	C4	C5	C6	Σ
#1	10	4	0	2	0	4	20
#2	9	2	0	3	2	1	17
#3	8	2	0	3	2	1	16
#4	8	0	2	2	2	4	18

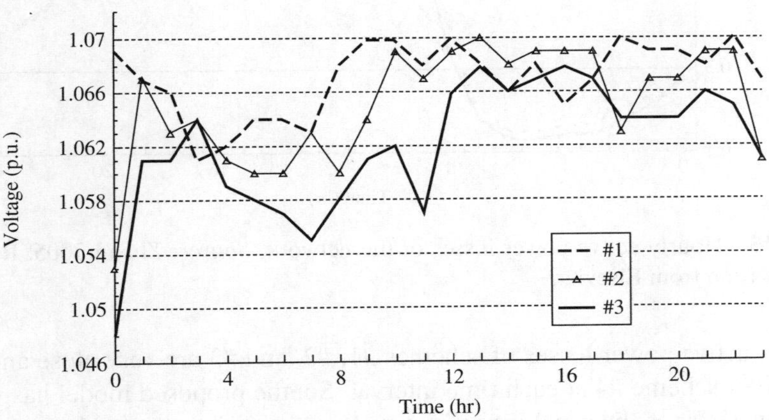

Figure 6.14 Response curves of voltage at bus 1. *Source*: Zhang 2005. Reproduced with permission from Elsevier.

The response curves of voltages at bus 1 are shown in Figure 6.14. It is many times that the voltages come close to or reach the upper limit 1.07 in scheme #1 and scheme #2. However, owing to the reduced limit bandwidths, the voltages are always not as near their secure upper limits 1.07 in scheme #3. The lowest/highest voltages of these six load buses in the 24 time intervals are listed in Table 6.5, where V_l and V_h represent the lowest and highest voltages of these buses in the whole day. Those of scheme #1 are too high, and those of scheme #3 are correct.

Figure 6.15 describes the curves of the hourly active power losses of the network under different control schemes. Although the curve of scheme #1 is the lowest one,

Table 6.5 Nodal voltage distributions for various schemes.

Bus no.	#1		#2		#3		#4	
	V_l	V_h	V_l	V_h	V_l	V_h	V_l	V_h
1	1.061	1.07	1.053	1.07	1.048	1.068	1.037	1.067
2	1.056	1.07	1.053	1.07	1.051	1.067	1.033	1.07
3	1.045	1.07	1.050	1.07	1.047	1.068	1.0197	1.051
4	1.047	1.07	1.058	1.069	1.052	1.068	1.033	1.068
5	1.053	1.07	1.014	1.05	1.007	1.048	1.001	1.041

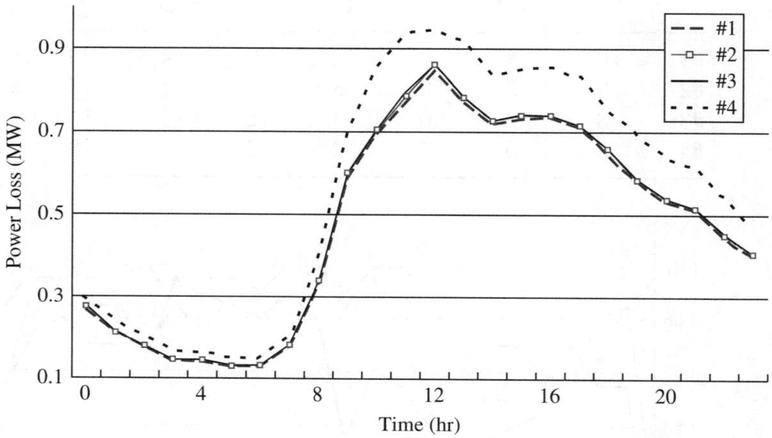

Figure 6.15 Hourly active power losses of the network. *Source*: Zhang 2005. Reproduced with permission from Elsevier.

the hourly active power losses of schemes #1, #2, and #3 are very close and are less than those of scheme #4 at each time interval. So, the proposed model has very little side effects as far as the capability of power loss saving is concerned.

Tinney *et al.* [225] give a hypothetical curve to describe the objective function f_1 as a function of the number of controls used to solve an OPF problem, which is shown in Figure 6.16. On this curve, point A corresponds to the smallest set of actions n_A that can produce a feasible solution. As more controls are used, the power loss of the system decreases. At point D, all of the n_D controls are used, and the power loss is the absolute minimum, but the number of controls is the largest. Point D is the solution obtained by scheme #1, and point A is the solution by scheme #4. They are not so suitable for real-time ORPD application.

When the cost of adjusting control devices is considered, as shown in Figure 6.16, f'_D represents the sum of power loss f_1 and CACD. Here, it is assumed that each control has the same UAC, so the curve CACD is directly proportional to n. Then, point B is the minimum of f'_D, which corresponds to the solution of schemes #2 or #3. It is

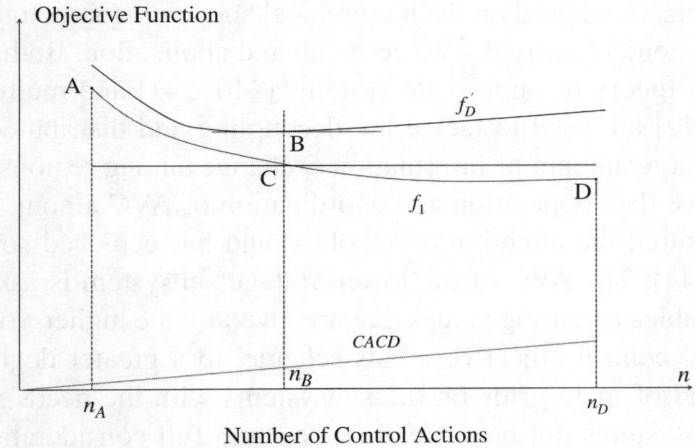

Figure 6.16 Relationship of objective functions with the number of controls. *Source*: Zhang 2005. Reproduced with permission from Elsevier.

Table 6.6 Nodal voltage distributions.

Scheme	#1	#2	#3	#4
Time (s)	0.483	0.480	0.486	0.508

obvious that the number of controls at point B is less than n_D, and the corresponding power loss $f_1((n_B)$ (at point C) is only a little higher than that obtained by scheme #1, $f_1(n_D)$ (at point D), and is much lower than the solution of scheme #4, $f_1(n_A)$ (at point A). That is why the curves of power loss of schemes #1, #2, and #3 are very close and are less than that of scheme #4.

The tests are performed on an Intel PIV 2.0 GHz. The average CPU time listed in Table 6.6 is only for the main ORPD procedure, which does not include the input or output time. The solution speed of the proposed algorithm is fast enough.

6.3 Game Theory Model of Multi-agent Volt/VAR Control

Voltage control is at the heart of power system operation to ensure system security and improve power quality [234–239], with automatic voltage control (AVC) [234, 235] as its primary mean. AVC usually aims at a reasonable distribution of reactive power to reduce active power losses and less operating times to guarantee the system's security, while there is an intense conflict between these two aspects. Since it is extremely complex to pursue reactive power optimization [236–240] for large-scale power systems, and not realistic to obtain massive real-time information comprehensively in large power grids to execute optimization computation of whole networks, decoupled and distributed optimization based on the hierarchical and zoning structure of power systems is the main trend of current AVC research and application. Both decomposition and coordination theory of large-scale systems [240, 241] and multi-agent systems (MASs) [233, 242] are used to tackle the decoupling and distributed optimization, depending on a large amount of information exchange among regions.

How to enhance the cooperation and coordination of AVC among regional power systems has captured the attention of scholars, and has achieved some progress in research [243, 244]. The AVC of the lower-voltage subsystem is asked to track the coordinated variables or setting values that are given by the higher-voltage main grid to meet the local control objectives. This scheme, to a greater degree, reflects the inflexible demand of main grids on the subsystems with the needs and the control capability of subsystems not being really taken into full consideration [244]. Various coordination strategies based on such a scheme are therefore based on experience rather than the mechanism, in which there are many subjective factors. As the coordination variables, the voltage amplitude and exchanging reactive power of the gateway are not only determined by the regional control, but also influenced by the control actions of the whole interconnected regions [243].

Some trends of power system development in recent years, such as the AC/DC hybrid interconnection, large receiving end, electromagnetic loop networks, have been strengthening inter-regional reactive power and voltage coupling continuously. Thus the volt/VAR control in any region will influence, to a certain extent, the reactive power distribution of other interconnected regions. This kind of influence will bring conflict and competition to the AVC strategies in extreme operating conditions, such as conditions with very heavy or very light power load. The reactive power is apparently overmuch in the valley condition as the HV/UHV/EHV transmission lines, especially cables, increase. The game theory [245–248] situation of volt/VAR control in multi-area power systems is becoming more and more obvious as a result of the increased power load difference between peak and valley.

Game theory is one of the essential attributes of the hierarchical and regional volt/VAR control in large-scale power systems, since the control purviews of different regional networks belong to different dispatch subjects and the AVC results interact with each other. In this chapter, game theory has been used to analyze the essences of the phenomenon that power system operators or AVC systems from different layers and regions contend for reactive resources for their own benefit in the volt/VAR control mismatching (VCM) condition. In order to get the game theory of AVC out of the "prisoners' dilemma" [245–248], a method via constructing a cooperative game mechanism was proposed to solve the VCM problem.

To analyze the Nash equilibrium of the AVC game quantitatively, the primary condition is that players' payoffs [245–248] under different strategy combinations can be accurately identified. The payoff function model of the AVC game was presented in Chapter 5 and used to analyze the game effect of two different control mechanisms quantitatively.

6.3.1 *Game Mechanism of Volt/VAR Control During Multi-level Power Dispatch*

6.3.1.1 VCM Problems in Large-scale Power Systems

In this section, a subsystem is defined as a power network that comprises a 220 kV substation and the 110 kV system connected to the 220 kV substation, while the main grid is defined as a power grid that includes the 500 kV and 220 kV power systems in a provincial region. To system operators of various levels and regional power systems, there are some objective differences in their pursuits of security, quality, and economy, and their awareness of networks and information available are also different, with modeling of outer systems remaining incomplete. The AVC actions of any networks therefore show strong individual rational characteristics: the expected control results are usually estimated through assuming decoupled regional power networks and the AVC systems try to control with strategies most favorable to their own ends. Unfortunately, AVC of adjacent regional networks interrupt each other. The results of reactive power "optimization" lacking coordination of control strategies at each layer not only fail to place themselves in the expected optimized state, but also cause poor quality of the voltage profile in the entire system. Thus the VCM problem has been produced, which is mainly manifested as follows.

1. When the loads are still increasing under heavy load operating conditions, the AVC systems of subsystems tend to claim reactive power from the transmission systems with more dynamic VAR sources in the first instance in order to reduce the operating times. It easily leads to the decline of the dynamic reactive power margin and voltage amplitude of the transmission grids, and causes the corresponding reduction of the voltage profile of subsystems. Such a vicious circle will not stop until the gateway reactive power constraints are violated. However, the global power systems are in an operating state far from the optimal one. When the loads are declining under light load operating conditions, there is also a similar systemwide problem of the voltage profile exceeding the upper limit of the constraint range.
2. The power generating capacity of power grids in those areas rich in distributed intermittent renewable energy power generations is always out of control, leading to considerable uncertainty of both direction and capacity of reactive power flow, and resulting in frequent voltage fluctuations in the lines. Especially in such areas where the power network was designed as a receiving-end system but later a large number of distributed generations were developed, it is difficult for the existing AVC strategy to adapt to this change and a good voltage quality in both substations and customer sides could hardly be ensured at the same time.
3. Owing to the application of a large number of DC transmissions and the situation of AC and DC hybrid/parallel transmission, the fluctuation range of inter-regional reactive power exchanging become very large when the interconnection operation mode changes. In situations where matched model and coordinated strategy between the AVCs are lacking, such sharp fluctuation always results in poor

quality of the whole systems and leads to voltage instability problems in extreme cases [243].

The results of voltage control mismatch may include: the irrational distribution of reactive power flow, great voltage differences among regions, systemwide exceeding of upper limit of voltage amplitude in holiday operating conditions, low-voltage qualification rate, insufficient dynamic reactive power reservation, large active power losses, and so on. These problems will threaten the safe operation of power grids.

6.3.1.2 Game Essence of Volt/VAR Control

A game is a description of strategic interaction that includes the constraints on the actions that the players can take and the players' interests, but does not specify the actions that the players do take [245].

Game theory considers that: first, players are rational, which means that everyone will maximize his/her self-interest under the constraints; second, they have conflicts in communication and cooperation and their behaviors interplay with each other, and the information is asymmetrical. Game theory is a method of studying mathematical models of conflict and cooperation between intelligent rational decision-makers [246]. Game theory has two major branches, noncooperative and cooperative game theory. Cooperative game theory is axiomatic, frequently appealing to Pareto optimality, fairness, and equity. Noncooperative game theory is economic in flavor, with solution concepts based on players maximizing their own utility functions subject to stated constraints [247].

Although some literature has been aware of the relevance between reactive power optimization and game theory, none points out the game nature of volt/VAR coordinated control. And there is also no literature or study which probes into the game modeling of large-scale power systems.

We find that using game theory to describe the volt/VAR control problem is entirely appropriate, according to the most classic model in game theory: The "prisoners' dilemma" model suggests that there is a conflict between individual and collective rationality [247]. A participator acts in the principle of maximizing his/her own interests, but it results in the damage to the interests of both. When regional AVCs are pursuing their respective optimizations, it can also result in the mismatch control problem in the whole systems. The mechanisms of these two problems are the same.

Since any regional voltage control will affect the reactive power distribution in adjacent regions to some extent, the AVC strategy in one region should consider how to coordinate with those in other adjacent regions. From the viewpoint of game theory, since competition and conflict consist in the AVC's utilization of reactive resources in the whole system, AVC strategies and payoffs are dependent on the other relevant AVC strategies and payoffs. Game theory is neither like reactive power optimization which depends on whole modeling of global networks nor like multi-agent systems which relies heavily on information exchange. However, it can reveal the essence of competition toward VAR resources and voltage control mismatch between each decision-maker.

6.3.1.3 Mathematical Model for Volt/VAR Control Game

Voltage control has distribution character in terms of physical structure and requirement on optimization, so game theory can be adopted to describe the nature of conflict, coupling, coordination, and competition. The game behaviors of voltage control can be defined as

$$G_{AVC} = \{S_1, S_2, \dots, S_n; F_1, F_2, \dots, F_n\} \tag{6.34}$$

where the game includes three elements:

1. Game players set $N = \{1, 2, \dots, n\}$, in which player i ($i \in N$) means the decision subject of voltage control, namely, AVC of the ith regional power networks.
2. Strategy set $S = \{S_1, S_2, \dots, S_n\}$, in which S_i is the pure strategy set that player i can choose, where s_i ($s_i \in S_i$) is a specific pure strategy, such as transformer tap position changing, reactive compensation switching, and generator voltage regulating.
3. Payoff function set $F = \{F_1, F_2, \dots, F_n\}$, in which the n-tuple function $F_i(s_1, s_2, \dots, s_n)$ is the award given to player i at the outcome of the game. Each AVC operator's payoff function can be calculated by using the secure, qualitative, and economic indices of power systems or be indicated according to the objective function value of each AVC.

When the game enters into a stable state, players are in equilibrium. Making a change on the part of any player in strategies would lead that player to earn less than remaining in the current strategy. In this state, namely, Nash equilibrium, no-one has the incentive to unilaterally change its action to deviate from the current situation, the mathematical description of which is as follows.

In the normal form of the game (6.34), the strategy combination $\{s_1^*, s_2^*, \dots, s_n^*\}$ is the Nash equilibrium if, to each player i, $i = (1, 2, \dots, n)$, s_i^* is the optimal strategy against $\{s_1^*, s_2^*, \dots, s_{i-1}^*, s_{i+1}^*, \dots, s_n^*\}$, that is, when satisfying

$$F_i\{s_1^*, s_2^*, \dots, s_{i-1}^*, s_i^*, s_{i+1}^*, \dots, s_n^*\} \geq F_i\{s_1^*, s_2^*, \dots, s_{i-1}^*, s_i, s_{i+1}^*, \dots, s_n^*\},$$
$$\forall\ s_i \in S_i \tag{6.35}$$

6.3.1.4 Analysis of Volt/VAR Control Mismatch by Using Game Theory

The frame of AVC game analysis was presented through an example in which the power systems were operating in a typical VCM scene under heavy load condition.

In this scenario, the main grid's dynamic VAR margin was being reduced as the loads were becoming heavier and heavier so that the main grid's voltage profile came down while some nodal voltages in subsystems were lower than the tolerance band. AVC systems have to respond under such situations.

(i) Qualitative Nash Equilibrium for VCM

In China, the jurisdiction scopes of provincial main grids include all the 500 kV substations, 220 kV buses and the power plants connected to these buses, while the jurisdiction scopes of subsystems comprise all the 220 kV and 110 kV substations (excluding 220 kV buses) and the power plants connected to these buses.

In this example, the set of voltage regulation game participants is

$$N = \{1, 2\} = \{\text{main grid, subsystems}\}$$

Assuming that there are no power plants participating in voltage control in the receiving-end subsystem, the strategy spaces of the main grid and subsystem to improve the voltage profile will mainly be

$$\begin{aligned} \mathbf{S}_1 &= \{S_{1C}, S_{1N}\} \\ &= \{\text{to further generate reactive power generation,} \\ &\quad \text{not to change reactive power generation}\} \end{aligned}$$

and

$$\begin{aligned} \mathbf{S}_2 &= \{S_{2C}, S_{2T}\} \\ &= \{\text{to switch in capacitor banks,} \\ &\quad \text{to increase tap positions of 220 kV main transformers}\} \end{aligned}$$

The nodal voltage magnitudes and gateway power factors should be in tolerance bands in the existing typical mechanism of AVC. Here, "gateway" means the join of the main grid and subsystem.

Hence payoffs $\{F_i\}$ can be calculated by strategy cost and gains, which is given in Table 6.7, and the normal form of the noncooperative game can be accordingly shown in Table 6.8.

Table 6.7 Costs and gains of the game.

Participant	Strategy	Cost	Avail for itself under this strategy	Avail for the other under this strategy
Main grid	S_{1C}	2	0	5
	S_{1N}	0	0	0
Subsystem	S_{2C}	4	5	0
	S_{2T}	2	5	−5

Table 6.8 Payoffs from noncooperative game of AVC in normal form

		Subsystem strategy	
		S_{2C}	S_{2C}
Main grid strategy	S_{1C}	$(-2, 6)$	$(-7, 8)$
	S_{1N}	$(0, 1)$	$(-5, 3)$

Experiential values without units are adopted in qualitative analysis in Table 6.7. Since the cost of changing tap position of the transformer is greater than that of switching in a capacitor bank, the cost of S_{2C} is assumed to be 4 and that of S_{2T} to be 2. No matter for which player, benefit of the control strategy that can banish the disqualification of voltage profile is enacted to be 5. For simplicity of analysis, the enhancing VAR generation of the main grid, of which the cost is about 2, is exactly what can eliminate the disqualification of voltage profile in the subsystem because the voltage profile in the subsystem can be increased by further generating reactive power in the main grid. Under these assumptions, the costs and profits of various strategies are itemized as follows.

1. S_{1C} gives a benefit of 0 to the main grid, since the voltage profile of the main grid has not exceeded the limits. However, it gives 5 to the subsystem by reason of its dispelling of the voltage profile disqualification.
2. S_{2C} gives a benefit of 5 to the subsystem by virtue of its dispelling of the voltage profile disqualification, but that to the main grid is only 0 since the voltage profile of the main grid has not exceeded the limits.
3. S_{2T} can increase the voltage profile of the subsystem, because of which the profit of the subsystem is 5, but at the cost of aggrandizing the reactive power load. To balance the incremental VAR load, the subsystem has to exact more reactive power from the main grid, making the main grid punished for exceeding the limit. Therefore, S_{2T} gives a profit of -5 to the main grid.

The payoffs of the AVC game in existing typical mechanism are shown in Table 6.8, which is called the normal form or game matrix in game theory. The numbers in parentheses denote the payoffs of the two players under a certain strategy combination. For instance, if the adopted strategy combination is $\{S_{1C}, S_{2C}\}$, as can be checked in Table 6.7, the main grid benefits 0 from this combination while its cost is 2; the payoff of the main grid under this combination is $0 + 0 - 2 = -2$, according to the above. The payoff of the subsystem that is $5 + 5 - 4 = 6$ can be calculated in the same way. Thus, the payoff combination of the players under $\{S_{1C}, S_{2C}\}$ is $(-2, 6)$. The payoff combinations under the other three strategy combinations can be calculated by similar means.

The Nash equilibrium for the game matrix showed in Table 6.8 can be solved by a method named iterated elimination of dominated strategies [245]. For the main grid, if the selected strategy of the subsystem is S_{2C}, the main grid will evidently select S_{1N} due to the fact that its payoff of selecting S_{1N} is 2 more than that of selecting S_{1C}; else if the subsystem selects S_{2T}, the best response of the main grid is likewise to select S_{1N} since its payoff of selecting S_{1N} is 2 more than that of selecting S_{1C} too. Accordingly S_{1N} is the main grid's dominated strategy – the best choice no matter which strategy is adopted by the subsystem. We can conclude that S_{2T} is the dominated strategy of the subsystem in the same way. That means the strategy set $\{S_{1N}, S_{2T}\}$ therefore is the solution, namely the Nash equilibrium of this game with payoff of $(-5, 3)$ to the players.

The total payoff of the whole system is as little as $-2\ (= -5 + 3)$ according to the equilibrium; however, it would increase to $4(= -2 + 6)$ if $\{S_{1C}, S_{2C}\}$ is chosen. That means the latter is better for the whole systems. From the viewpoint of power systems, $\{S_{1C}, S_{2C}\}$ means more reactive power generations to balance the reactive loads to improve the voltage profiles and save active power losses. Obviously, the low-efficiency equilibrium in this situation is neither the global optimization of the whole systems nor the highest gains for both players – the players' dominated strategies deteriorate the AVC efficiency of the whole systems. This outcome demonstrates that Nash equilibrium cannot always guarantee a collective rationality effect to the game in a noncooperative mechanism as in the "prisoners' dilemma" model. The strategies of AVC in such a mechanism could simply increase the tap positions of transformers to attempt to improve the nodal voltage magnitudes of the subsystem, but would result in more reactive power load demands and more reactive power flow transmission far from the main grid to the subsystem. The voltage profiles of the whole networks would be worse therefore, yielding the VCM phenomenon.

(ii) Qualitative Nash Equilibrium for Cooperative Game

A cooperative mechanism can be built to solve the VCM problem. Cooperative game theory is concerned primarily with coalitions – groups of players – who coordinate their actions and pool their winnings. Consequently, one of the problems here is how to divide the extra earnings (or cost savings) among the members of the formed coalition [248]. The objective of AVC cooperative game research of large-scale power systems is to augment the payoff of each player by establishing a cooperative mechanism and consequently increase the profit of the whole systems.

A 220 kV substation is the connection between the main grid and subsystem, and its gateway voltage often drops under heavy load condition. It is helpful for the security and loss saving of the whole systems to improve gateway voltages. Both sides of the AVC game players consequently have the responsibility to advance the 220 kV nodal magnitude in the heavy load condition. In other words, an increment of voltage should be taken into consideration as part of the avails of both players, and the avails listed in Table 6.7 can be revised as shown in Table 6.9. The normal type of cooperative AVC game is shown in Table 6.10 as the result.

Table 6.9 The costs and gains of the game considering gateway voltage.

Participant	Strategy	Cost	Avail for itself under this strategy	Avail for the other under this strategy
Main grid	S_{1C}	2	3	8
	S_{1N}	0	0	0
Subsystem	S_{2C}	4	8	3
	S_{2T}	2	3	−7

Table 6.10 Payoffs from the cooperative game of AVC.

		Subsystem strategy	
		S_{2C}	S_{2C}
Main grid	S_{1C}	(4, 12)	(−6, 9)
strategy	S_{1N}	(3, 4)	(−7, 1)

In a cooperative game, the examination mechanism demands the AVC operators pay attention to voltage quality of the gateways (220 kV buses), which means that, if a strategy can promote voltage magnitudes of the gateways, it will augment the profits of each player, or it will not. Accordingly, the costs and profits of various strategies in a cooperative mechanism are given as:

1. S_{2C} brings benefits of 3 to the main grid and 8 to the subsystem because it has a supporting effect on voltage magnitudes of 220 kV buses.
2. S_{2T} also has incremental effect on voltage magnitudes of 220 kV buses, so the profits of the main grid and subsystem given by S_{2T} are increased to 3 and 8.
3. S_{2T} has decreasing effect on voltage magnitudes of the gateway. As a result, it cut the benefits of the main grid and subsystem down to 3 and −7.

In Table 6.10, Nash equilibrium is $\{S_{1C}, S_{2C}\}$ with payoff (4, 12), which can attain a high benefit of 16 for the whole systems. This high-efficiency equilibrium means the coordinated optimization of the main grid and subsystem.

It is concluded that cooperative mechanisms are helpful to attain equilibrium with high efficiency and unify the individual and collective rationality. We can judge whether a certain mechanism is good or not to the whole systems' optimization through analyzing Nash equilibrium. It is the key to solving the VCM problem and getting the AVC game out of the "prisoners' dilemma" to form a cooperation. What is more, how to model the payoff function exactly is another focus since equilibrium is related to the payoff function model.

6.3.2 *Payoff Function Modeling of Multi-agent Volt/VAR Control*

6.3.2.1 Concept and Formulation

The payoff function is a mathematical function describing the award given to a single player at the outcome of a game [245, 246]. In the game of AVC, the payoff function should consider the relevant operation indices to which AVC operators pay very close attention in a certain control mechanism and often includes the control avail and cost. AVC's avail consists of the values of voltage quality enhancement, power loss saving, voltage security improvement, and so on. The costs of readjusting transformer tap position and reactive compensation comprise the cost of the payoff function, which are not merely the physical cost as in the literature [249], but also the mental scenario cost that can reflect the player's awareness. Especially when the number of capacitors and switching times of transformer taps are limited, the cost in payoff function is several times larger than the physical one considering the mental cost. Under these descriptions, we can formulate the payoff function as follows:

$$F = f\left(\sum_{v=10}^{500} \alpha_v \Delta V_v\right) + g(\Delta P_{loss}) + h(\delta_{VSM}) + k(C_R) + \beta \tag{6.36}$$

where ΔV_v is the summation of voltage magnitude increments of the buses in voltage level v $(v = 10, 110, 220, 500\text{kV}, \dots)$; α_v is a weight coefficient that reflects the influence of ΔV_v to each player on its award; ΔP_{loss}, δ_{VSM}, and C_R are the power loss saving, the increment of voltage security margin, and the total cost of adjusting the control devices in the power network corresponding to each player, respectively, when an action set is executed; f, g, h and k are functions that convert different forms of variables into a uniform dimension and they can usually be linearized; and β is a penalty function whose value will be 0 if the examination indices, such as nodal voltage magnitude and power factor of gateway, are qualified, else it will be a large negative value to force $F \leq 0$.

The values of ΔV_v, ΔP_{loss}, and δ_{VSM} in (6.36) are related to a certain strategy set $\{s_1, s_2, \dots, s_n\}$. This means that F_i (the payoff of player i) is a function of strategy set space $S = \{s_1, s_2, \dots, s_n\}$.

In practical operation and control of power systems, the concrete forms of payoff functions are dependent on the power system operators' cognizance. For example, if the operators do not care or cannot quantitatively calculate the power loss, $g(\Delta P_{loss}) = 0$; and if the operators want to decrease the operating times to guarantee the system's security, $k(C_R)$ is several times higher than the theoretical value C_R. Therefore, (6.36) is just a general formula whose concrete form depends on actual situation.

We can assume an active item vector $I = \{f, g, h, k\}$ to denote the indices with which the players are concerned in a certain mechanism. For instance, $I = \{f, k\}$ means that the players care about voltage amplitude increment ΔV_v and readjusting cost C_R in some mechanism. From (6.36), we can figure out that the payoff is also related to weight coefficients like α_v, so a weight coefficient vector $W = \{\alpha_v\}$ ($v = 10, 110, 220, 500\text{kV}, \ldots$) could be assumed in the same way.

In summary, the payoff function of the AVC game in a large-scale power system is not only a function of the strategy set, but also related to both game mechanisms and weights of relevant variables, described as follows:

$$F_i = u(S, I_i, W_i) \tag{6.37}$$

If the AVC strategy is discrete, we can number all the strategy combinations, and then define the payoff of player i under the jth combination as F_{ij}.

6.3.2.2 Explanation of the Model

Under a certain game mechanism, AVC operators' awareness of system operation indices is determinate, i.e. active item vector I is definite. Thus, weight coefficient W_i is the only independent variable of F_{ij}, which can be described as

$$F_{ij} = u(W_i) \tag{6.38}$$

If strategy set $s_1^*, s_2^*, \ldots, s_n^*$ is the Nash equilibrium of AVC game and its number is j^* in the strategy set M, according to (6.35), F_{ij} satisfies the following condition:

$$F_{ij*} \geq F_{ij}, \forall \quad j \in N \tag{6.39}$$

Since F_{ij} is a function of W_i, there are two possibilities in the above-mentioned AVC game analysis process.

1. No matter which values the weight coefficients take in their spaces, the Nash equilibrium is definite, which means that the equilibrium would not change as the weight coefficients vary. However, the Nash equilibrium might be a global optimal solution or a mismatch one for the whole system.
2. The Nash equilibrium changes as the weight coefficients vary, so we can transfer the equilibrium from a mismatch "dilemma" solution to an optimal solution by changing the weight coefficients. This goal can be achieved via solving (6.39). This conclusion's guiding significance for engineering lies in the fact that we can change the emphasis of the examination indices of volt/VAR control to coordinate the AVC operated by different operators and avoid VCM.

For the above case 1, we can transfer the Nash equilibrium from a bad solution to an optimal solution by changing the construction of vector I, while we can achieve this goal by changing the values of W_i for case 2. In either case, cooperative mechanisms should be established to form a cooperative game.

Assuming that $\{s_1', s_2', \ldots, s_n'\}$ is already known as the optimal solution for the AVC in some scenario, our aim is to make it the Nash equilibrium. The first step is to assign the combination's number to j^* in (6.39) and solve it. Then the solution, value ranges of the weight coefficients, will correspond to the relevant cooperative constraints.

6.3.2.3 Application to an Actual Power System

(i) The Power System for Simulation

A 500 kV/220 kV/110 kV regional power system of China Southern Grids is used as the simulation object in this section, whose model is shown in Figure 6.17, where the 220 kV substation A and its 110 kV network belong to the subsystem operator, while the 500 kV substation and 220 kV power plants belong to the main grid. For the sake of simplicity, all the power supplied by the 220 kV substations except A is equivalently and correspondingly connected to their 110 kV and 10 kV bus. The base case load flow is fitted according to the peak load of a given operating year of this system. This simulation is run in the BPA power system analysis program.

(ii) Two Game Mechanisms

1. *Mechanism 1: the existing typical mechanism*. The detailed information of the concerned items of volt/VAR control participants under mechanism 1 is described in Table 6.11. Then, in mechanism 1, payoff functions F_{1j} for the main grid and F_{2j}

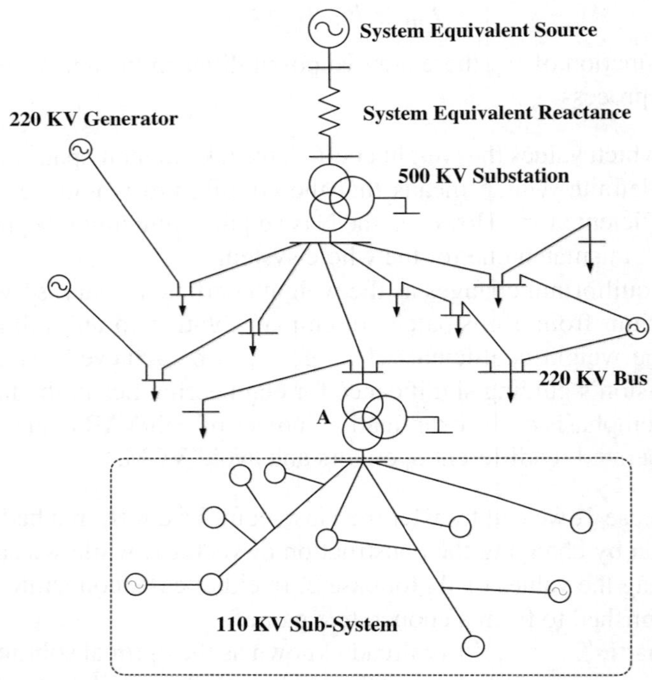

Figure 6.17 Model for simulation. *Source*: Zhang 2005. Reproduced with permission from Elsevier.

for the subsystem under some strategy combination j are given as follows:

$$F_{1j} = k_1(C_{R1}) + \beta \tag{6.40}$$

$$F_{2j} = k_2(C_{R2}) + \beta \tag{6.41}$$

where the subscript 1 stands for the main grid while 2 stands for the subsystem. Apparently, the control objective of mechanism 1 is to eliminate exceeding the voltage profile and power factor limits without meeting the needs of energy saving and optimization.

2. *Mechanism 2: focus on increment of gateway voltage magnitude*. Just like what is described in section 6.3.1.4(ii), mechanism 2 demands that the operators consider the increment of gateway voltage magnitude in operation, especially in the subsystem. What is more, considering the needs of voltage security, dispatch departments should be concerned with not only the voltage qualification but also the voltage quality. This demand needs to be incarnated in the payoff functions of this mechanism, which are set as follows:

$$F'_{1j} = f(\alpha_{220}\Delta V_{220} + \alpha_{500}\Delta V_{500}) + k_1(C_{R1}) + \beta \tag{6.42}$$

$$F'_{2j} = f(\alpha_{10}\Delta V_{10} + \alpha'_{200}\Delta V_{200}) + k_2(C_{R2}) + \beta \tag{6.43}$$

(iii) Strategy Combination Analysis

In this simulation example, the adopted control strategies and their theoretical adjusting costs, which are calculated based on the method proposed in [249], are presented in Table 6.12. The systemwide voltage profile is low because the base case power flow is in a peak load condition. All the 220 kV generators are running according to rated power factors, and therefore only the capacitors of the 500 kV substation are available for the main grid. For the subsystem, the control measures available are switching in capacitor banks and changing the tap position of transformers in both

Table 6.11 Detailed information of concerned items

Item	Active or not	Reason
β	Yes	Power system operators pay much attention to qualification of voltage profile
h	No	It is still too difficult to consider voltage security margin in real-time operation
g	No	Real-time power flow calculation base is still not accurate enough in a large-scale power system such as China Southern Grid
f	No	Operators will not be commanded to improve the voltage quality while the voltage profile is qualified
k	Yes	Operators hope to reach the control target with the lowest cost

Table 6.12 Detailed information of concerned items.

Player	Strategy symbol	Specific strategy	Adjusting cost/$
Subsystem	S_{2T}	Change the tap positions of transformers in A with one step	8.958
	S_{2C}	Switching in two capacitor banks in A (20 MVAR)	9.560
Main grid	S_{1N}	Not change VAR generation	0.000
	S_{1C}	Switching in a capacitor bank in 500 kV substation (20 MVAR)	7.166

Table 6.13 Control effects of AVC.

Strategy combination	S_{1C}, S_{2C}	S_{1C}, S_{2T}	S_{1N}, S_{2C}	S_{1N}, S_{2T}
Serial number j	1	2	3	4
ΔV_{500}/kV	0.500	0.200	0.200	0.000
ΔV_{220}/kV	8.400	3.700	4.100	−0.700
ΔV_{10}/kV	1.057	1.259	0.815	1.018
ΔV_{10}/kV	1.057	1.259	0.815	1.018
ΔP_{loss}/kV	20 960	20 972	20 963	20 988
Cost/$	15.726	16.124	8.560	8.958
Qualified or not of main grid	Yes	Yes	Yes	Yes
Qualified or not of subsystem	Yes	Yes	Yes	Yes

the 220 and 110 kV substations. In this example, the subsystem's control actions are set in the 220 kV substation, not only for the sake of simplicity but also for a more evident effect on the main grid. In addition, the principles of setting specific control strategies are that the costs of the two strategies of the subsystem are equivalent and the compensated VAR capacities of two players are also equivalent.

Once the strategy combinations are ascertained, the corresponding power flows can be calculated. Power flows corresponding to different strategy combinations are compared and analyzed, with the concrete data being shown in Table 6.13.

As can be checked in Table 6.13, if $\{S_{1C}, S_{2C}\}$ is adopted, the voltage profiles of both main grid and subsystem have a larger advance, which is helpful to enhance security and decrease the power loss of the whole system. Therefore, $\{S_{1C}, S_{2C}\}$ is the best strategy combination for coordinative optimal volt/VAR control of the four combinations. Conversely, if $\{S_{1N}, S_{2T}\}$ is chosen, the global power loss is maximal and the 220 kV gateway voltage magnitudes are even lower than those in the initial power flow. Consequently, $\{S_{1N}, S_{2T}\}$ is the worst strategy combination although its adjusting cost is less than the others.

Table 6.14 Payoffs of the game in mechanism 1 in normal form.

		Subsystem strategy	
		S_{2C}	S_{2C}
Main grid	S_{1C}	(F_{11}, F_{21})	(F_{12}, F_{22})
strategy	S_{1N}	(F_{13}, F_{23})	(F_{14}, F_{24})

Thus, in this case, if a mechanism makes $\{S_{1N}, S_{2T}\}$ the Nash equilibrium of the AVC game, this mechanism will be considered to have a noncooperative effect. If another mechanism makes $\{S_{1C}, S_{2C}\}$ the Nash equilibrium of the AVC game, it can be deemed as a cooperative mechanism.

(iv) Nash Equilibrium Analysis

1. *Game in mechanism 1.* The normal form of the AVC game in mechanism 1 is shown in Table 6.14. Equations (6.40) and (6.41) are used to calculate the payoffs of both the players quantitatively. First of all, a linear model is adopted to tackle the adjusting cost for simplicity, namely $k(C_R) = \gamma C_R$, where the mental cost is γ times the physical one. Second, substitute the variables of (6.40) and (6.41) with the data in Table 6.12 and Table 6.13. Third, considering that all the examination indices are qualified and the setting control strategies have little effect on the 500 kV bus, β and ΔV_{500} are ignored. Then, the result is

$$\begin{cases} F_{11} = F_{12} = -7.166\gamma_1 \\ F_{13} = F_{14} = 0.000 \\ F_{21} = F_{23} = -9.560\gamma_2 \\ F_{22} = F_{24} = -8.958\gamma_2 \end{cases} \tag{6.44}$$

   As can be calculated from (6.44), $F_{11} < F_{13}$, $F_{12} < F_{14}$, $F_{21} < F_{22}$, and $F_{23} < F_{24}$, which means S_{1N} is the dominated strategy of the main grid in mechanism 1 while S_{2T} is the subsystem's.

   This result demonstrates that the Nash equilibrium is independent of the values of the weight coefficients. In other words, such a mechanism might lead the AVC game to be in an inferior status, namely the so-called VCM problem and the "prisoners' dilemma", as a result of the operators taking some measures unilaterally which go against optimization of the whole system to pursue their own qualification with the minimal costs. Without the consciousness of coordination and cooperation, a VCM complex of operators competing with each other for VAR resource is ineluctable.

Table 6.15 Payoffs of the game in mechanism 2 in normal form.

		Subsystem strategy	
		S_{2C}	S_{2C}
Main grid strategy	S_{1C}	(F'_{11}, F'_{21})	(F'_{12}, F'_{22})
	S_{1N}	(F'_{13}, F'_{23})	(F'_{14}, F'_{24})

2. *Game in mechanism 2*. The normal form of the AVC game in mechanism 2 is shown in Table 6.15. A linear model is used to tackle f and k in the same way, that is $f(x) = x$ and $k(x) = \gamma x$. Substitute the variables of (6.42) and (6.43) with the data in Table 6.12 and Table 6.13. Then the payoffs can be calculated to be as follows:

$$\begin{cases} F'_{11} = 8.400\alpha_{220} - 7.166\gamma_1 \\ F'_{12} = 4.100\alpha_{220} - 7.166\gamma_1 \\ F'_{13} = 3.700\alpha_{220} \\ F'_{14} = -0.700\alpha_{220} \\ F'_{21} = 8.400\alpha'_{220} + 0.815\alpha_{10} - 8.958\gamma_2 \\ F'_{22} = 4.100\alpha'_{220} + 1.018\alpha_{10} - 9.560\gamma_2 \\ F'_{23} = 3.700\alpha'_{220} + 1.057\alpha_{10} - 8.958\gamma_2 \\ F'_{24} = -0.700\alpha'_{220} + 1.259\alpha_{10} - 9.560\gamma_2 \end{cases} \tag{6.45}$$

We cannot reach the Nash equilibrium by solving (6.45) as the dominated strategies of both the main grid and the subsystem vary according to the weight coefficients. In order to make $\{S_{1C}, S_{2C}\}$ the Nash equilibrium, the first step is to set j^* of (6.39) to 1, from which we can get

$$\begin{cases} F'_{11} > F'_{13} \\ F'_{12} > F'_{14} \\ F'_{21} > F'_{22} \\ F'_{23} > F'_{24} \end{cases} \tag{6.46}$$

and

$$\begin{cases} 4.700\alpha_{220} - 7.166\gamma_1 > 0 \\ 4.300\alpha'_{220} - 0.203\alpha_{10} - 8.958\gamma_2 > 0 \end{cases} \tag{6.47}$$

Inequalities (6.47) is the solution of (6.46) and its intervals are located in the upper left side of the lines in Figure 6.18 and Figure 6.19. For the main grid, if its weight

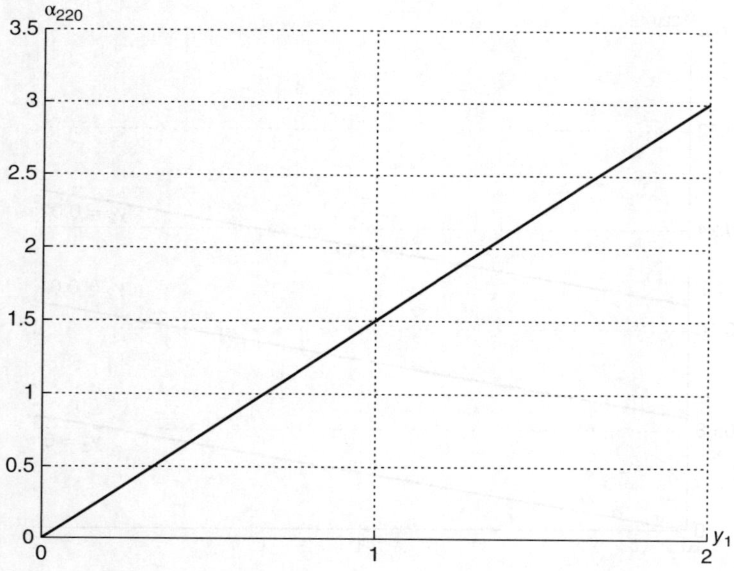

Figure 6.18 Results for main grid.

coefficient point (γ_1, α_{220}) is located in the upper left side of the line in Figure 6.18 (in other words, the value of γ_1 is 0.656 times less than that of α_{220}), which is consistent with (6.46), S_{1C} is the dominated strategy. On the contrary, if (γ_1, α_{220}) is located in the opposite side of the line, S_{1N} will be the dominated strategy, which is considered to be adverse to the global optimization according to section 6.3.2.2. In conclusion, after considering the benefit of elevating the voltage profile, we still cannot turn the Nash equilibrium into an optimal solution to the whole system directly. However, we can, at least, solve the VCM problem to achieve the coordinative optimization in the whole system by changing the values of the weight coefficients.

6.4 Volt/VAR Control in Distribution Systems Using an Approach Based on Time Interval

At present, research in the field of volt/VAR control for distribution systems can be divided into two categories: off-line setting control and real-time control. Research in off-line setting control aims to find dispatch schedules for switching capacitors and OLTC setting at substations for the day ahead according to optimization calculations based on load forecasts for the day ahead, while research in real-time control aims to control the aforementioned devices based on real-time measurements and experiences. The second category of control requires a higher level of distribution system automation and more hardware and software support. Until recently, the

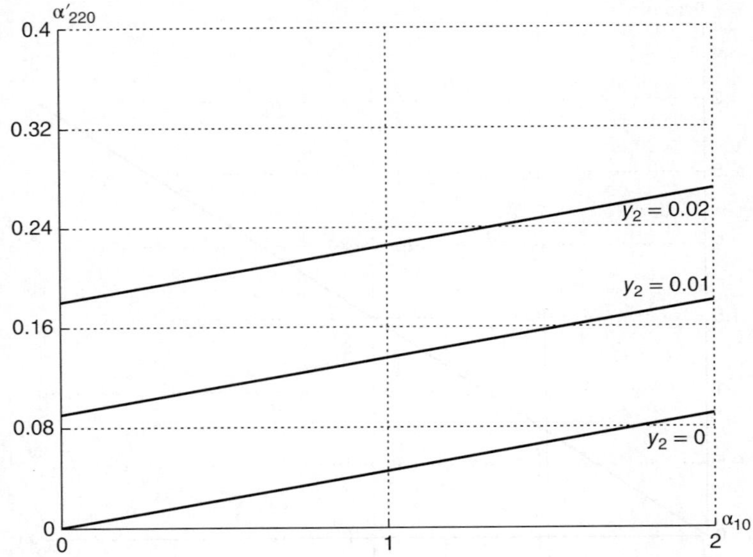

Figure 6.19 Results for main subsystem.

majority of distribution systems did not reach such standards. Furthermore, it is very difficult for real-time control to consider the overall load change as well as the constraints of maximum allowable switching operations for a number of volt/VAR control devices.

In the off-line setting control category, dynamic programming is often applied [186, 250–252], but because of the heavy computational burden such approaches are very time-consuming even when simplifications are made. When dynamic programming is employed to determine the optimal dispatch of capacitors on a feeder of a distribution system [250] or optimal volt/VAR control in a distribution substation [251], the burden is acceptable because of the relatively small searching space. However, to find the optimal dispatch schedule for OLTC settings at a substation and all switching capacitors across the whole distribution system for the day ahead by dynamic programming [252] requires a very large search space that is computationally time-consuming. To reduce the computation burden, Liu and Qiu [186] decompose the problem into two subproblems, one at the substation level and the other at the feeder level. Dynamic programming and fuzzy logic control algorithms are utilized to solve the two subproblems, respectively.

This section presents a volt/VAR control strategy to solve the off-line setting control problem by coordinating all volt/VAR control devices in a distribution system. The aim is to find the optimal dispatch schedule for all such devices in a distribution system based on the forecasted hourly loads of each bus, so that the total energy loss can be minimized. The constraints taken into account include the maximum allowable number of switching operations in a day for OLTC setting and all capacitor switching, as well as the nodal voltage limits.

To simplify the control of OLTC setting at substations, this section proposes a time-interval control strategy that is based on load forecasts. The daily load at the secondary bus of substations is divided into several continuous load levels using a genetic algorithm. OLTC setting can occur between different load levels, but no OLTC setting can occur during a load level.

The optimal dispatch of all volt/VAR control devices is a multi-phase decision-making problem. For each hour, it is a discrete and nonlinear problem. Therefore, using traditional mathematical methods can be very complex, while dynamic programming entails a heavy computational burden. In this section a genetic algorithm is employed that uses a special encoding method to avoid such problems.

6.4.1 Problem Formulation

With the development of a distribution management system (DMS), loads along each feeder bus and substation secondary bus can be obtained for the day ahead by employing short-term load forecasting techniques.

Generally, voltages at the primary bus of a substation change slightly over a day and are therefore assumed to have a constant value in this section. OLTC setting is based on the change of load. To minimize energy losses more effectively, the voltage at the secondary bus of a substation can vary within its limits and is not kept close to a specified value, which differs from the procedure used elsewhere [186, 251, 252]. All volt/VAR control devices are constrained by a maximum number of switching operations that can be performed during a day. To avoid control engineer inconvenience, the maximum number of switching operations for a capacitor along a feeder is two, which is less than that of a capacitor at a substation secondary bus.

The following objective function is used to minimize the energy loss for the day ahead:

$$J = min \sum_{i=1}^{24} (\Delta P_i + \Delta P_{i+1}) \Delta T_i / 2 \tag{6.48}$$

where ΔT_i is the period between time i and $i + 1$ (1 h in this section), ΔP_i and ΔP_{i+1} are the real power losses at times i and $i + 1$, respectively, and $(\Delta P_i + \Delta P_{i+1})\Delta T_i/2$ is the approximate energy loss in a period.

The objective function is subject to the standard power balancing equality constraints as well as the following additional inequality constraints:

$$V_{min} < V_{m,i} < V_{\max} \tag{6.49}$$

$$\sum_{i=1}^{24} |TAP_i - TAP_{i-1}| \le MK_T \tag{6.50}$$

$$\sum_{i=1}^{24} |C_{m,i} \oplus C_{m,i-1}| \le MK_{Cm} \tag{6.51}$$

where $V_{m,i}$ is the voltage of node m at time i, V_{min} and $V_{\max}$ are nodal voltage limits, TAP_i is the tap position at hour i, MK_T is the maximum switching operations for OLTC, $C_{m,i}$ is the status of capacitor m (ON or OFF) at time i, and MK_{Cm} is the maximum switching operations for capacitor m.

6.4.2 Load Level Division

It is difficult to specify the controlling parameters when applying automated techniques to control OLTC at a substation level. It should also be noted that, because of the probabilistic nature of load forecasting, it could be construed as inaccurate to determine a dispatch schedule of OLTC settings based only on load forecasting [186, 251, 252]. However, from inspection of two typical load curves [253] shown in Figure 6.20 and Figure 6.21, it can be seen that several apparent load levels exist during a day. Therefore, if the start and end times of each load level are found, the tap position may only move when the load transits from one load level to another and is then kept fixed during each continuous load level. This method not only takes into account the overall daily load change, but also removes unnecessary movements. If the error of the load forecast is not too great, a dispatch schedule of load levels found in this way that last several hours can be put into the operation for the day ahead directly.

Now the question is how to recognize the location of each load level. First, the number of load levels S in a day is assumed known; it can be based on the load forecast, MK_T, and control engineer experience. After that, the genetic algorithm (GA) is employed to determine the start and end times of each load level. The fitness function is

$$F = F_{\max} - min \sum_{i=1}^{S_n} \sum_{j=1}^{K_i} [(P_{ij} - PA_i)^2 + (Q_{ij} - QA_i)^2] \tag{6.52}$$

subject to

$$\sum^{S_n} K_i = 24 \tag{6.53}$$

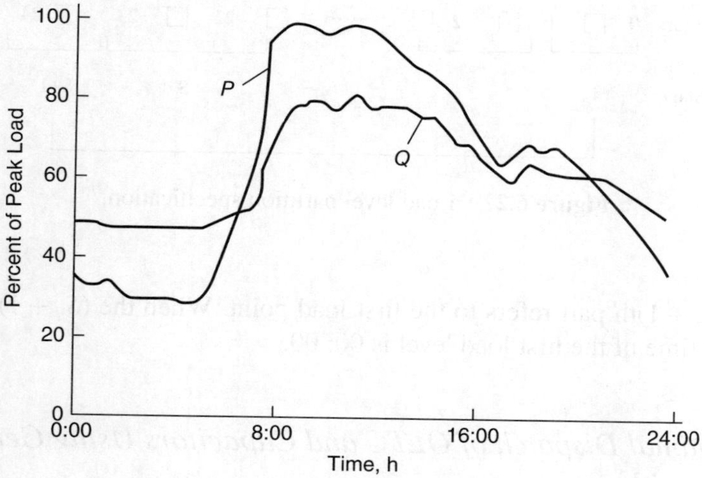

Figure 6.20 Typical daily load curve (1).

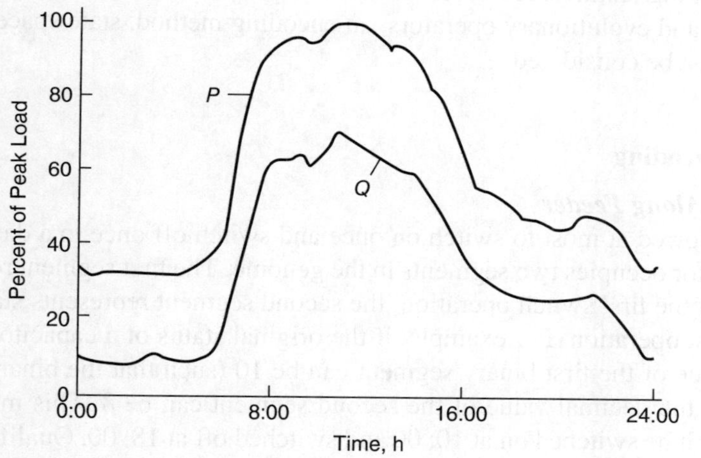

Figure 6.21 Typical daily load curve (2).

where F_{max} is constant (converts fitness function to standard form), P_{ij} and Q_{ij} are the active and reactive powers of the jth load point of the ith load level, PA_i and QA_i are the average active and reactive powers of the ith load level, and K_i is the number of load points of the ith load level.

There are 24 forecasted load points representing each hour of the following day; each load level lasts $25 - S_n$ hours at most and 1 h at least. The method for handling this problem is shown in Figure 6.22. It should be noted that there are S_n variants and a duration of S load levels. The start time of the first load level is not always the start time of the forecasted load, usually 00: 00. Dividing the 24 load points into $S_n + 1$

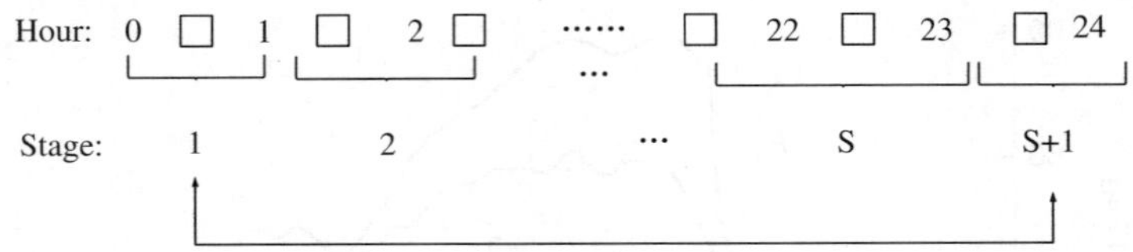

Figure 6.22 Load level partition specification.

parts, the $(S_n + 1)$th part refers to the first load point. When the $(S_n + 1)$th part lasts 0 h, the start time of the first load level is 00: 00.

6.4.3 Optimal Dispatch of OLTC and Capacitors Using Genetic Algorithm

A GA is suitable for solving a problem with discrete, nonlinear, and multi-phase decision-making features. However, to choose a proper selection strategy, controlling parameters, and evolutionary operators, an encoding method, state space, and fitness function must be considered.

6.4.3.1 Encoding

Capacitors Along Feeder

They are allowed at most to switch on once and switch off once in a day. Therefore, each capacitor occupies two segments in the genome. The first segment represents the start time of the first switch operation; the second segment represents status duration after the first operation. For example, if the original status of a capacitor is OFF, the decimal value of the first binary segment can be 10 (such that the binary segment is 01010) and the decimal value of the second segment can be 8. This means that the capacitor will be switched on at 10: 00 and switched off at 18: 00. Qualified segments are those for which the sum of these two values is not greater than 24.

Capacitors at Secondary Bus

Their switching numbers are unknown, but they must be constrained by the maximum number of switches that are allowable each day. Each capacitor holds one segment in the genome. The length of the segment is 24 bits. If the value of the ith bit is 0, it denotes that the capacitor is OFF at time i. Qualified segments should satisfy the constraint of (6.51).

OLTC

The operational characteristic is that the tap position can be different at different load levels and remains constant during each load level. Hence, it occupies s segments in the genome. The lengths of each segment are equal and is dependent on the number of tap positions. The OLTC in the test system has 17 tap positions; the segment length can be 4 (assuming that the tap position difference between two neighboring load levels is no greater than 15). Qualified segments should be subject to

$$\sum_{i=1}^{s_n} |TAP_i - TAP_{i-1}| \leq MK_T \tag{6.54}$$

Here TAP_i represents the tap position during the ith load level.

6.4.3.2 Analysis of Computational Burden

According to the aforementioned encoding method, the genome length will be

$$L = LU_1 S + 2LU_2 C_f + 24C_s \tag{6.55}$$

where LU_1 and LU_2 are the segment length of each OLTC and capacitors along feeders, respectively, and C_f and C_s are the number of capacitors along feeders and at the secondary bus of the substation, respectively.

If there are five and two capacitors along the feeders and at the secondary bus of the substation, respectively, let $LU_1 = 4$, $LU_2 = 5$, and $S_n = 4$, then L_u, the size of the searching space, is 2^{114}. If dynamic programming is applied, there will be $2^7 \times 17$ states at each load point. If $N = 24$, the size of the searching space would be $(2^7 \times 17)^{24}$, which is more than 2^{264}. Even if four tap positions are chosen at each stage, the size of the search space is still very large (2^{216}). It should also be emphasized that, to test each state of the genome using the method presented in this paper, a load flow must be calculated 24 times. However, it should also be noted that many states in the searching space violate the constraints of (6.50) and (6.51), and therefore load flow calculation is unnecessary for them. In this way the computational burden can be greatly reduced.

6.4.3.3 Fitness Function

The evolutionary objective of the GA is the minimization of energy loss while satisfying all of the constraints. If any states violate the constraints of (6.50) and (6.51), a small fitness function value will be returned, skipping the calculation of energy loss. So, considering the constraint of (6.49), the fitness function is

$$F = \max\left[F_{max} - \left(w_1 J_p + w_2 \sum_{n=1}^{24}\sum_{m=1}^{M} \Delta U_{m,n}\right)\right] \tag{6.56}$$

where J_p is the percentage energy loss, $\Delta U_{m,n}$ is the value of voltage violation of node m at hour n, and w_1 and w_2 are coefficients of energy loss and voltage violation, respectively.

6.4.3.4 Selection Strategy and Evolutionary Operators

A canonical GA has been adopted, which can be described as follows. During the evolutionary process the individuals with the best fitness function values enter directly into the next generation. A tournament selection model has been adopted to choose the parents for crossover. The children are generated by one-point crossover of their parents and the mutation probability is fixed with regard to the mutation operator.

6.4.3.5 Computational Procedure

A schematic flowchart of the computational procedure is shown in Figure 6.23. The input data include network parameters, forecasted loads, a specified number of load levels S, and parameters of volt/VAR control devices, including original status and parameters for the GA. The GA and load flow calculation are two independent modules.

6.4.4 *Example*

The distribution system in Figure 6.24 is used to demonstrate the effectiveness of the proposed method. The study system has been developed from the distribution system in Civanlar and Grainger [254]. For research purposes, the branch between node 0 and 1 is changed from a line to a transformer branch equipped with OLTC. Two

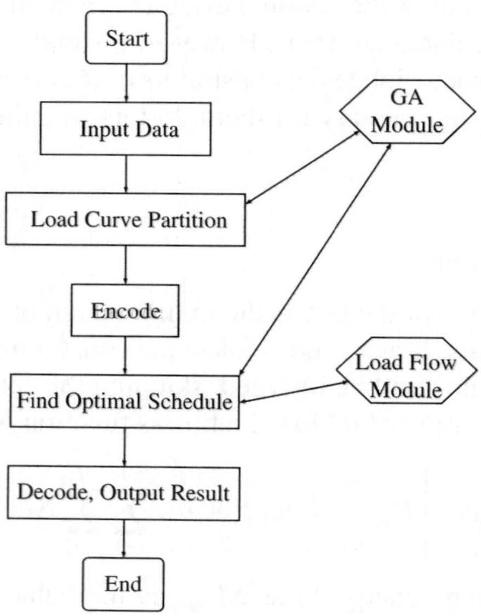

Figure 6.23 Flowchart of time-interval base volt/VAR control algorithm.

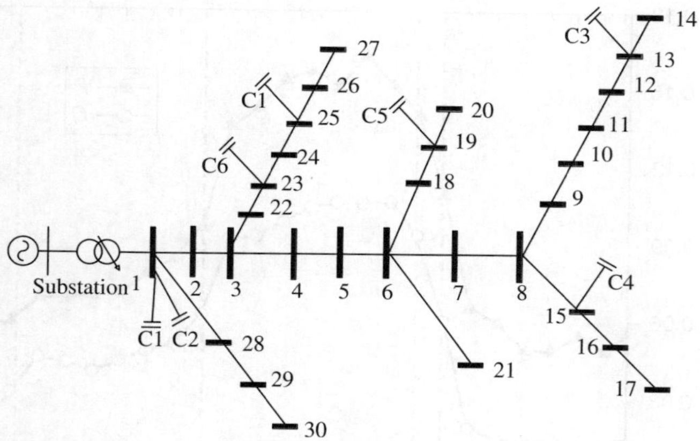

Figure 6.24 One-line diagram of test distribution system.

Table 6.16 Capacitor data for distribution system.

Capacitor no.	C_1	C_2	C_3	C_4	C_5	C_6	C_7
Location (bus no.)	1	1	13	15	19	23	25
Capacitance (kVAR)	900	900	600	600	300	900	900

switched capacitors (C_6 and C_7) are added at node 1. The OLTC has 17 tap positions ($[-8, -7, \ldots, 0, 1, \ldots, 7, 8]$). It can change the voltage from −5% to +5%. Table 6.16 presents the detailed data of the capacitors. The impedance of the transformer between nodes 0 and 1 is (0.0178 + j0.3471) per unit and the base power is 52.9 MVA. The length of each branch line is set to 1 mile. Let $MK_T = 30$ and let $MK_C = 2$ and 6 for capacitors along feeders and at the secondary bus of a substation, respectively. The upper and lower limits of voltage for each bus are 1.05 per unit and 0.95 per unit, respectively. The voltage at the primary bus of a substation is 1.0 per unit. See Shenkman [253] for the maximum reactive load of the bus.

The load model has a great influence on the optimal result. All loads are linked via lines without transformers in the test system. Therefore, if a load model of constant power is adopted, the proposed method must lead to high voltage of the whole system under the objective function (6.48). In simulation tests, 50% of the load of each bus is constant power load and the other 50% is constant impedance load.

We assume that all the loads change during a day according to the load profile shown in Figure 6.21, but that they all vary randomly by ±15% around the nominal level, both real and reactive parts. In this way, loads for all 24 hours are obtained. For the case of $S_n = 4$, the load profiles at buses 1 and 4 are shown in Figure 6.25, where the dash-dot lines indicate the boundaries between load levels. Again, we assume that

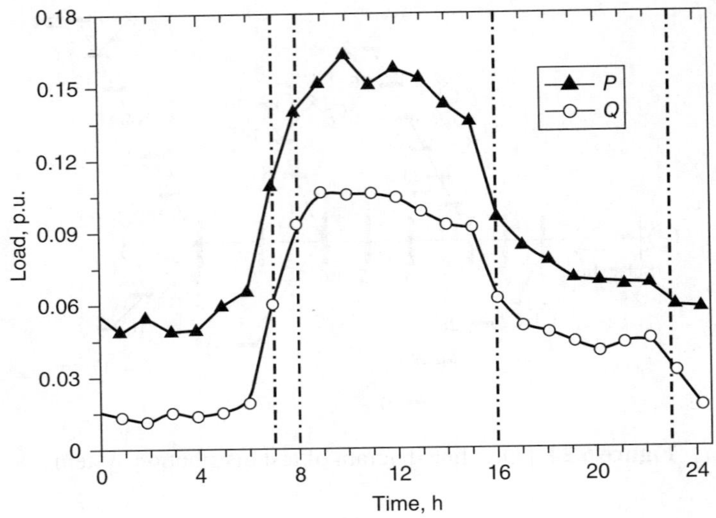

Figure 6.25 Four-load level partition results.

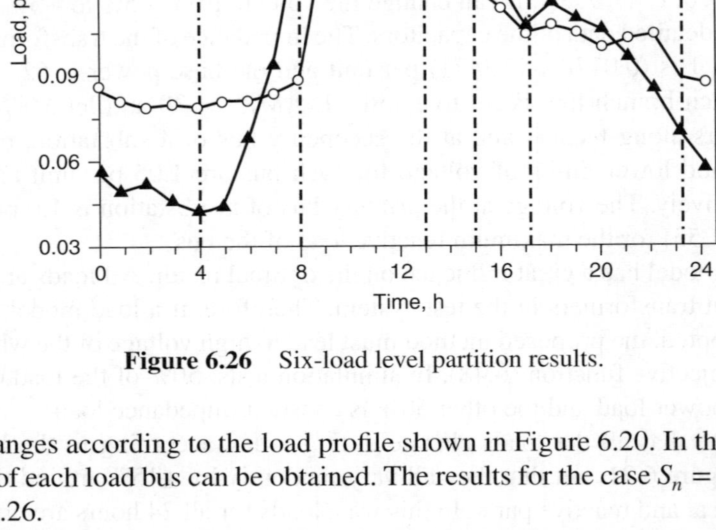

Figure 6.26 Six-load level partition results.

all load changes according to the load profile shown in Figure 6.20. In the same way, 24 h loads of each load bus can be obtained. The results for the case $S_n = 4$ are shown in Figure 6.26.

Table 6.17 shows one optimal dispatch result of OLTC and capacitors based on the loads and load levels shown in Figure 6.25. The original position of the tap is 0, and

Table 6.17 Optimal dispatch schedule for day ahead (for capacitors: 0 = OFF; 1 = ON).

Hour	TAP	C_1	C_2	C_3	C_4	C_5	C_6	C_7
0	−1	1	1	0	1	0	0	0
1	−1	1	1	0	1	0	0	0
2	−1	1	1	0	1	0	0	0
3	−1	1	1	0	1	0	0	0
4	−1	1	0	0	1	0	0	0
5	−1	0	0	0	1	0	0	1
6	−1	0	1	0	1	0	0	1
7	−1	1	0	1	1	0	1	1
8	+2	1	0	1	1	1	1	1
9	+4	1	1	1	1	1	1	1
10	+4	1	1	1	1	1	1	1
11	+4	1	1	1	1	1	1	1
12	+4	1	1	1	1	1	1	1
13	+4	1	0	1	1	1	1	1
14	+4	1	0	1	1	1	1	1
15	+4	1	0	1	1	1	1	1
16	+4	0	0	1	1	1	1	1
17	0	1	0	1	1	1	1	1
18	0	1	0	1	1	1	0	1
19	0	1	0	0	1	1	0	1
20	0	1	0	0	1	1	0	1
21	0	1	0	0	1	1	0	1
22	0	0	0	0	1	1	0	1
23	0	0	0	0	1	1	0	1

the original status of each capacitor is OFF (0). The number of switching operations for OLTC in the whole day is 10. Capacitors C_1 and C_2 switch six times in a day; some of the capacitors along the feeder switch twice, and some switch only once; C_4 is on for the whole day.

The voltage at node 14 is the lowest in the test system. The voltage change is shown in Figure 6.27. Before optimization, all the capacitors are OFF and the tap position is fixed at 0 for the whole day. After optimization, the schedule of Table 6.17 is carried out to simulate the effects. The voltage profile at node 14 is greatly improved.

The partial DP curve plotted in Figure 6.27 illustrates the optimized solution that has been obtained using the partial dynamic programming method [252]. The same objective function as used in Liang and Cheng [252] has also been adopted for this study. At each stage all the $2^7 \times 17$ possible states are calculated and 2^9 optimal paths are saved. In this system the voltage of node 14 relies heavily on capacitor C_3 because the objective function aims to keep the secondary bus voltage close to a specified value (in this case 1.0 p.u.). With regard to the results obtained to plot the partial DP

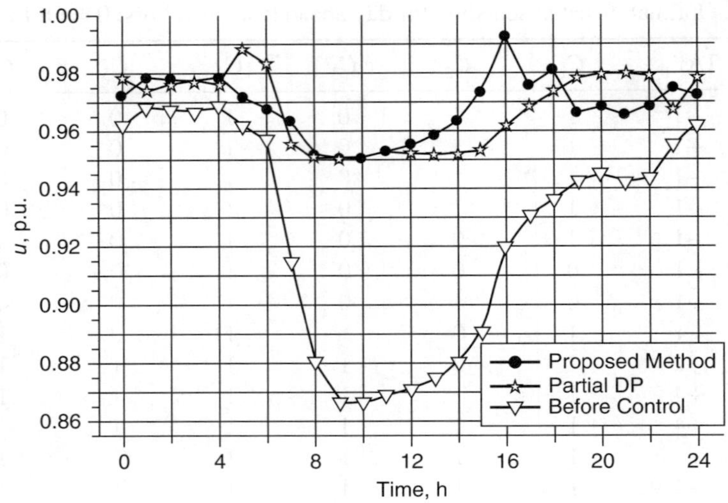

Figure 6.27 Voltage change of bus 14 over a day.

curve, capacitor C_3 is ON from 05: 00 to 23: 00, which leads to higher voltage at the light load hours and lower voltage at the heavy load hours when compared against the proposed method.

A real power loss comparison is shown in Figure 6.28. During the peak load hours, the real power loss is greatly decreased, which meets the volt/VAR control aim of this method. From Figure 6.28 one can see that the proposed method is more effective than the partial DP method with regard to reducing real power losses.

Now let us study the influence of the maximum number of allowable switching operations MK_C for capacitors at the secondary bus of the substation. For the load profile shown in Figure 6.21 and according to the four levels of load partitioning shown in Figure 6.25, let MK_C =2, 4, 6, 8, and 10, respectively. Table 6.18 lists the tap positions across the four load levels ($TAP_i, i = 1, \ldots, 4$), the energy loss (J_e) and the minimum voltage for the day ahead (U_{min}) for different values of MK_C. One can see that the minimum voltage is less than 0.95 when $MK_C = 2$; $TAP_4 > TAP_3$ whilst the load value of the fourth load level is less than that of the third load level when $MK_C = 2$ and 4. The OLTC schedules are almost identical when $MK_C = 6$, 8, and 10. Therefore, for the test system, $MK_C = 6$ is optimal as we need to find a balance between improvement of voltage profile, loss reduction, schedule complexity, and capacitor depreciation cost.

With regard to the load profile shown in Figure 6.20, its change is more complicated than the one shown in Figure 6.21. If we change S and keep the values of MK_T andMK_C constant, we can study the influence on the OLTC schedule. Letting S_n range from 2 to 7, the results are shown in Table 6.19. As the peak value of the load is so heavy, the maximum tap position value is +8 for all S_n. As S_n increases, there are more and more tap positions during a whole day. At the same time this means that the complexity

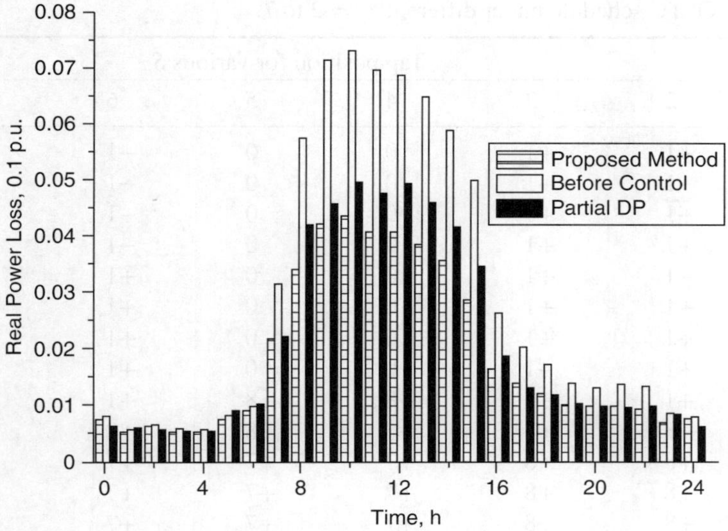

Figure 6.28 Comparison of real power losses.

Table 6.18 Influence of maximum allowable switching operations for capacitors.

MK_c	2	4	6	8	10
TAP_1	0	0	−1	0	0
TAP_2	+2	+2	+2	+2	+2
TAP_3	+5	+4	+4	+4	+4
TAP_4	+7	+6	0	0	0
J_1 (kWh)	2526.5	2433.6	2428.1	2430.7	2430.8
U_{min}	0.947	0.951	0.952	0.952	0.953

to schedule or control OLTC increases. Therefore, considering the results shown in Table 6.19, S_n=5 should be appropriate for the load profile shown in Figure 6.20.

Based on the load level partition result shown in Table 6.19 for $S_e = 5$, a dispatch schedule of OLTC is obtained. Figure 6.29 illustrates the tap movements in a day for both the proposed and partial DP methods. It is important to note that tap positions at 0 indicate no movement and that the cumulative tap movement in a day is 16 for both methods. However, the OLTC schedule obtained from the proposed method is simpler and easier to implement in practice when compared against the schedule obtained from the partial DP method.

The proposed method has been implemented using the C++ programming language. Population size and evolutionary generation of the GA is 30 and 4500, respectively. The average computing time for the method is about 3 min running on a P4

Table 6.19 OLTC schedule under different $S_e = 2$ to 7.

Hour	Tap position for various S_e					
	2	3	4	5	6	7
0	+1	+1	0	0	−1	−1
1	+1	+1	0	0	−1	−1
2	+1	+1	0	0	−1	−1
3	+1	+1	0	0	−1	−1
4	+1	+1	0	0	+1	+1
5	+1	+1	0	0	+1	+1
6	+1	+1	0	0	+1	+1
7	+1	+1	0	0	+1	+1
8	+1	+1	+8	+8	+1	+1
9	+8	+8	+8	+8	+8	+8
10	+8	+8	+7	+7	+7	+7
11	+8	+8	+7	+7	+7	+7
12	+8	+8	+7	+7	+7	+7
13	+8	+8	+7	+7	+6	+7
14	+8	+8	+7	+7	+6	+5
15	+8	+8	+7	+3	+6	+5
16	+8	+8	+7	+3	+6	+5
17	+8	+8	+7	+3	+6	+5
18	+8	+1	+1	+1	+1	+1
19	+8	+1	+1	+1	+1	+1
20	+8	+1	+1	+1	+1	+1
21	+8	+1	+1	+1	+1	+1
22	+1	+1	+1	+1	+1	0
23	+1	+1	+1	+1	+1	0

1.8 GHz/512 MB RAM, while the partial DP method requires about 40 min computing time on the same processor.

6.4.5 *Summary*

A new volt/VAR control approach to distribution systems is proposed in this chapter. It includes two steps: (i) dividing the forecasted load of the next day into several load levels; and (ii) determining the optimal dispatch schedule of all volt/VAR control devices. A GA is employed in both steps and a specifically developed encoding method is used in step (ii). OLTC settings remain unchanged in each load level. Simulations show that the control of OLTC settings via this method is both simple and effective. By using a valid encoding method, the searching space of this optimization problem is reduced considerably, especially when compared to a dynamic programming approach. Test

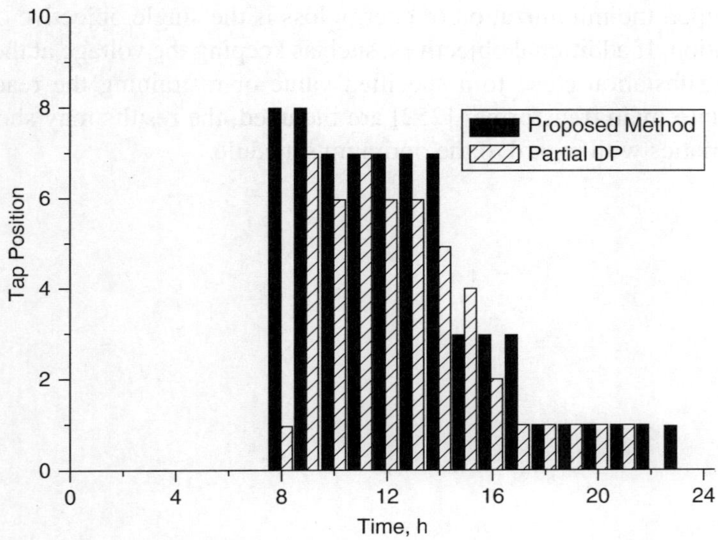

Figure 6.29 OLTC schedule of the next day under five load levels.

results indicate that the proposed method improves the voltage quality, whilst also reducing the energy loss significantly. However, the following conclusions should be noted.

1. Most load profiles have several apparent load levels during a day. For complicated load curves, a large number of load levels should be chosen.
2. In the current method, the number of load levels is obtained from the load forecast, system conditions, and experience. Future research can develop mathematical methods to determine the optimum number of load levels.
3. In this chapter, 24 load points representing each hour are chosen. If 48 or 96 load points are necessary, the allowable time interval between two switching operations of volt/VAR control devices must be considered. Furthermore, a more effective encoding method would be needed for capacitors at the secondary bus of a substation.
4. Voltage at the primary bus of a substation is a constant value in this chapter, as it usually changes little during a day. If its change cannot be ignored, it must also be considered during the load level division.
5. The tap position may differ greatly from two neighboring load levels. So when the load is altering from one level to another, the OLTC settings should be modified gradually.
6. In the simulations, the number of switching operations for OLTC over the whole day is much less than the maximum allowable number. This leaves a large margin with regard to operational changes of the day ahead.

7. In this paper, the minimization of energy loss is the single objective of volt/VAR optimization. If additional objectives, such as keeping the voltage at the secondary bus of a substation close to a specified value or restraining the reactive power flowing into main transformer [252] are included, the results may show different characteristics with regard to the optimum schedule.

7

Modeling and Analysis of Electricity Markets

7.1 Oligopolistic Electricity Market Analysis Based on Coevolutionary Computation

In recent years different equilibrium models have been used in the analysis of strategic interaction between participants in an electricity market, including the oligopoly models of Cournot, Bertrand, and Stackelberg, supply function equilibrium (SFE), and collusion [255, 256]. Among them the Cournot and SFE models are the most extensively used models for analyzing pool-based electricity markets.

The Cournot oligopoly model assumes that strategic firms employ quantity strategies: each strategic firm decides its quantity to produce, while treating the output level of its competitors as a constant. Hogan [257] and Cardell *et al.* [258] apply the Cournot quantity approach to a single-period market trading. The market model developed exploits the standard approach to interpreting a market equilibrium as defining the first-order conditions for a related optimization problem. Borenstein and Bushnell [259] simulate the California electricity market after deregulation as a static Cournot market with a competitive fringe. The model indicates the potential for significant market power in high demand periods. They discuss the weaknesses of concentration measures as a viable measure of market power in the electricity industry, and propose a market simulation approach based on Cournot–Nash concept [260]. The Cournot equilibrium in a transmission-constrained network is investigated in Cunningham *et al.* [261].

The general SFE model was introduced by Klemperer and Meyer [262] and first applied to electricity market analysis by Green and Newbery [263], in which each firm chooses as its strategy a "supply function" relating its quantity to its price. Green and Newbery [263] assume that each firm submits a smooth supply schedule, relating amount supplied to marginal price and look for the noncooperative Nash

Power System Optimization – Large-scale Complex Systems Approaches, First Edition.
Haoyong Chen, Honwing Ngan and Yongjun Zhang.

equilibria of the spot market, which implies a high markup on marginal cost and substantial deadweight losses. Green [264] further models the effect of three policies that could increase the amount of competition in the electricity spot market in England and Wales by the SFE approach. Hobbs *et al.* [265] formulate the problem of calculating SFE in the presence of transmission constraints as a mathematical program with equilibrium constraints (MPEC). Rudkevich *et al.* [266] present an analysis that estimates the price of electricity dispatched and sold through a PoolCo, using a closed-firm mathematical formula derived from the analytical concept of SFE models. Baldick [255] uses an example from the literature to compare Cournot and supply function equilibrium models of bid-based electricity markets with and without transmission constraints. He has demonstrated that the parameterization of the supply function model has a significant effect on the calculated results. Day and Hobbs [267] present a conjectured supply function (CSF) model of competition among power generators on a linearized DC network.

The common feature of these models is to find the market equilibrium under certain assumptions. Some models are successful in studying oligopoly behavior in real electricity markets, but they are generally difficult in obtaining closed-form solutions when considering the practical issues in electricity market, such as nonconvexity and discontinuity of cost function and inter-temporal scheduling of generators. Even worse, market equilibrium may not be unique in some circumstances [255, 258–260, 262]. Thus a new method needs to be explored to analyze market equilibrium in more realistic and complicated conditions.

There exist at least four distinct approaches to answer the problems concerning strategic interaction in an electricity market [265]. The first is empirical analysis of existing markets. The second approach is to model and solve the analytical market equilibria. The other two relatively new approaches are laboratory experimentation [268] and agent-based market simulation [269, 270].

A recent fast-developing area rests on the application of coevolutionary computation [66] to electricity market analysis. Coevolutionary computation is an extension of conventional evolutionary algorithms (EAs). It models an ecosystem consisting of two or more species. Multiple species in the ecosystem coevolve and interact with each other and result in the continuous evolution of the ecosystem [66, 67, 271].

Price [82] uses a coevolutionary genetic algorithm (GA) to model several standard industrial organization games, including a simple model of an electricity pool. He demonstrates that coevolutionary computation has a potential role in applied work requiring detailed market simulation through several examples. Cau [83] develops a co-evolutionary approach to study the dynamic behavior of participants over many trading intervals. Son and Baldick [272] proposed a hybrid coevolutionary programming approach for Nash equilibrium search in games with local optima. The transmission-constrained electricity market examples are studied.

The coevolutionary computation model can be regarded as a special form of the agent-based computational economics (ACE) model. ACE is a computational study of economies modeled as dynamic systems of interacting agents [273]. Each participant in the market under investigation is represented by an agent in the model. The ability of ACE to capture the independent decision-making behavior and interactions of individual agents provides a very good platform for the modeling and simulation of electricity markets. The coevolutionary computation model has many advantages compared to other agent-based models in that it uses the well-developed GA (or other EAs) for agent "strategic learning". GAs are found to be effective analogues of economic learning [274].

A "cooperative coevolutionary algorithm" for unit commitment is first presented by the author, which can be thought of as a simulation of perfect competitive electricity market with price-taking generating units [68]. The coevolutionary computation based approach is systematically extended to analysis of oligopolistic electricity markets. This chapter presents a new unified framework of coevolutionary computation applied to the electricity market (CCEM) for both the one-shot and the repeated games of oligopolistic electricity markets and shows its high computational efficiency and great potentials in practical application. The standard Cournot model and the newly presented Pareto improvement model are combined together to form an integrated framework to analyze both the one-shot and the repeated electricity market games.

7.1.1 *Market Model Formulation*

This section starts from the Cournot oligopoly model. In the standard static Cournot game, the firms make their quantity decisions at the same time, and each firm behaves independently to maximize its own profit [275]. It is well known that the Cournot oligopoly has a Nash equilibrium in which every firm has its maximum profit while assuming that other firms have fixed their outputs. It is apparent that, when all firms have reached such a point, none has any incentive to change unilaterally, and so the situation is viewed as the market equilibrium. However, the situation will be quite different in the presence of repeated games. Because the Nash equilibrium is generally Pareto-inefficient, all firms can increase their payoff by jointly modifying their decisions. Thus a set of "Pareto improvement" solutions exist [275], which can be the results of "tacit collusion". Although explicit collusion is illegal, the danger of tacit collusion always exists [276]. In this section, a new model is formulated to consider this situation. For comparison, a "cooperative" collusion model is also formulated, in which all firms simply work together as a pure monopoly.

7.1.1.1 Standard Cournot Model

Suppose there are I firms, and each firm has its cost,

$$C_i(q_i) = a_i q_i^2 + b_i q_i + c_i, \quad i = 1, \ldots, I \tag{7.1}$$

$$\underline{q}_i \le q_i \le \overline{q}_i \tag{7.2}$$

where q_i is the quantity generated by firm i; a_i, b_i, and c_i are the coefficients of the firm's cost function, with $a_i > 0$, $b_i \geq 0$, and $c_i \geq 0$; and $\underline{q}_i$ and $\overline{q}_i$ are the lower and upper limits of the quantity, respectively.

If it is assumed that there is negligible transmission loss, then the aggregate demand Q will be equal to the total output of all the firms in the market, as shown below:

$$Q = \sum_{i=1}^{I} q_i \tag{7.3}$$

The market price p depends on Q and their relationship is represented by the inverse linear demand function in (7.4) or the inverse constant elasticity demand function in (7.5): linear demand

$$p = A - B \cdot Q \tag{7.4}$$

constant elasticity demand

$$p = \lambda Q^{1/\epsilon} \tag{7.5}$$

where A and B are positive coefficients of the linear demand function, λ is a positive constant, and ϵ is the elasticity, which will typically be negative.

The Cournot model is criticized for the problem of its sensitivity to the elasticity of market demand, which is difficult to specify in the electricity market. To overcome this problem, the constant elasticity demand model is also adopted in this chapter, in addition to the conventional linear demand model. Different demand elasticity can be simulated by varying the value of ϵ to obtain results nearer to the practical situation.

The profit of the firm i is

$$\pi_i(q_1, \ldots, q_I) = pq_i - C_i(q_i) \tag{7.6}$$

For simplicity of notation, let

$$\boldsymbol{q} = (q_1, \ldots, q_I) \tag{7.7}$$

and

$$\boldsymbol{\pi} = (\pi_1, \ldots, \pi_I) \tag{7.8}$$

So, if a vector quantity $\boldsymbol{q}^* = (q_1^*, \ldots, q_I^*)$ is a Cournot–Nash equilibrium, it must satisfy

$$\begin{aligned} \pi_i(\boldsymbol{q}^*) &= \pi_i(q_1^*, \ldots, q_I^*) \\ &= \max_{q_i} \pi_i(q_1^*, \ldots, q_{i-1}^*, q_i, q_{i+1}^*, \ldots, q_I^*) \end{aligned} \tag{7.9}$$

Therefore, the following first-order conditions must be satisfied:

$$\left.\frac{\partial \pi_i(q_1^*, \ldots, q_{i-1}^*, q_i, q_{i+1}^*, \ldots, q_I^*)}{\partial q_i}\right|_{q_i = q_i^*} = 0, i = 1, \ldots, I \tag{7.10}$$

For the linear demand function in (7.4), condition (7.10) can be explicitly expressed as (7.11) and the Cournot equilibrium can be easily obtained:

$$2(B + a_i)q_i^* + B\left(\sum_{j\neq i} q_j^*\right) = A - b_i \tag{7.11}$$

However, for the constant elasticity demand function in (7.5), the first-order condition will be a complicated nonlinear equation set.

7.1.1.2 Pareto Improvement Model

The Cournot–Nash equilibrium is the result of the one-shot noncooperative game. In the real electricity market, the auction is repeated daily in a pool-based electricity market. This raises the question as to what the outcomes will be in the repeated games. The results are given by numerous folk theorems in game theory [277]. The most important one concerning the oligopoly game is presented by Friedman [278]. Since the Cournot–Nash equilibrium is generally not Pareto-optimal, it is not suitable to use as a viable outcome in the repeated oligopoly game. Under some assumptions, the Pareto-optimal solutions can be the noncooperative equilibria in infinitely repeated games (supergames) [277, 278]. These equilibria can be achieved by "tacit collusion" among the firms. Since there are many Pareto-optimal solutions that can be the noncooperative equilibria of an infinitely repeated game, a particular solution is dependent on any additional conditions imposed on the game. This section will only focus on searching for the general Pareto improvement solutions of the Cournot oligopoly. The definition for Pareto optimality is given in Definition 7.1 below. The Pareto improvement model is given after Definition 7.2 below, which defines Pareto dominance.

Definition 7.1 (Pareto optimality) *A point* $\boldsymbol{q}^* = (q_1^*, \ldots, q_I^*)$ *is Pareto-optimal if, for every possible* $\boldsymbol{q} = (q_1, \ldots, q_I)$ *and* $i = 1, \ldots, I$, *either*

$$\pi_i(\boldsymbol{q}) = \pi_i(\boldsymbol{q}^*), i = 1, \ldots, I \tag{7.12}$$

or there is at least one $i \quad (1 \leq i \leq I)$ *such that*

$$\pi_i(\boldsymbol{q}) < \pi_i(\boldsymbol{q}^*) \tag{7.13}$$

Definition 7.2 (Pareto dominance) *A vector* $\boldsymbol{\sigma} = (\sigma_1, \ldots, \sigma_I)$ *is said to dominate* $\boldsymbol{\tau} = (\tau_1, \ldots, \tau_I)$ *(denoted by* $\boldsymbol{\sigma} > \boldsymbol{\tau}$ *) if and only if*

$$\sigma_i \geq \tau_i, i = 1, \ldots, I \tag{7.14}$$

and there is at least one $i \quad (1 \leq i \leq I)$ *such that*

$$\sigma_i > \tau_i \tag{7.15}$$

For clarity, we denote Cournot–Nash equilibrium by $\boldsymbol{q}^{*(n)} = (q_1^{*(n)}, \ldots, q_I^{*(n)})$ and Pareto-optimal solutions by $\boldsymbol{q}^{*(p)} = (q_1^{*(p)}, \ldots, q_I^{*(p)})$ hereafter.

The objective of the Pareto improvement model is to find the Pareto improvement set (denoted by Ω) of the Cournot oligopoly model (7.1)–(7.6), i.e. the set of Pareto-optimal solutions with some producers better off and none of the producers worse off compared to the outcome at the Cournot–Nash equilibrium, which can be written as

$$\begin{aligned} &\Omega = \{\boldsymbol{q}^{*(p)} \mid \pi_i(\boldsymbol{q}^{*(p)}) \geq \pi_i(\boldsymbol{q}^{*(n)})\} \\ &\overline{\exists} \quad \boldsymbol{q} = (q_1, \ldots, q_I), \pi(\boldsymbol{q}) \succ \pi(\boldsymbol{q}^{*(p)}) \end{aligned} \tag{7.16}$$

where $\overline{\exists}$ means "do not exist".

The Pareto improvement model (7.1)–(7.6) and (7.16) forms a classical multi-objective optimization problem with inequality constraints. Apparently, at a solution of the multi-objective optimization problem, no producer can achieve more profit without decrease of other producers' profits, and all producers' profits are more than those at the Cournot–Nash equilibrium, so we call it the "Pareto improvement solution". Based on the theory in Tirole [277] and Friedman [278], any such solutions can be the result of the repeated oligopoly game.

7.1.1.3 Collusion Model

Different from tacit collusion achieved in repeated games, explicit collusion is generally illegal under antitrust law. The collusion model is formulated in this section for comparison.

By collusion, the firms work together to extract as much total profit as they can from the market [275]. Therefore the objective of the collusion model is

$$\max_{(q_1, \ldots, q_I)} \sum_{i=1}^{I} \pi_i(q_1, \ldots, q_I) \tag{7.17}$$

Equations (7.1)–(7.6) and (7.17) make up the collusion model, which is an ordinary multi-variable optimization problem. The solutions of the model can be denoted by $\boldsymbol{q}^{*(c)}$.

7.1.2 *Electricity Market Analysis Based on Coevolutionary Computation*

This section introduces a new framework of electricity market analysis based on coevolutionary computation (CCEM).

7.1.2.1 General Framework of CCEM

The above market models can be illustrated by Figure 7.1. Each producer submits its optimal trading strategy to the Power Exchange (PX). The trading strategy is different for different market models. Here for the standard Cournot and Pareto improvement models with quantity decision participants, the trading strategy is the quantity to be generated. PX then calculates the market price using (7.3) and (7.4) or (7.5) above according to the demand characteristics and market rules. Each producer can calculate its profit with the market price and its trading strategy according to (7.6). Figure 7.1 suggests an agent-based simulation method for electricity market analysis, in which each participant in the market under investigation is represented by an agent in the model. Each agent makes its decisions based on the knowledge of itself and the environment. The agents interact with each other through the system model. Here the system model consists of the market rules and demand function.

This section presents an agent-based simulation approach to the electricity market based on coevolutionary computation. Coevolutionary computation is developed from traditional evolutionary algorithms (EAs), which simulates the coevolutionary mechanism in nature and adopts the notion of an ecosystem. The agents simulate the species in an ecosystem. Multiple species coevolve and interact with each other and result in the continuous evolution of the ecosystem. The species are genetically isolated—individuals only mate with other members of their species. They interact with one another within a shared domain model. Coevolutionary computation is a relatively new area in evolutionary computation research. Its theory and applications are still rapidly developing [66, 271, 67].

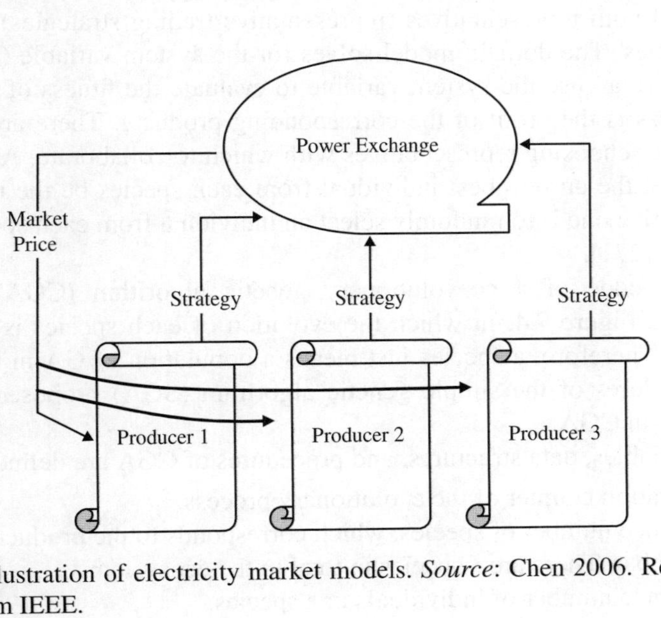

Figure 7.1 Illustration of electricity market models. *Source*: Chen 2006. Reproduced with permission from IEEE.

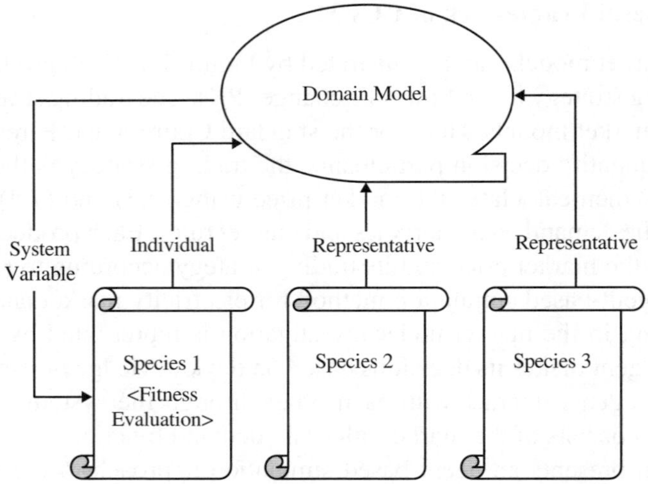

Figure 7.2 Framework of cooperative coevolutionary model. *Source*: Chen 2006. Reproduced with permission from IEEE.

The basic coevolutionary computation model is shown in Figure 7.2 [67], which is an analogue of Figure 7.1. Each producer in Figure 7.1 is represented by a species in the ecosystem. Each species evolves a bundle of individuals, which represent the candidate trading strategies of the corresponding producer. PX is modeled with the domain model. Each species is evolved through the repeated application of a conventional EA. Figure 7.2 shows the fitness evaluation phase of the EA from the perspective of species 1. To evaluate an individual (trading strategy) from species 1, collaboration is formed with representatives (representative trading strategies) from each of the other species. The domain model solves for the system variable (market price). Then species 1 can use the system variable to evaluate the fitness of its individual. Here the fitness is the profit of the corresponding producer. There are many possible methods for choosing representatives with which to collaborate. An obvious one is simply to let the current best individual from each species be the representative, and an alternative one is to randomly select an individual from each species to be the representative [271].

The pseudo-code of a coevolutionary genetic algorithm (CGA) is given in Figure 7.3 and Figure 7.4, in which the evolution of each species is handled by a standard GA. Therefore, a species just means a population of GA in this algorithm and the procedures of the simple genetic algorithm (SGA) proposed in Goldberg [166] are used in CGA.

The key variables, data structures, and procedures of CGA are defined as follows:

k Generation counter of the evolutionary process.

i Sequence number of species, which corresponds to the producer sequence number. There are I species in total in the ecosystem.

n Sequence number of individuals in a species.

k = 0

for each species i do

Begin

initialize the species population Pop_0^i

evaluate fitness of each individual $Ind_0^{i,n}$ *in* Pop_k^i

choose a representative Rep_0^i *from*

end

while termination condition = false do

begin

for each species i do

begin

reproduction from Pop_{k+1}^i *to get* $Mate_k^i$

crossover and mutation from $Mate_k^i$ *to get* Pop_{k+1}^i

evaluate fitness of each individual $Ind_{k+1}^{i,n}$ *in* Pop_{k+1}^i

choose a representative Rep_{k+1}^i *from* Pop_{k+1}^i

end

k = *k* + 1

end

Figure 7.3 Pseudo-code of CGA. *Source*: Chen 2006. Reproduced with permission from IEEE.

Chromosome — A 16-bit binary string (two bytes) is used to encode the quantity q_i generated by the corresponding producer i (namely the producer's trading strategy) for each species i, which is called the chromosome.

Fitness — The fitness value of a chromosome in species i is the profit of the corresponding producer i when its generating quantity q_i equals the decoded value of the chromosome.

Individual — The nth individual $Ind_k^{i,n}$ of species i in the kth generation is composed of a chromosome, its decoded value, and fitness value.

for each individual in do

begin

form a collaboration with representatives from other species,

which is $\left(\mathrm{Re}\,p_k^1, \mathrm{Re}\,p_k^2, \ldots, Ind_k^{i,n}, \ldots, \mathrm{Re}\,p_k^I\right)$

decode the collaboration and form a tentative vector of

generating quantities $\left(\hat{q}_1, \hat{q}_2, \ldots, \hat{q}_i, \ldots, \hat{q}_I\right)$ *of all producers*

calculate market price with (3), (4-1) or (4-2)

calculate profit of producer i with (5), *which is used as the fitness value of*

end

Figure 7.4 Pseudo-code of fitness evaluation procedure. *Source*: Chen 2006. Reproduced with permission from IEEE.

Population	The population Pop_k^i of species i in the kth generation contains a number of individuals $Ind_k^{i,n}$, which is evolved by GA.
Initialization	The chromosomes of each species are randomly initialized to integers between 0 and 65 535 (two bytes). The initial population of species i is denoted by Pop_0^i.
Reproduction	A simple roulette wheel selection or tournament selection is used to choose members for the mating pool $Mate_k^i$ from Pop_k^i for species i in the kth generation. Fitter individuals have a higher number of offspring in the succeeding generation.
Crossover	The crossover operator takes two parent chromosomes and generates two offspring chromosomes by swapping some parts of their strings. A two-point crossover operator is used, which means that two crossover positions are selected and the two parent chromosomes swap the strings between them with a certain probability.
Mutation	A bit-wise mutation operator is used, in which each bit of the chromosome randomly changes from 1 to 0 or vice versa with certain probability.
	The $(k+1)$th generation population Pop_{k+1}^i of species i is formed from the kth generation mating pool $Mate_k^i$ through repeated application of crossover and mutation operators.
Choice of representative	Representative Rep_k^i of species i in the kth generation is chosen from Pop_k^i with the aforementioned method.
Fitness evaluation	The process of evaluating the fitness of individuals in Pop_k^i is shown in Figure 7.4.
Termination condition	A simple termination condition is used, namely that the number of generations reaches its preset upper limit.

GA is used to evolve each species in our model, which is an effective analogue of economic learning. Riechmann [274] shows that economic learning via GA can be described as a specific form of an evolutionary game, which is a stochastic process repeatedly turning one population of individuals into another. Each repeated turn of GA consists of several stochastic processes [274]: reproduction, which is interpreted as learning by imitation; crossover, which is interpreted as learning by communication; mutation, which is interpreted as learning by experiment; evaluation, which is interpreted as playing the role of the market as an information-revealing device; and selection, which decreases the number of unsuccessful trading strategies.

The ability of the coevolutionary computation model to facilitate the individual participant's strategic learning and the interactions among the participants provides a very good platform for the modeling and simulation of electricity markets.

7.1.2.2 CCEM for Standard Cournot Model

For the Cournot model (7.1)–(7.11), the above CGA is used for simulation. In a Cournot oligopoly, a producer i only needs to optimize its own profit π_i expressed by (7.6), with the quantity q_i as the decision variable. The simple roulette wheel selection is used in the reproduction operator of each species. A "greedy" method is used for selecting representatives [271]. In this method the current best individual from each species is selected as the representative so as to facilitate the fast convergence of the simulation.

From the calculation process above, we can see that the species are coordinated by the market price p. When producer i changes its quantity q_i to gain more profit in (7.6), it will change the market price according to the inverse demand function in (7.4) or (7.5), and in turn changes the profits of other producers. Other producers will behave in the same way. The adjustment process will continue until no-one can get more profit by changing its quantity without changes of the quantities of other producers—in other words, the market reaches Cournot–Nash equilibrium.

7.1.2.3 CCEM for Pareto Improvement Model

The market simulation with the Pareto improvement model (7.1)–(7.6) and (7.16) is actually a multi-objective optimization problem. The above CGA is used for simulation. A producer i should optimize the Pareto-optimal profits (see Definition 7.1) for all producers in the market, with its own quantity q_i as the decision variable. Different from CCEM for the standard Cournot model, a tournament selection is used in the reproduction operator of each species.

The main techniques used for multi-objective optimization are taken from Deb et al's NSGA-II [279]. The main features of NSGA-II include: a fast nondominated sorting procedure, a fast crowded distance computation procedure, and a crowded tournament selection operator. The fast nondominated sorting procedure sorts a population into different nondomination levels. The fast crowded distance computation procedure is to estimate the density of solutions surrounding a particular solution in the population. The crowded tournament selection operator guides the selection process of the algorithm toward a uniformly spread-out Pareto-optimal front. The crossover and mutation operators of NSGA-II are the same with standard GA, while the selection operator is different. A binary tournament selection operator is used and the selection criterion is based on the crowded comparison operator. The details of NSGA-II can be found in Deb *et al.* [279].

The techniques of NSGA-II are applied to the evolution of each species, and thus the CCA in Figure 7.3 and Figure 7.4 becomes a multi-objective coevolutionary computation model.

7.1.3 Example

To validate the models and algorithms described in this chapter, the test case with three producers from Cunningham *et al.* [261] is used. The data of producers' costs in table I therein [261] are reproduced here in Table 7.1. Numbers 1, 2, and 3 are the sequence numbers of the producers, respectively.

To illustrate the ability of CCEM to handle the nonlinear models and simulate different market situations, a constant elasticity demand function is used in simulation besides the linear demand function used by Cunningham *et al.* [261]. The demand function parameters are listed in Table 7.2: ϵ is the elasticity of demand, which is typically negative. The demand with elasticity greater than 1 in absolute value is

Table 7.1 Producers' cost data.

Producer's cost function	No.	a_i	b_i	c_i
$C_i(q_i) = a_i q_i^2 + b_i q_i + c_i$,	1	0.007 859	1.360575	9 490.366
$0 \leq q_i \leq 2000$	2	0.010 526	−2.07807	11 128.95
	3	0.006 478	8.105354	6 821.482

Table 7.2 Demand function parameters.

Simulation cases	Inverse demand function	Parameters	
		A or λ	B or ϵ
Case 1: linear	$p = A - B \cdot Q$	106.1176	0.0206
Case 2: inelastic	$p = \lambda Q^{1/\epsilon}$	3.52×10^8	−0.5
Case 3: elastic	$p = \lambda Q^{1/\epsilon}$	9969.7	−1.5

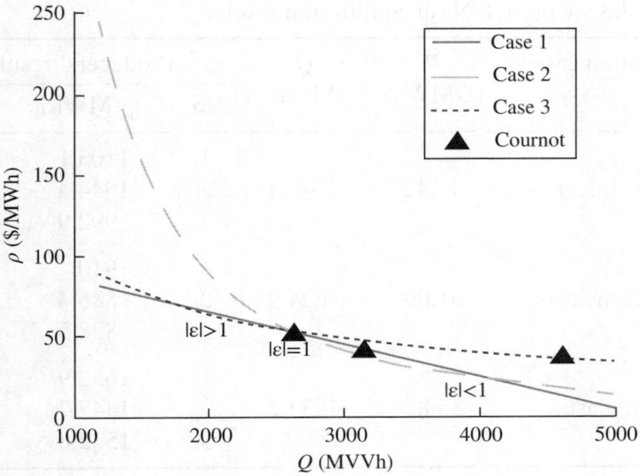

Figure 7.5 Market demand and price of CCEM for standard Cournot model. *Source*: Chen 2006. Reproduced with permission from IEEE.

an elastic demand, and the elasticity less than 1 in absolute value is for an inelastic demand. The demand curves used for simulation are shown in Figure 7.5. The solid line is the linear demand curve; the dashed line is an inelastic demand with the elasticity −0.5; the dotted line is an elastic demand with the elasticity −1.5. We choose the values of parameter λ so that the three different demand curves intersect at the point with $|\epsilon| = 1$ on the linear demand curve.

7.1.3.1 CCEM for Standard Cournot Model

We apply the coevolutionary computation algorithm described in section 7.1.2.2 for the standard Cournot model to the test case. The evolution process runs for a maximum 150 generations, with the population size 100. The elitist strategy preserving a single copy of the best individual is adopted. The crossover probability p_c is set to 0.9, and the mutation probability p_m is set to 0.06.

The Cournot–Nash equilibrium with linear demand function can be easily obtained by solving the linear equation system (7.11). However, the first-order conditions for Cournot–Nash equilibrium with constant elasticity demand function are generally nonlinear equation systems and difficult to solve. This section uses the test cases with constant elasticity demand functions to demonstrate the ability of CCEM for solving nonlinear Cournot models. The simulation results are listed in Table 7.3. The demand curves and market demands and prices at Cournot–Nash equilibrium are also shown in Figure 7.5. The solid triangles denote the market aggregate demands and prices at the Cournot–Nash equilibria. They are on different demand curves, which denote the corresponding demand functions used for simulation. To verify the simulation results, an iterative Nash equilibrium search algorithm [272] has been

Table 7.3 Cournot–Nash equilibrium results.

Simulation cases	P (£/MWh)	Q (MWh)	Producers' results		
			No.	q_i (MWh)	π_i ($)
			1	1103.1	25 141
Case 1: linear	41.42	3140.4	2	1044.4	22 823
			3	992.9	19 874
			1	910	29 253
Case 2: inelastic	51.09	2624.9	2	886.4	27 726
			3	828.5	24 340
			1	1652.9	26 082
Case 3: elastic	35.87	4633.2	2	1447.7	21 751
			3	1532.6	20 518

used to calculate the Cournot–Nash equilibria. The results are the same as those in Table 7.3 obtained using CCEM.

From the simulation results, we can observe that CCEM can find the Cournot–Nash equilibrium in all cases. The Cournot–Nash equilibrium with inelastic demand has higher market price and less generating output, while the Cournot–Nash equilibrium with elastic demand has lower market price and more generating output. These results are consistent with experiences from real markets.

The variations of prices with respect to the generation number of the simulation in typical runs are shown in Figure 7.6. We can see that the market prices converge to

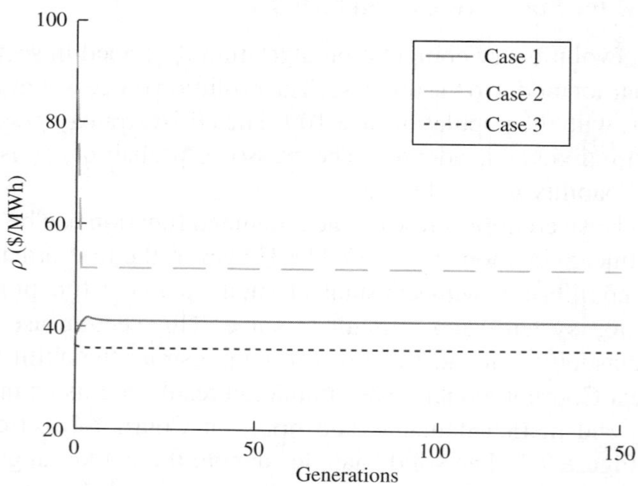

Figure 7.6 Variation of market price in coevolution process. *Source*: Chen 2006. Reproduced with permission from IEEE.

the Cournot–Nash equilibrium rapidly in all three cases. CCEM has high efficiency in the market simulations.

In addition, simulations using different settings of p_c and p_m are also performed to observe the influence of the parameters of the algorithm. The results are listed in Table 7.4, where d_1, d_2, and d_3 are the Euclidean distances between the simulation results and the actual Cournot–Nash equilibrium quantities for the linear, inelastic, and elastic demand cases, respectively. Since the results are stochastic, the values of d_1, d_2, and d_3 are averaged over 10 runs. The values of p_c and p_m vary from 0.0 to 1.0. The other algorithm parameters are the same as those in the above simulations. From the results, it can be observed that the algorithm is more sensitive to p_m than to p_c. The simulation results are the best when p_m is between 0.05 and 0.1 and p_c is between 0.5 and 0.9. The simulation results may imply that, since the decision space is simple (one variable for each producer), mutation is the main operator to explore new search spaces.

Table 7.4 Cournot–Nash equilibrium results.

p_c	p_m	d_1	d_2	d_3	p_c	p_m	d_1	d_2	d_3
	0	32.28	21.25	32.57		0	0.39	1.02	0.51
	0.001	6.23	1.34	11.36		0.001	0.2	0.27	0.26
	0.005	2.07	0.66	0.53		0.005	0.14	0.1	0.07
0	0.01	1.94	0.23	0.39	0.7	0.01	0.05	0.04	0.05
	0.05	0.06	0.05	0.06		0.05	0.06	0.04	0.05
	0.1	0.05	0.05	0.06		0.1	0.05	0.05	0.05
	0.5	0.24	0.14	0.13		0.5	0.25	0.16	0.14
	1	26.45	11.84	18.2		1	0.05	0.06	1.66
	0	1.94	0.29	1.28		0	0.29	0.16	0.15
	0.001	0.69	0.94	0.69		0.001	0.49	0.12	0.12
	0.005	0.3	0.17	0.2		0.005	0.14	0.08	0.06
0.2	0.01	0.06	0.18	0.19	0.9	0.01	0.07	0.04	0.05
	0.05	0.05	0.05	0.05		0.05	0.05	0.04	0.05
	0.1	0.05	0.05	0.07		0.1	0.05	0.05	0.05
	0.5	0.21	0.13	0.18		0.5	0.23	0.16	0.14
	1	0.07	0.07	4.39		1	0.07	0.07	0.63
	0	0.18	0.42	0.11		0	0.17	0.15	0.09
	0.001	0.52	0.17	1.05		0.001	1	0.13	0.2
	0.005	0.85	0.08	0.09		0.005	0.24	0.08	0.06
0.5	0.01	0.06	0.09	0.06	1	0.01	4.81	0.09	0.05
	0.05	0.06	0.05	0.05		0.05	0.05	0.05	0.05
	0.1	0.06	0.05	0.05		0.1	0.06	0.04	0.05
	0.5	0.28	0.18	0.17		0.5	0.16	0.19	0.16
	1	0.07	0.05	1.13		1	0.07	0.06	0.64

In addition, we use a more realistic five-firm example system [280] to further validate CCEM. The example system is based on the cost data for the five strategic firms industry in England and Wales subsequent to the 1999 divestiture. The firms' cost data are reproduced here in Table 7.5. The linear demand function [280] is used in the simulation, where $A = 350$ and $B = 10$. The simulation results are listed in Table 7.6. The iterative Nash equilibrium search algorithm proposed in Son and Baldick [272] is also used to verify the simulation results. This study shows that CCEM has potential in practical electricity market analysis.

7.1.3.2 CCEM for Pareto Improvement Model

We use the algorithm described in section 7.1.2.3 to perform the simulation. The algorithm parameters are the same as those in the standard Cournot model simulations in the last section.

The basic simulation results are listed in Table 7.7. Since the Pareto improvement solutions are not unique, the average and standard deviation values of market prices, market aggregate demand (sum of all producers' output), producer quantities, and profits are calculated. The data in Table 7.7 are the average values with the standard deviation values in brackets. For the cases with linear and elastic demand, the market prices are highly concentrated, with the average of 60.84/MWh and 53.41/MWh, respectively. These average prices are higher than the prices at the Cournot–Nash

Table 7.5 Producers' cost data.

Producer's cost function	No.	a_i	b_i	c_i
	1	1.3435	12	0
	2	2.3075	12	0
$C_i(q_i) = a_i q_i^2 + b_i q_i + c_i$	3	0.8945	8	0
	4	0.965	8	0
	5	2.3075	12	0

Table 7.6 Cournot–Nash equilibrium results.

Simulation cases	P (£/MWh)	Q (GWh)	Producers' results		
			No.	q_i (GWh)	π_i ($\times 10^3$)
			1	5.391	6.869
			2	4.679	5.615
Five-firm system	80.4	26.96	3	6.142	8.56
			4	6.069	8.413
			5	4.68	5.615

Table 7.7 Pareto improvement results.

Simulation cases	P(£/MWh)	Q(GWh)	Producers' results		
			No.	q_i (GWh)	π_i ($\times 10^3$)
	60.84	2198	1	780.0(95.0)	31 928(3812)
Case 1: linear	(2.53)	(123)	2	741.2(93.2)	29 533(3843)
			3	676.8(87.8)	25 766(3460)
	0.104		1	0.031(0.000)	
Case 2: inelastic	(0.016)		2	0.031(0.006)	
			3	0.042(0.015)	
	53.41	2559	1	897.3(101.0)	30 658(2870)
Case 3: elastic	(2.34)	(168)	2	842.7(63.7)	28 050(2336)
			3	819.3(111.4)	25 695(2968)

equilibria. The producer quantities are distributed, with the average values less than the Cournot–Nash equilibrium quantities. Although the individual producer quantities are distributed, the sums of all producers' output are concentrated, with the average values much less than that of the Cournot–Nash equilibrium. The producer profits are distributed, with the average values more than the Cournot–Nash equilibrium profits. The results show that the producers raise their profits by collectively withholding the quantities supplied and increasing the market price. Although the profits can be allocated among the producers quite differently, the market prices are similar.

For the case with the inelastic demand, the situation is quite different. From Table 7.7 we can see that the producer quantities are driven to near zero, and the market price gets impossibly high. In turn, the producer profits get impossibly high. This is because the demand is inelastic, the producers can raise the market price to an abnormal level by collectively withholding their outputs. This situation can also be observed in real markets.

The influence of the algorithm parameters on simulation results has also been studied. Since the influence is similar to that of the standard Cournot model in Table 7.4, the results are not shown for brevity.

For comparison, the results of the collusion model in section 7.1.1.3 are listed in Table 7.8, where the producers work cooperatively to maximize their total profit. The model can be solved by any ordinary nonlinear optimization method. For the cases with linear and elastic demand, the market prices and demands are near to the average values of the Pareto improvement results in Table 7.7. For the cases with the inelastic demand, the producers' quantities are driven near zero, and the market price and producers' profit become infinitely large. In fact, for the cases with linear and elastic demand, it is easy to prove that the collusion results are Pareto-optimal. However, the collusion results do not necessarily belong to the Pareto improvement solutions. For the case with linear demand, we can see that, although the total profits of all producers are the highest in the collusion result, the profit of producer 3 is lower than the

Table 7.8 Collusion results.

Simulation cases	P (£/MWh)	Q (MWh)	Producers' results		
			No.	q_i (MWh)	π_i ($\times 10^3$)
Case 1: linear	60.56	2211.5	1	867.9	35 968
			2	811.3	32 762
			3	532.3	19 265
Case 2: inelastic	Infinite	Infinit-esimal	1	Infinitesimal	Infinite
			2	Infinitesimal	Infinite
			3	Infinitesimal	Infinite
Case 3: elastic	52.03	(2652.4)	1	1016.8	33 907
			2	922.5	29 830
			3	713.0	21 205

Cournot–Nash equilibrium in Table 7.3. This result cannot be achieved without any explicit agreement on division of total profit among the producers.

Using the producers' profits as the coordinates of a point in a three-dimensional space, the Pareto improvement solutions can be visualized as in Figure 7.7. Only the Pareto improvement solutions with the linear demand function are shown. The Pareto front is determined by an iterative Monte Carlo (IMC) search method and the crosses identify the Pareto-optimal profit points. The big circle designates the collusion profits of each producer. The small circles designate the Pareto improvement profits. The

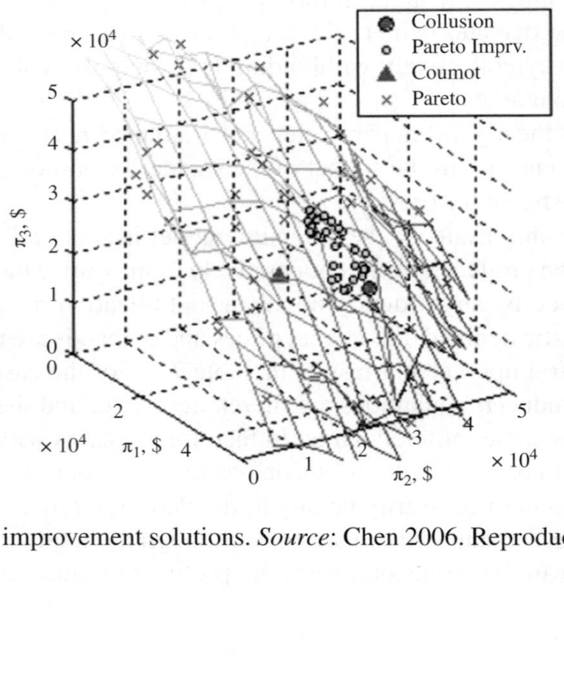

Figure 7.7 Pareto improvement solutions. *Source*: Chen 2006. Reproduced with permission from IEEE.

solid triangle designates the Cournot–Nash equilibrium profits, which is under the Pareto front. We can see that the collusion result lies on the Pareto front but outside of the Pareto improvement solutions. The Pareto improvement profits are fairly concentrated. The Pareto improvement profits obtained by CCEM coincide with the Pareto front determined by the IMC method and thus the correctness can be justified. The same conclusion can be reached in the elastic demand case.

7.1.4 *Summary*

This section presents a coevolutionary computation model for electricity market analysis (CCEM). The coevolutionary computation model can be looked at as a special form of the agent-based computational economics model. Each participant in the market is represented by a species in the model and the interactions among market participants are embodied in the coevolutionary process of the species.

The standard Cournot model and the newly presented Pareto improvement model are combined together to form an integrated framework to analyze both the one-shot and the repeated electricity market games. The linear and constant elasticity demand functions are used in simulations.

It has been found that CCEM can find the Cournot–Nash equilibrium in the standard Cournot model and the Pareto improvement solutions in the Pareto improvement model with high efficiency. It can handle the nonlinear market models which are difficult to handle by conventional methods. The Cournot model is often criticized for the problem of sensitivity to the specification of market demand, and thus the framework presented in this section using different demand functions can greatly help to overcome this difficulty.

7.2 Supply Function Equilibrium Analysis Based on Coevolutionary Computation

In recent years, different equilibrium models have been used in the analysis of the strategic interaction between participants in electricity markets; among them, the supply function equilibrium (SFE) and Cournot models are the most extensively used models for analyzing pool-based electricity markets [255, 281].

The Cournot oligopoly model assumes that strategic firms employ quantity strategies: each strategic firm decides its quantity to produce, treating the output level of its competitors as a constant [281]. The Cournot model often suffers from the problem of sensitivity to the specification of the market demand.

The SFE model offers a more realistic view of electricity markets. The general SFE model was introduced by Klemperer and Meyer [262] and first applied to electricity market analysis by Green and Newbery [263], in which each firm chooses its strategy as a "supply function" relating its quantity to its price. Green and Newbery [263] assume that each firm submits a smooth supply schedule, relating amount supplied to marginal price, and look for the noncooperative Nash equilibria of the spot market, which implies a high markup on marginal cost. Green [264] further models the effect of three policies that could increase the amount of competition in the electricity spot market in England and Wales. The linear supply function model with asymmetric firms is introduced in the analysis. Each firm is supposed to have quadratic costs. The marginal costs are assumed to be zero at zero output. Accordingly, the constant item of the linear supply function is supposed to be zero [264]. However, Green's analysis allows for nonzero constant terms that are the same for all firms. Green [282] analyzed the electricity contract market in England and Wales. He models competition in the spot market with supply functions and linear marginal costs. Conjectural variations are used to model different degrees of competition in the contract market.

Rudkevich *et al.* [266] model a pool-based electricity market as a noncooperative game with a number of identical profit-maximizing generating firms. They use the analytical solution of SFE and relax the convexity and differentiability conditions to allow for the realistic stepwise supply curves to be studied. Rudkevich [283] further presents a stylized model of the learning process through which power generating firms can adjust their supply bidding strategies in order to achieve a rational profit-maximizing equilibrium behavior in the form of SFE. He also assumes that the constant item of the linear marginal cost function of every player is zero.

Baldick and Hogan [280] consider a supply function model of an electricity market where strategic firms have capacity constraints. They show that, if firms have heterogeneous cost functions and capacity constraints, then the differential equation approach to finding the equilibrium supply function may not be effective because it produces supply functions that fail to be nondecreasing. They analyze the nondecreasing constraints and characterize piecewise continuously differentiable equilibrium. To find stable equilibria, they numerically solve for the equilibrium by iterating in the function space of allowable supply functions. Baldick *et al.* [284, 285] consider an SFE model of interaction in an electricity market, assuming a linear demand function and considering a competitive fringe and several strategic players with capacity limits and affine marginal costs. They assume that bid rules allow affine or piecewise affine nondecreasing supply function by firms and extend the results of Green and Rudkevich concerning the linear SFE solution. The capacity constraints are introduced and an *ad hoc* approach to constructing piecewise affine supply curves is proposed. Baldick [255] uses an example from the literature to compare Cournot and supply function equilibrium models of pool-based electricity markets with and without transmission constraints. He demonstrates that the parameterization of the supply function model has a significant effect on the calculated results.

Hobbs *et al.* [265] formulate the problem of calculating SFE in the presence of transmission constraints as a mathematical program with equilibrium constraints (MPEC). They use a single pricing-period model. Each generating firm submits bids to an independent system operator (ISO), choosing its bids (manipulating the intercept of the bid functions) to maximize profits subject to anticipated reactions by rival firms. The single-firm model is formulated as an MPEC with a parameter-dependent spatial price equilibrium problem as the inner problem. In the multi-firm model each player solves an MPEC of the single-firm type alternately to find the Nash equilibria. Day and Hobbs [267] present a conjectured supply function (CSF) model of competition among power generators on a linearized DC network. CSF models differ from other approaches in that they represent each of the generating firm's conjectures regarding how rival firms will adjust sales in response to price changes. The market participants' profit maximization and the market clearing conditions define a mixed complementarity problem (MCP). Both the variable slope with fixed intercept and variable intercept with fixed slope affine CSF models are considered. Day and Bunn [286] present an agent-based computational model that models firms as seeking optimal supply functions in response to the bids of the other firms. The approach involves iterations in the supply function space and is consistent with the analytical method for the simplified circumstance but has the potential to model much more realistic situations with respect to discontinuous supply functions, nonlinear demand curve, and asymmetric firms.

The SFE model is successful in studying oligopoly behavior in electricity markets, but it is generally difficult to obtain the solution with the differential equation approaches, especially when considering the practical issues in the electricity market, such as nonconvexity of the cost function. Even worse, market equilibrium may not be unique especially when the SFE model is applied to the single pricing-period market with no uncertainty of demand [255, 262, 263]. Thus a new method needs to be explored to analyze market equilibrium in more realistic and complicated conditions.

Coevolutionary computation models for electricity market equilibrium analysis have been presented [281, 272], which model an ecosystem consisting of two or more species. Multiple species in the ecosystem coevolve and interact with each other and result in the continuous evolution of the ecosystem. The coevolutionary computation model is found to be an effective and powerful approach for electricity market simulation and analysis. This section focuses on the simulation of SFE models based on coevolutionary computation and shows its high computational efficiency and great potential in practical application.

The coevolutionary computation model offers an alternative way to obtain the solution of the differential equation describing the equilibrium by simulation of strategic best-response behavior of the market participants. The simulation corroborates results in Baldick [255] and Baldick and Hogan [280].

7.2.1 Market Model Formulation

The notation used in this section will be as follows:

i	generating firm index;
I	number of generating firms;
q_i	quantity generated by firm i;
$C_i(q_i)$	cost function of firm i;
p	market price;

$S_i(p)$	supply function of firm i;
$D(p,t)$	aggregate demand function at pricing period t;
$D_0(t)$	underlying load characteristic for each period t;
π_i	profit of firm i.

This section uses the general model in Baldick *et al.* [285], which we quote here for reference. Suppose there are I firms in the market, and each firm has a strictly convex quadratic cost function:

$$C_i(q_i) = \frac{1}{2}c_i q_i^2 + a_i q_i, \quad i = 1, \dots, I \tag{7.18}$$

where q_i is the quantity generated by firm i; and c_i and a_i are the coefficients of the firm's cost function, with $c_i > 0$ and $a_i \geq 0$. Therefore the marginal cost function is affine for each firm:

$$\frac{\mathrm{d}C_i(q_i)}{\mathrm{d}q_i} = c_i q_i + a_i, \quad i = 1, \dots, I \tag{7.19}$$

Assume that the supply function of each firm is $S_i(p), i = 1, \dots, I$, where p is the market price. Market rules require that supply functions be nondecreasing. When there is negligible transmission loss, if the market clearing condition is satisfied, the aggregate demand $D(p,t)$ at pricing period t will be equal to the total output of all the firms in the market as shown below:

$$D(p,t) = \sum_{i=1}^{I} S_i(p) \tag{7.20}$$

We use the following form of the demand curve:

$$D(p,t) = D_0(t) - \gamma p \tag{7.21}$$

The underlying load characteristic is specified by $D_0(t)$ and, for each period t, the demand is linear in p with slope $\mathrm{d}D/\mathrm{d}p = -\gamma$. The coefficient γ is assumed to be positive.

An equilibrium consists of a set of supply functions, one for each firm, such that each firm is maximizing its profits, given the supply functions of the other firms. Suppose firm i produces the residual demand of the other firms (that is, total demand less the other firms' supply) to meet the market clearing condition. Then its profit π_i at each period can be written as

$$\pi_i(p,t) = p\left(D(p,t) - \sum_{j \neq i} S_j(p)\right) - C_i\left(D(p,t) - \sum_{j \neq i} S_j(p)\right) \tag{7.22}$$

If p^* is an equilibrium price at pricing period t and the nondecreasing constraints on supply functions are not binding at the equilibrium [280], it must satisfy

$$\pi_i(p^*,t) = \max_p \pi_i(p,t), \quad i = 1, \dots, I \tag{7.23}$$

Profit function (7.22) can be differentiated with respect to price:

$$\frac{d\pi_i(p,t)}{dp} = D(p,t) - \sum_{j\neq i} S_j(p) + \left[p - C_i'\left(D(p,t) - \sum_{j\neq i} S_j(p)\right)\right] \times \left(\frac{dD(p,t)}{dp} - \sum_{j\neq i}\frac{dS_j(p)}{dp}\right) \tag{7.24}$$

Setting this derivative to zero, we obtain the first-order condition for profit maximization:

$$S_i(p) = [p - C_i'(S_i(p))]\left(-\frac{dD}{dp} + \sum_{j\neq i}\frac{dS_j(p)}{dp}\right) \tag{7.25}$$

Any solution to the coupled differential equations (7.25) for $i = 1, \ldots, I$ such that $S_i(p)$ is nondecreasing is an SFE. These equations do not involve the load characteristic $D_0(p,t)$ but depend on the demand slope dD/dp. However, in the case where the profit function (7.22) is not concave, the first-order condition (7.25) can no longer be satisfied.

We consider two different forms of supply function in this section: the affine supply function and the piecewise affine supply function.

7.2.1.1 Affine Supply Function Model

We assume that the market rules specify that the supply function of each firm is also affine:

$$S_i(p) = \frac{1}{\tilde{c}_i}(p - \cap_i), \quad i = 1, \ldots, I \tag{7.26}$$

The parameters $\tilde{c}_i$ and $\cap_i$ are chosen by firm i subject to the requirement that $\tilde{c}_i > 0$.

For the affine marginal cost function (7.19), affine supply function (7.26), and affine demand function (7.21), we obtain

$$\frac{1}{\tilde{c}_i}(p - \cap_i) = \left[p - c_i\left(\frac{1}{\tilde{c}_i}(p - \cap_i)\right) - a_i\right]\left(\gamma + \sum_{j\neq i}\frac{1}{\tilde{c}_i}\right), \quad i = 1, \ldots, I \tag{7.27}$$

There are possibly $2 \times I$ variables, namely, $\tilde{c}_i, \cap_i, i = 1, \ldots, I$, for the I equations, and thus when the single pricing period is considered and each firm i can choose the parameters $\tilde{c}_i$ and $\cap_i$ arbitrarily, the solution will not be unique. Different restrictions on the parameterization of supply function are reported in the literature.

Baldick [255] summarizes the parameterization of supply function into four categories as follows (the notation is different from that in Baldick *et al.* [285]): (1) *c* parameterization, where firm i can choose $\tilde{c}_i$ in (7.26) to be an arbitrary positive value, but is required to specify a fixed and pre-chosen value of $\cap_i$; (2) *a* parameterization, where firm i can choose $\cap_i$ in (7.26) arbitrarily, but is required to specify a fixed and pre-chosen value of $\tilde{c}_i$; (3) ($c \propto a$) parameterization, where firm i can choose $\tilde{c}_i$ and $\cap_i$ subject to the condition that $\tilde{c}_i$ and $\cap_i$ have a fixed linear relationship, i.e. there is a fixed pre-chosen $k > 0$ and firm i must then choose $\tilde{c}_i$ and $\cap_i$ so that $\tilde{c}_i = k \cdot c_i$ and $\cap_i = k \cdot a_i$; and (4) (c, a) parameterization, where firm i can choose $\tilde{c}_i$ and $\cap_i$ arbitrarily.

It should be noted that if we assume that each firm must specify a single supply function that applied to all pricing periods over an extended length of time, we will have

$$\frac{1}{\tilde{c}_i} = \left(1 - \frac{c_i}{\tilde{c}_i}\right)\left(\gamma + \sum_{j \neq i} \frac{1}{\tilde{c}_j}\right), i = 1, \ldots, I \tag{7.28}$$

and

$$\tilde{a}_i = a_i \tag{7.29}$$

The proof can be found in Baldick *et al.* [285].

All the above four parameterization cases of supply function are considered in this section.

7.2.1.2 Piecewise Affine Supply Function Model

To consider more complicated conditions in real electricity markets, we take a numerical approach to finding the SFE, which is allowed to be piecewise affine. For comparison, we use the similar treatments stated in Baldick and Hogan [280].

The supply function is defined as a piecewise affine nondecreasing function with break-points evenly spaced between $(\underline{p} + 0.1)$ and $(\overline{p} - 0.1)$, where $\underline{p}$ and $\overline{p}$ are the lower and upper limits of the market price, respectively. No generation capacity constraint is considered. An illustration of firm i's piecewise affine supply function is given in Figure 7.8. The kth $(1 \leq k \leq K)$ linear section of firm i's supply function existing in the price range $p_{i,k-1}$ to $p_{i,k}$ and the capacity range $q_{i,k-1}$ to $q_{i,k}$ can be

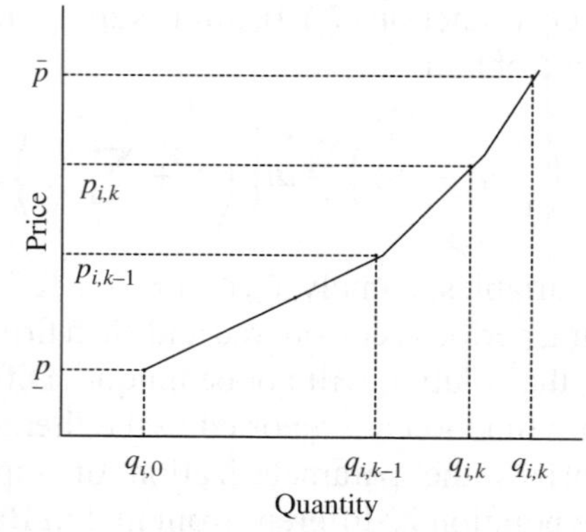

Figure 7.8 Firm i's piecewise affine supply function. *Source*: Chen 2006. Reproduced with permission from IEEE.

defined as

$$q = q_{i,k-1} + \frac{1}{\beta_{i,k}}(p - p_{i,k-1})$$
$$= q_{i,k-1} + \frac{q_{i,k} - q_{i,k-1}}{p_{i,k} - p_{i,k-1}}(p - p_{i,k-1}) \quad (7.30)$$

where $\beta_{i,k}$ is the slope of the kth linear section.

This section numerically estimates the equilibrium supply functions of any shapes with piecewise affine and continuous functions having a large number of pieces.

7.2.2 Coevolutionary Approach to Analyzing SFE Model

Coevolutionary computation is a relatively new area in evolutionary computation research. Its theory and applications are still rapidly developing [66, 271]. Coevolutionary computation was first applied to power system generation scheduling by Chen and Wang [68] and has been successfully applied to analysis of the oligopoly electricity market [281]. Son and Baldick [272] proposed a hybrid coevolutionary programming approach for Nash equilibrium search in games with local optima. The general framework of coevolutionary computation is introduced in Chen *et al.* [281] and here we focus only on the special considerations for analysis of the SFE model.

The basic coevolutionary computation model used in this section is shown in Figure 7.9 [66]. It is actually a special form of agent-based simulation approach to the electricity market. Coevolutionary computation is developed from traditional evolutionary algorithms (EAs), which simulate the coevolutionary mechanism in nature and

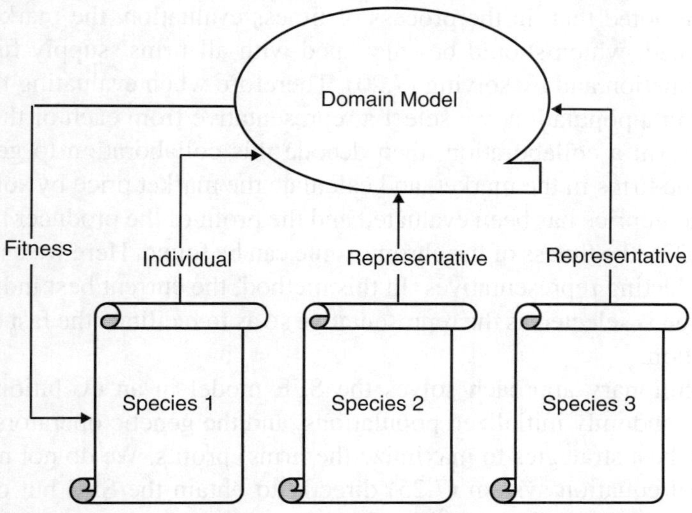

Figure 7.9 Framework of cooperative coevolutionary model. *Source*: Chen 2006. Reproduced with permission from IEEE.

adopt the notion of an ecosystem. The agents simulate the species in an ecosystem. Multiple species coevolve and interact with each other and result in the continuous evolution of the ecosystem. The species are genetically isolated—individuals only mate with other members of their species. They interact with one another within a shared domain model. In this section, a firm (agent) in the market is represented by a species in the ecosystem, which is evolved by a conventional genetic algorithm (GA). Figure 7.9 shows the fitness evaluation phase of GA from the perspective of species 1. To evaluate individuals from one species, collaborations are formed with representatives from each of the other species. The implementation details of the coevolutionary approach are presented in Chen *et al.* [281].

For the SFE model, one firm in the market is represented by one species, and thus there are I species in the ecosystem. Each firm i needs to optimize its own profit π_i expressed by (7.22).

7.2.2.1 Coevolutionary Approach for Affine Supply Function Model

For the affine supply function model, in different parameterization cases, different parameters (slope or intercept) of the supply function are chosen as the variables to be optimized (which are called "strategic variables"). For example, in the c parameterization case, the slope of the supply function $\tilde{c}_i$ is chosen as the strategic variable. We can see that in (c, a) parameterization, we have two strategic variables, and in the other cases we have only one strategic variable. We use a binary code chromosome to encode the strategic variables, and use the profit of the firm as the fitness of the chromosome directly. In fact, the GA is used to solve the parameter optimization problem in the supply function space.

It should be noted that, in the process of fitness evaluation, the market price p is to be determined, which should be calculated with all firms' supply functions and the demand function and by solving (7.20). Therefore when evaluating the fitness of an individual in a population, we select a representative from each of the other populations and form a collaboration, then decode this collaboration to get all supply functions of the firms in the market, and calculate the market price by solving (7.20). When the market price has been evaluated and the profit of the producer has been calculated by (7.22), the fitness of the chromosome can be found. Here we use a "greedy" method for selecting representatives. In this method, the current best individual from each population is selected as the representative so as to facilitate the fast convergence of the simulation.

The coevolutionary approach solves the SFE model in an evolutionary process starting from randomly initialized populations, and the genetic operators are used to search for the best strategies to maximize the firms' profits. We do not need to solve the differential equation system (7.25) directly to obtain the SFE but can get it by simulating the strategic behavior of the firms in markets.

7.2.2.2 Coevolutionary Approach for Piecewise Affine Supply Function Model

The algorithm framework of the coevolutionary approach for the piecewise affine supply function model is similar to that for the affine supply function, the main differences being the coding structure of the chromosome and the starting conditions. From section 7.2.2.1 we know that to fully describe a piecewise affine supply function, we should keep the quantity and price values of each break-point. Since the price break-point values are specified as in section 7.2.2.1, we only need to keep the quantity break-point values. For convenience of handling, we keep the quantity value at the price $\underline{p}$ and the slope reciprocal values of each linear section of the piecewise affine supply function instead of the quantity values of each break-point. That is, the coding structure of a chromosome is as shown in Figure 7.10. The chromosome consists of $K + 1$ binary genes, each of which encodes a corresponding variable value.

With the affine supply function model, the evolutionary process can start from randomly initialized populations; but with the more complicated piecewise affine supply function model, the evolutionary process needs to start from an estimate of the equilibrium supply function, which is called a "starting function" in Baldick *et al.* [284]. As there [284], we encode the competitive, affine, and Cournot starting functions, respectively, into the initial representatives of each species as the initial conditions of the evolutionary process.

7.2.3 Example

We use the five-firm example system in Baldick and Hogan [280] to validate the coevolutionary approach described in this section. The example system is based on the cost data for the five strategic firms industry in England and Wales subsequent to the 1999 divestiture. The firms' cost data are reproduced in Table 7.9. No generating capacity constraints are considered in the case study.

A demand slope of $\gamma = 0.1$ GW/(£/MWh) and a load duration characteristic of

$$D_0(t) = 10 + 25(1 - t), 0 \le t \le 1 \tag{7.31}$$

with quantities measured in GW are used in Baldick and Hogan [280], where the time argument t is normalized to go from 0 to 1 and $t = 0$ corresponds to peak load conditions. We use the same demand function form in our simulation and suppose

1	2	3	...	K+1
$q_{i,0}$	$1/\beta_{i,1}$	$1/\beta_{i,2}$	...	$1/\beta_{i,K}$

Figure 7.10 Coding structure of chromosome for piecewise affine supply function model. *Source*: Chen 2006. Reproduced with permission from IEEE.

Table 7.9 Firms' cost data from table 2 in Baldick and Hogan [280].

Firm's cost function	Firm no. i	c_i (£/MWh^2)	a_i (£/MWh)
	1	2.687	12
	2	4.615	12
$C_i(q_i) = \frac{1}{2}c_i q_i^2 + a_i q_i$	3	1.789	8
	4	1.93	8
	5	4.615	12

that one trading day equals 48 pricing periods, and viz. one pricing period spans $1/48$ time interval.

7.2.3.1 Simulation with the Affine Supply Function Model

We simulate the competitive case, the Cournot case, and the four different parameterization cases mentioned in section 7.2.1.1. In the competitive case, all firms behave competitively by bidding their true costs; while in the Cournot case, all firms employ quantity strategies. All cases excluding the (c, a) parameterization are simulated at a single pricing period. In the simulation of the (c, a) parameterization case, two different situations are considered. One is the single pricing-period simulation, in which the supply function is applied to a single pricing period; the other is the multiple pricing-period simulation, in which a single supply function is applied to all pricing periods.

Theoretically, the SFE does not depend on the load characteristic but does depend on the demand slope $\mathrm{d}D/\mathrm{d}p$. However, if there are multiple SFEs, the SFE that can be found by the numerical framework depends on the load characteristic. Here we use the load when $t = 0$, namely,

$$D(p, 0) = D_0(0) - \gamma p = 35 - \gamma p$$

for the single pricing-period simulation.

A 16-bit binary code chromosome is used to encode a strategic variable in each GA population. The evolutionary process starts from a randomly initialized population and runs for a maximum 150 generations, with the population size 100. A two-point crossover and a bit-flip mutation genetic operators are used. The elitist strategy preserving a single copy of the best individual is adopted. The crossover probability is set to 0.9, and the mutation probability is set to 0.06. The computational time of each run is within 1 s on a Pentium 4 PC with 2.8 GHz CPU.

The simulation results are listed in Table 7.10. The simulation results show that, in the single pricing-period simulations, the market prices in all four SFE cases are higher than in the competitive case and lower than in the Cournot case; the profits

Table 7.10 Simulation results of affine supply function model.

Test cases	P (£/MWh)	Firms' results				
		Firm no.	$\tilde{c}_i$ (£/MWh^2)	$\cap_i$ (£/MWh)	q_i (GWh)	π_i ($\times 10^3$£)
		1	2.687	12	5.536	0.858
		2	4.615	12	3.223	0.499
compet.	26.87	3	1.789	8	10.55	2.074
		4	1.93	8	9.78	1.923
		5	4.615	12	3.223	0.499
		1	—	—	5.391	6.869
		2			4.679	5.615
Cournot	80.4	3	—	—	6.142	8.56
		4			6.069	8.413
		5			4.68	5.615
		1	3.522	12	5.919	1.59
		2	5.386	12	3.87	0.961
c param.	32.85	3	2.69	8	9.237	3.192
		4	2.817	8	8.819	3.002
		5	5.386	12	3.87	0.961
		1	2.687	15.66	5.897	1.423
		2	4.615	14.13	3.765	0.848
a param.	31.5	3	1.789	14.63	9.43	2.96
		4	1.93	14.15	8.993	2.777
		5	4.615	14.13	3.765	0.848
		1	3.115	13.91	5.902	1.521
		2	5.025	13.07	3.827	0.914
$(c \propto a)$ param.	32.3	3	2.342	10.47	9.321	3.1
		4	2.477	10.27	8.893	2.913
		5	5.025	13.07	3.827	0.914
		1			5.391	6.869
		2			4.68	5.615
(c, a) param. single period	80.4	3	not unique	not unique	6.141	8.56
		4			6.069	8.413
		5			4.68	5.615
		1	3.519	12.01	—	31.99
		2	5.389	12		19.33
(c, a) param. multiple period	—	3	2.692	7.986		74.16
		4	2.815	8.02		69.74
		5	5.381	12.01		19.33

of the firms are more than the profits in the competitive case but less than the profits in the Cournot case. The results depend greatly on the method of parameterization, except for the (c, a) parameterization case; the results of other parameterization cases are close.

In the single-period $(c; a)$ parameterization simulation case, the slope and intercept of the supply function of each firm are not unique, but the market clearing price and the firms' outputs are unique and equal to the results of the Cournot case. In this case the slope of the supply function is a very large positive value and the intercept is a very large negative value. This is easy to understand since, as the slopes become large, the supply functions of the firms become less elastic and the firms behave like a Cournot quantity decision-maker. Such a limiting SFE is a focal equilibrium under (c, a) parameterization because it is mutually beneficial to all the strategic players [255]. The simulation results demonstrate that players can converge to this focal equilibrium without any explicit coordination between them.

On the other hand, in the multiple-period (c, a) parameterization simulation case, the slope and intercept of the supply function of each firm become unique and equal to the results of single-period c parameterization with minor computational errors. This result comes from the fact that, in the c parameterization case, we suppose the firms honestly bid $\cap_i = a_i$, which is an equilibrium condition for multi-period bidding. The seemingly simple result is important since we impose no strategic hypothesis other than the general genetic operators on the firms' strategic learning, and only the multi-period bidding requirement settles the supply functions down on a unique equilibrium. This is a key notion of SFE theories [262–264, 285].

The simulation results show that the coevolutionary approach is highly efficient in affine supply function model analysis. Rudkevich [283] presents an alternative model of the learning process in which each firm starts from the perfectly competitive supply function and adjusts the slope of the supply function through a stylized formula and using the market observed data. Rudkevich proves that firms can achieve an SFE through this learning process. Our model differs from Rudkevich's in that: first, it does not need to start from the pre-assumed supply functions rather than from any randomized supply functions; and second, it does not need an assumed learning model but facilitates the strategic learning by the general evolutionary operations. Therefore the simulation of strategic learning behaviors in the market can be more realistic with the coevolutionary approach.

7.2.3.2 Simulation with the Piecewise Affine Supply Function Model

As in Baldick *et al.* [284], the piecewise affine supply function has 40 break-points evenly spaced between $(\underline{p} + 0.1)$ and $(\overline{p} - 0.1)$. A price cap $\overline{p} = 40$ £/MWh and a price minimum of $\underline{p} = 12$ £/MWh are used in our simulation. As mentioned before, the evolutionary process needs to start from an estimate of the equilibrium supply function in the piecewise affine supply function model simulation. We should define a set of starting functions as the initial estimates of the supply functions. Three different starting functions are used [284]:

1. "competitive", where the supply functions are the inverses of the marginal cost functions, namely $\tilde{c}_i = c_i$ and $\cap_i = a_i$ in (7.26);
2. "affine SFE", where the supply functions are given by (7.26), with the coefficients satisfying (7.28) and (7.29);
3. "Cournot", where quantities and prices under Cournot competition are calculated for each $t \in [0, 1]$ and a supply function drawn through the resulting price–quantity pairs, namely $\tilde{c}_i = c_i + 1/\gamma$ and $\cap_i = a_i$ in (7.26).

An 8-bit binary code chromosome is used to encode a variable in each GA population. The evolutionary process starts from the representatives encoding the starting functions and runs for a maximum 100 generations, each generation containing 15 nested generations of evolution of each species. The other algorithm parameters are the same as those in the affine supply function model simulation. The simulation results are shown in Figure 7.11, Figure 7.13, Figure 7.14, Figure 7.15, Figure 7.16, Figure 7.18, Figure 7.19, Figure 7.20, Figure 7.21, and Figure 7.22. Since the results of the coevolutionary approach are stochastic, we run the algorithm for 20 times in each simulation and calculate the average results. The average computational time of each time run is 20 s on a Pentium 4 PC with 2.8 GHz CPU.

The affine supply functions are shown in Figure 7.11. In this figure and all subsequent figures: firm 1 is shown as a solid line; firms 2 and 5 have identical costs and capacities and are shown superimposed as a dash-dot line; firm 3 is shown as a solid line; and firm 4 is shown as a dotted line. The coevolutionary approach starting from

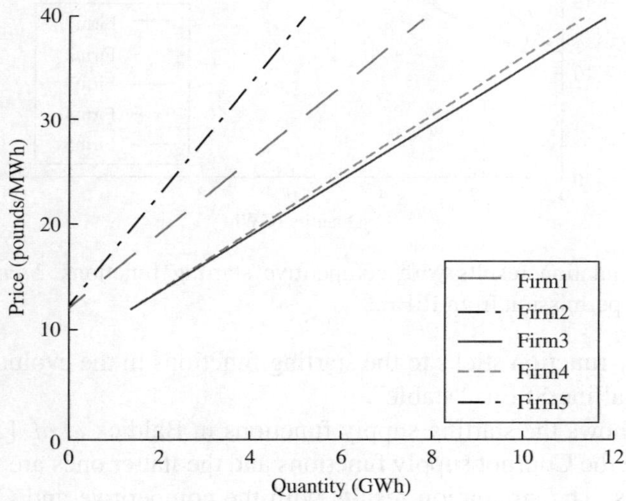

Figure 7.11 Affine supply functions. *Source*: Chen 2006. Reproduced with permission from IEEE.

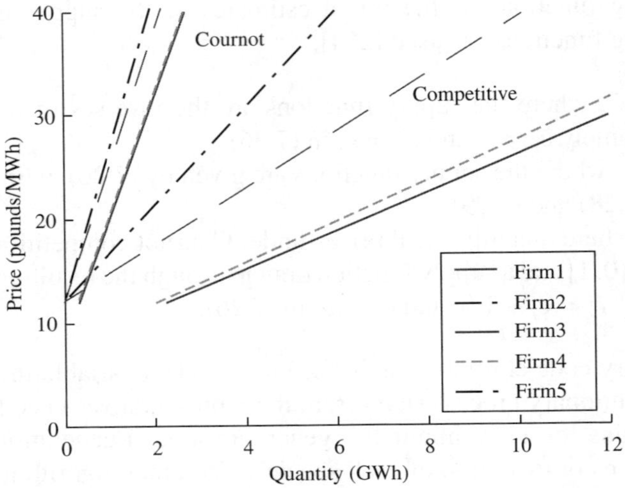

Figure 7.12 Starting supply functions. *Source*: Chen 2006. Reproduced with permission from IEEE.

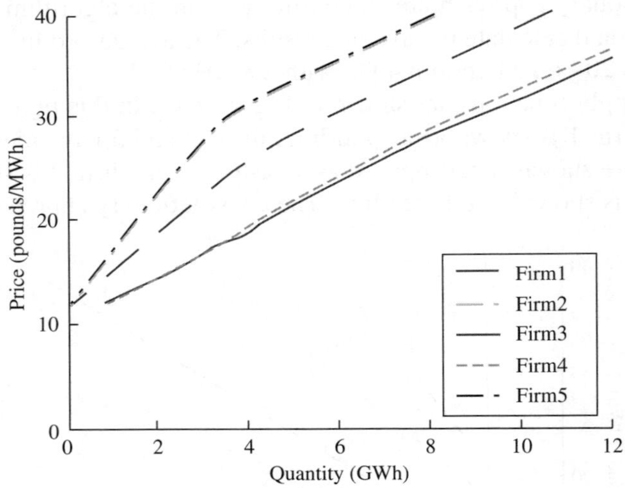

Figure 7.13 Simulation results with competitive starting functions. *Source*: Chen 2006. Reproduced with permission from IEEE.

the affine supply function sticks to the starting functions in the evolutionary process. That means the affine SFE is "stable".

Figure 7.12 shows the starting supply functions in Baldick *et al.* [284], where the steeper lines are the Cournot supply functions and the flatter ones are the competitive supply functions. The simulation results with the competitive and Cournot starting functions are shown in Figure 7.13 and Figure 7.14, respectively. Simulation results show that generally the coevolutionary approach can find the equilibrium supply

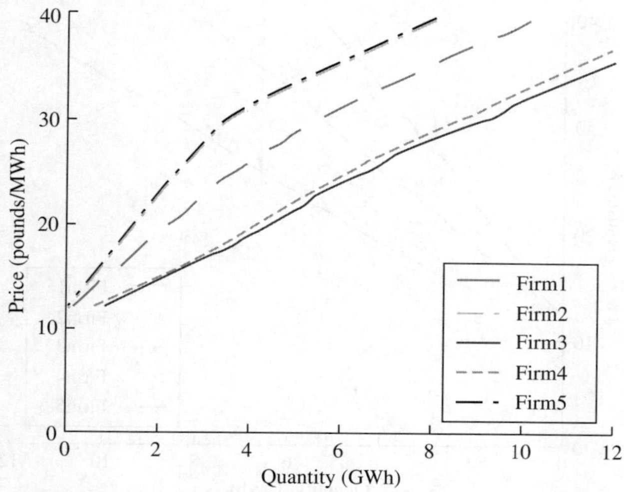

Figure 7.14 Simulation results with Cournot starting functions. *Source*: Chen 2006. Reproduced with permission from IEEE.

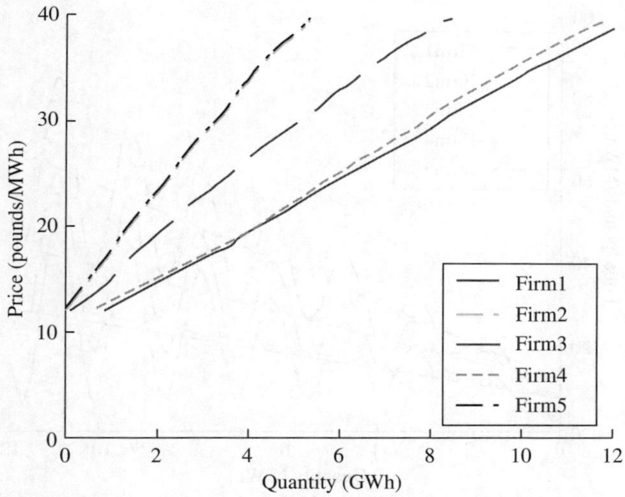

Figure 7.15 Simulation results with competitive starting functions and load duration characteristic (7.32). *Source*: Chen 2006. Reproduced with permission from IEEE.

functions with minor random fluctuations and the average over a number of runs will "smooth down" the fluctuations and find the equilibrium supply functions. The segments of supply functions with prices below about 30 £/MWh coincide well with the affine supply functions, but the segments with prices above 30 £/MWh deviate from the affine supply functions. The reason is that the peak load is 35 GW and the market prices will not exceed a few more than 30 £/MWh, so the supply function segments with prices above 30 £/MWh are not used through the whole clearing

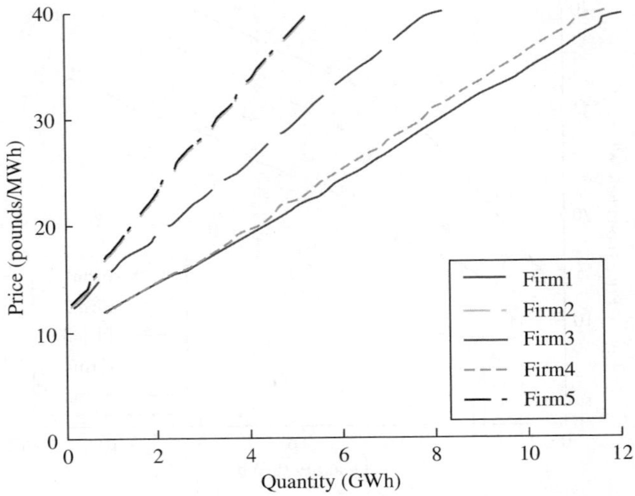

Figure 7.16 Simulation results with Cournot starting functions and load duration characteristic (7.32). *Source*: Chen 2006. Reproduced with permission from IEEE.

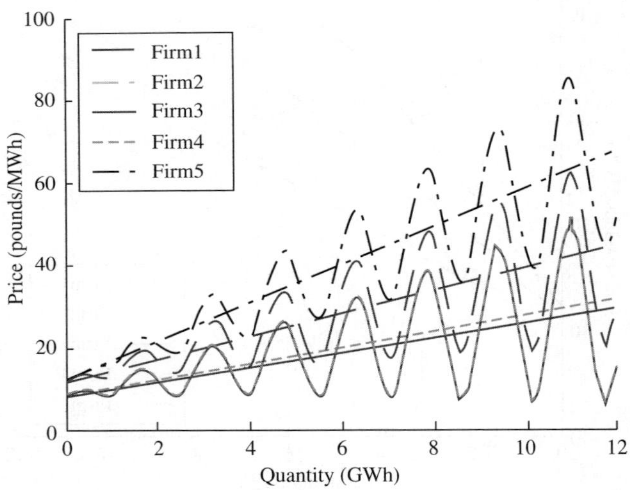

Figure 7.17 Marginal cost functions of cost functions (7.33). *Source*: Chen 2006. Reproduced with permission from IEEE.

process, and then their simulation results are just the averages over a number of random runs. To verify this assertion, we change the load duration characteristic to

$$D_0(t) = 10 + 40(1 - t), \quad 0 \le t \le 1 \tag{7.32}$$

Namely, the peak load is increased to 50 GW. The simulation results are shown in Figure 7.15 and Figure 7.16, from which we can see that the equilibrium supply functions coincide well with the affine SFE. Figure 7.11, Figure 7.13, and Figure 7.14

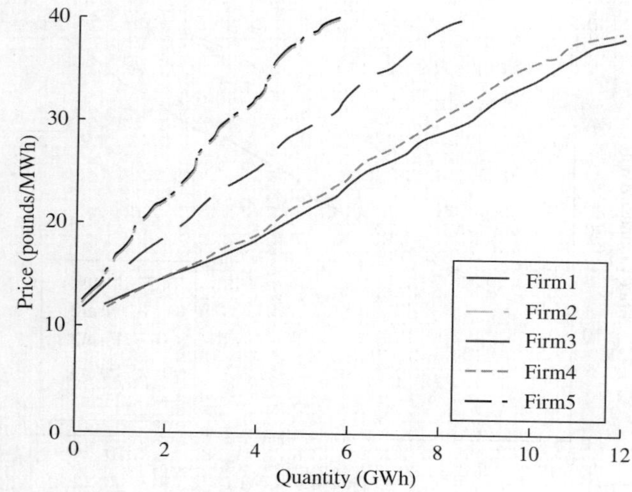

Figure 7.18 Simulation results with affine starting functions and cost functions (7.33). *Source*: Chen 2006. Reproduced with permission from IEEE.

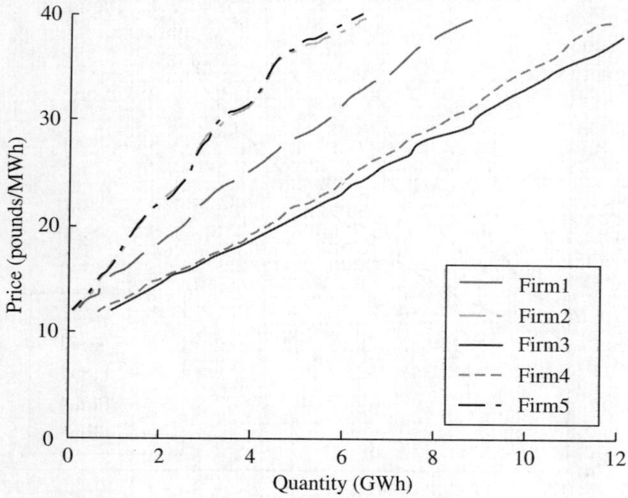

Figure 7.19 Simulation results with competitive starting functions and cost functions (7.33). *Source*: Chen 2006. Reproduced with permission from IEEE.

correspond to figures 31, 28, and 34, respectively, in Baldick *et al.* [284]—the results are similar.

To test the robustness of the proposed approach, we modify the quadratic cost function by superimposing a sinusoidal component. The cost function (7.18) is changed to

$$C_i(q_i) = \frac{1}{2}c_i q_i^2 + a_i q_i + \frac{1}{2}q_i \sin(4q_i), \quad i = 1, \dots, I \tag{7.33}$$

which is nonconvex.

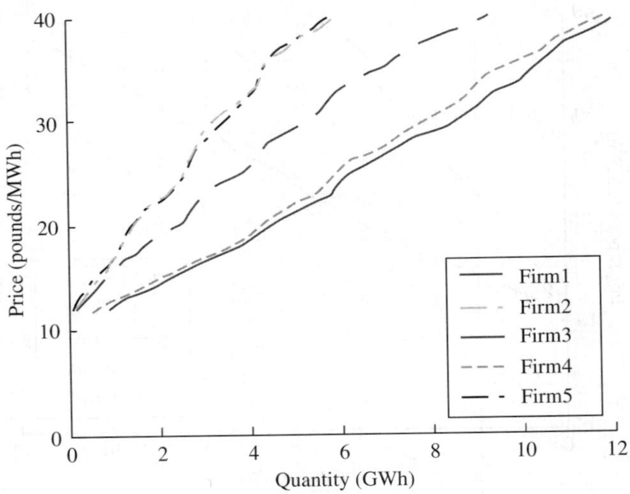

Figure 7.20 Simulation results with Cournot starting functions and cost functions (7.33). *Source*: Chen 2006. Reproduced with permission from IEEE.

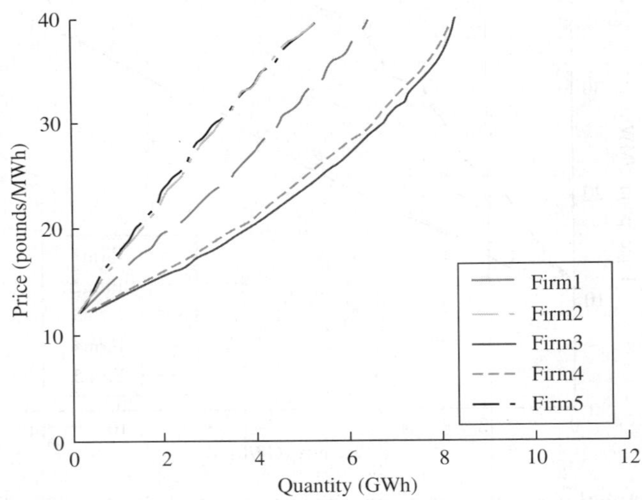

Figure 7.21 Simulation results with competitive starting functions and $\gamma = 0.02$. *Source*: Chen 2006. Reproduced with permission from IEEE.

Although the cost is changed slightly, the marginal cost is changed significantly, which is illustrated in Figure 7.17. The straight lines in Figure 7.17 are the original affine marginal cost functions (7.19) and the sinusoidal curves designate the new marginal cost functions derived from (7.33). Since the cost function is nonconvex and the profit function (7.22) is nonconcave, the first-order condition (7.25) can no longer be satisfied. There is no analytical solution in this case.

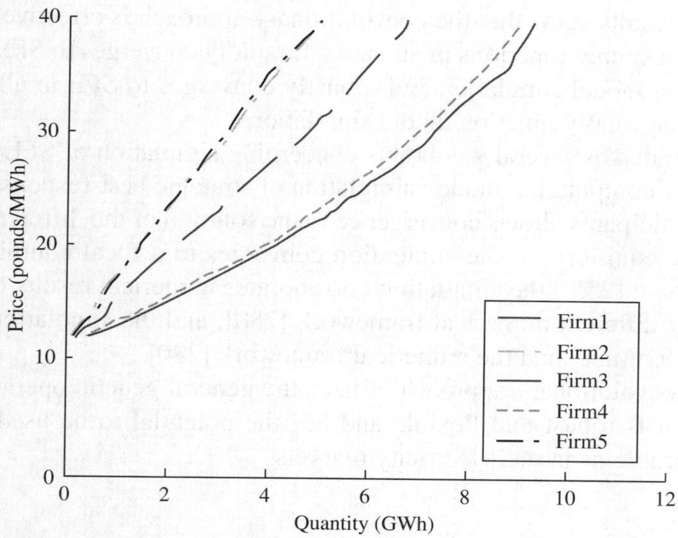

Figure 7.22 Simulation results with Cournot starting functions and $\gamma = 0.02$. *Source*: Chen 2006. Reproduced with permission from IEEE.

The simulation results are shown in Figure 7.18, Figure 7.19, and Figure 7.20. Since the affine SFE is not stable any more in this case, it is also used as starting functions and the result is shown in Figure 7.18. The simulation results with competitive starting functions and Cournot starting functions are shown in Figure 7.18 and Figure 7.19, respectively. We can see that the equilibrium supply functions converge to the affine SFE superimposed with sinusoidal fluctuations. The coevolutionary approach is robust to alteration of the firms' cost functions.

Figure 7.21 and Figure 7.22 show simulation results with reduced demand elasticity, in which coefficient γ in demand function (7.21) is reduced to 0.02. The "hockey-stick" shaped equilibrium supply functions emerge in the results, which imply that the firm can find a strategy to increase the slopes of last segments in the supply function to increase market prices and its profits.

The simulation results show that the coevolutionary approach is effective and powerful in SFE model analysis.

7.2.4 *Summary*

This section presents a coevolutionary approach to analyzing supply function equilibrium models of an oligopolistic electricity market. The affine and piecewise affine supply function models are considered. Different parameterization cases of the affine supply function model are analyzed. The piecewise affine and continuous supply functions that have a large number of pieces are used to numerically estimate the equilibrium supply functions of any shapes. Cases with different peak loads, non-quadratic and nonconvex costs, and different demand elasticities are simulated.

Simulation results show that the coevolutionary approach is effective and can find the equilibrium supply functions in all cases. It rapidly converges to SFE in the affine supply function model simulation and robustly converges to SFE in all cases of the piecewise affine supply function model simulation.

This section makes several key points concerning simulation of SFE based on the coevolutionary computation model: simulation of strategic best-response behavior of the market participants allows convergence to the solution of the differential equation describing the equilibrium; the simulation converges to a focal equilibrium as discussed elsewhere [255]; the simulation corroborates numerical results obtained with a qualitatively different numerical framework [280]; and the simulation takes much less computation time than the numerical framework [280].

Since the coevolutionary approach utilizes the general genetic operators in strategic learning, it is robust and flexible and has the potential to be used to solve the equilibrium problems in real electricity markets.

7.3 Searching for Electricity Market Equilibrium with Complex Constraints Using Coevolutionary Approach

In recent years, the electricity industry throughout the world has been undergoing a huge restructuring process. Competition, which has been introduced in order to increase social welfare and to improve market efficiency, is displacing government regulation as the major factor to determine electricity price. The electricity market is more akin to an oligopoly at the current stage, and therefore the market participant has market power to influence the price by strategic actions. A crucial issue related to strategic bidding is the analysis of market equilibrium, which is the steady state of a participant's strategic behavior. Nash equilibrium (NE) not only is used to understand the optimal strategy to maximize their own profit for a market participant, but also is very useful to forecast and monitor the potential performance of a market structure before the market is introduced for market regulators [256].

Different incomplete competition market models have been used to analyze market equilibrium; among them, the Cournot and supply function equilibrium (SFE) models are the most extensively used for analyzing the pool-based electricity market [255]. The classic Cournot and SFE models which neglect network constraints are widely studied. The general SFE model was developed by Klemperer and Meyer [262] for calculating the participant's maximum profit equilibria with uncertain demands, and was first applied by Green and Newbery [263] to analyze the competitive strategic behavior in the British electricity market. Baldick *et al.* [284] focused on a linear supply function equilibrium (LSFE) model for interaction in an electricity market, assuming a linear demand function, and consider several strategic participants with constrained capacity and affine marginal costs.

Recently, calculating the Nash equilibrium of an electricity market considering network constraints has become a research hotspot. Some iterative NE search methods that use repeated participant profit maximization based on local optimization algorithms have been applied to complicated electricity market. Hobbs *et al.* [265] formulated the single-firm model as a mathematical program with equilibrium constraints (MPEC) with a parameter-dependent spatial price equilibrium problem as the inner problem, by calculating each firm's MPEC in turn using a penalty interior point algorithm, then seek a pure strategy NE of the multi-firm model. Weber and Overbye [26, 27] modeled the bidding problem as a two-level optimization problem in which strategic participants try to maximize their profit under the constraints that their dispatch and price are determined by optimal power flow, and then employed the modified Newton approach to solve it. Sensitivity functions of each participant's profit with respect to bidding strategies were developed to solve this type of two-level optimization problem [287]. The bidding process with congestion management has been formulated as a three-level optimization problem [288], a probabilistic methodology is then proposed to analyze the market equilibrium.

Correia *et al.* [289] developed a systematic procedure to search for multiple Nash equilibria in optimal feasible subregions defined by binding transmission constraints. The mixed nonlinear complementarity problem (NCP) is obtained by combining the optimization conditions of each strategic participant's optimization problem [290], and then the mixed nonlinear complementarity approach was proposed to solve it with a special nonlinear complementarity function. In order to overcome the "local NE trap" problem, Son and Baldick [272] suggested hybrid coevolutionary programming to solve games with local optima.

Techniques for searching for market equilibrium based on the definition of Nash equilibrium are also extensively employed. That is, the NE can be found as the intersection of players' best-response curves in the simple game. Cunningham *et al.* [261] applied the manual approach to seek NE by drawing best-response curves and searching for an intersection. An approach combining the graphical and analytical methods was shown [291] to calculate the optimal bidding strategies of suppliers. Liu and Wu [292] pointed out that the participants' decision space should be divided into subsets in the electricity market taking account of network constraints. The equilibrium was then solved by analyzing the possible intersection of best responses in different subsets. Xu and Baldick [293] characterized the residual demand derivative analytically from the solution of a specific weighted least-squares problem, and the strategic behaviors were then analyzed based on the best response constructed by the residual demand derivative.

A recent rapidly developing area rests on the application of coevolutionary algorithms to electricity market analysis. Price [82] employed the coevolutionary genetic algorithm to the study of a simple electricity market pool model, and showed its potential role as an analytical tool for general market study through several simulation examples. Cau and Anderson [83] investigated the coevolutionary genetic algorithm on a two-player electricity pool market game. A novel framework of electricity market analysis based on coevolutionary computation approach was proposed [281]; the standard Cournot model and the Pareto improvement model were used to show that the proposed method is highly efficient and can handle the nonlinear market model. Further, the method was applied to analyzing both the affine and piecewise affine supply function models, which are difficult to handle by conventional approaches [294]. A two-settlement electricity market, including a forward market and a spot market, was formulated as a two-stage game [295], and a coevolutionary computation algorithm was then employed to determine the market equilibrium. Price [82] and Zhang *et al.* [295] employed the coevolutionary algorithms to the study of the electricity market without taking account of network constraints; thus, the coevolutionary computation approach needs to be tested to analyze market equilibrium in more realistic and complicated conditions.

7.3.1 Market Model Formulation

The general electricity market model [26, 27] is used here. The market mainly includes an independent system operator (ISO), generating companies (GenCos), and market consumers. GenCos submit bids which consist of megawatt outputs along with corresponding prices, and market consumers submit bids that link their demands with associated prices; the ISO then uses a security-constrained economic dispatch to determine supplies, demands, and spot prices after collecting bids. The details of the market model are presented as follows.

7.3.1.1 Strategic Participants' Bids

Market strategic participants should not be restricted to either a single GenCo or a single consumer. Any combination of several GenCos and consumers could constitute an economic entity. Therefore, "strategic participant" means a single entity that controls the strategic bids of a set of GenCos and consumers to maximize its own welfare.

Suppose strategic participant t controls n_t GenCos and m_t consumers. Each GenCo has a strictly convex quadratic generation cost function

$$C_{ti}(P_{Gti}) = \frac{1}{2}a_{ti}P_{Gti}^2 + b_{ti}P_{Gti} \quad (i = 1, 2, \dots, n_t) \tag{7.34}$$

where P_{Gti} is the generation output of GenCo t_i; and a_{ti} and b_{ti} are the coefficients of the cost function of GenCo t_i. The corresponding marginal cost is calculated as:

$$\frac{\mathrm{d}C_{ti}(P_{Gti})}{\mathrm{d}P_{Gti}} = a_{ti}P_{Gti} + b_{ti} \tag{7.35}$$

which is a linear function of its generation output.

Bids are restricted to linear functions in the model, so the GenCos are assumed to compete by submitting their bids in the form of linear supply functions (LSFs)

$$P_{Gti}^R = (p_{ti} - b_{ti}^R)/a_{ti}^R \tag{7.36}$$

where P^R_{Gti} and p_{ti} are the submitted quantity and spot price of GenCo t_i; and a^R_{ti} and b^R_{ti} are the coefficients (strategic variables) of the LSF. Four parameterization techniques for strategic variables are usually used, including a, b, $(a \propto b)$, and (a, b) parameterizations [255]. Weber and Overbye [26] showed that the shape of the bidding curve is not vital to participants for a single market simulation. Therefore, the $(a \propto b)$ parameterization is chosen since its format is similar to the actual production cost functions and it can be easily implemented for only one variable. That is, GenCo t_i will choose a^R_{ti} and b^R_{ti} subject to the condition that the two variables have a fixed linear relationship, so the strategic parameter k_{Gti} is used to vary the bid from the true marginal cost function as

$$P^R_{Gti} = (p_{ti}/k_{Gti} - b_{ti})/a_{ti}, k_{Gti} > 0 \tag{7.37}$$

So the strategic parameters of all GenCos form the participant's strategic generation vector $\boldsymbol{k}_{Gt}$ with k_{Gti} as an element.

For each consumer, the true marginal benefit bid is defined as a linear function

$$P_{Dtj} = (p_{tj} - \beta_{tj})/(-\alpha_{tj}) \quad (j = 1, 2, \ldots, m_t) \tag{7.38}$$

where P_{Dtj} and p_{tj} are the demand and spot price of consumer t_j, and α_{tj} and β_{tj} are the coefficients of the benefit function. Therefore its corresponding benefit can be calculated as

$$B_{tj}(P_{Dtj}) = \int p_{tj}\, \mathrm{d}P_{Dtj} = (-0.5)\alpha_{tj}P^2_{Dtj} + \beta_{tj}P_{Dtj} \tag{7.39}$$

A strategic variable k_{Dtj} can be similarly used to construct its bid as follows:

$$P^R_{Dtj} = (p_{tj}/k_{Dtj} - \beta_{tj})/(-\alpha_{tj}) \tag{7.40}$$

Thus the participant's strategic consumer vector $\boldsymbol{k}_{Dt}$ can be constructed by all consumers' strategic variables.

Then based on (7.37) and (7.40) for each owning GenCo and consumer, respectively, the strategic participant determines the strategic vectors and submits the bids to the ISO to seek the maximum profit.

7.3.1.2 Market Clearing Model

After receiving the bids of all participants, the ISO solves an optimization problem to determine the supplies, demands, and locational marginal prices (LMPs) based on the optimal power flow. In the market clearing process, the bids are treated as the marginal cost and benefit of the bidder, so the optimization problem—which is to maximize a quasi-social welfare, subject to the generating capacity constraints and transmission capacity constraints—can be formulated as follows:

$$\max \quad \boldsymbol{\Gamma}_{ISO}(\boldsymbol{K}_G, \boldsymbol{K}_D) = \sum_{t=1}^{T} \boldsymbol{K}_{Dt}\boldsymbol{B}_t - \sum_{t=1}^{T} \boldsymbol{K}_{Gt}\boldsymbol{C}_t \tag{7.41}$$

$$\text{s.t.} \qquad \boldsymbol{H\theta} = \boldsymbol{P}_G - \boldsymbol{P}_D$$
$$F_{\min l} \le F_l \le F_{\max l}$$
$$P_{\min Gti} \le P_{Gti} \le P_{\max Gti}$$
$$(l = 1, 2, \ldots, L; \quad t = 1, 2, \ldots, T; \quad i = 1, 2, \ldots, n_t)$$

Here the following notation has been used:

$\boldsymbol{K}_G$	vector of strategic generation variables, with $\boldsymbol{K}_{Gt}$ as an element;
$\boldsymbol{K}_D$	vector of strategic consumer variables, with $\boldsymbol{k}_{Dt}$ as an element;
T	number of strategic participants;
$\boldsymbol{B}_t$	vector of benefit for participant t, with B_{tj} as an element;
$\boldsymbol{C}_t$	vector of cost for participant t, with C_{tj} as an element;
$\boldsymbol{H}$	susceptance matrix;
θ	vector of bus voltage angles;
$\boldsymbol{P}_G$	vector of bus generation outputs;
$\boldsymbol{P}_D$	vector of bus demands;
F_l	power flow on line l;
$F_{\min l}, F_{\max l}$	lower and upper flow limits on line l;
L	number of lines in the system;
$P_{\min Gti}, P_{\max Gti}$	lower and upper bounds of outputs for GenCo i controlled by participant t.

In (7.41), the first equality constraint is the DC power flow equation, and the LMP at each bus is the Lagrange multipliers of the corresponding power flow constraint. The second inequality constraint is the transmission line constraint; and the third is the generation capacity constraints for each GEN.

Once the power market is cleared, each GenCo will be paid according to its LMP times the corresponding awarded generation; so its profit is the amount of revenue received minus the cost of supplying it. The consumer is charged the spot price, and its profit is defined as the amount of benefit received from using the power minus the expense incurred in purchasing it.

7.3.1.3 Market Equilibrium Problem

Each rational participant maximizes its profits by choosing and adjusting its strategies based on the market situation, so the optimization problem of each participant is formulated as the following two-level mathematical program:

$$\max \quad R_t(\boldsymbol{k}_{Gt}, \boldsymbol{k}_{Dt}) = \sum_{i=1}^{n_t} (\lambda_{Gti} \times P'_{Gti} - C_{ti}(P'_{Gti}))$$

$$+\sum_{j=1}^{m_t}(B_{tj}(P'_{Dtj}) - \lambda_{Dtj} \times P'_{Dtj}) \tag{7.42}$$

$$\text{s.t.} \quad k_{\min Gti} \leq k_{Gti} \leq k_{\max Gti}$$

$$k_{\min Dtj} \leq k_{Dtj} \leq k_{\max Dtj}$$

where we use the following notation:

$\lambda_{Gti}, \lambda_{Dtj}$ LMP of the node located by GenCo i, consumer j of participant t, respectively;
P'_{Gti} awarded generation of GenCo i;
P'_{Dtj} awarded demand of consumer j;
$k_{\min Gti}, k_{\max Gti}$ lower and upper limits of the generation strategic variable k_{Gti};
$k_{\min Dtj}, k_{\max Dtj}$ lower and upper limits of the consumer strategic variable k_{Dtj}.

In (7.42), the first and second inequality constraints are the strategic variable limits, because the lower and upper limits of the electricity price are usually restricted by the market regulator who needs to be vigilant against anti-competitive acts of market participants. The third constraint shows that this two-level problem forms a nested optimization problem.

It is easy to see that the participant's profit is involved not only with its own strategy, but also with the strategies of its opponents. Therefore, the market model forms a noncooperative game, and market equilibrium means a point which satisfies every player's optimization condition given the other players' bids. That is, each player has no incentive to unilaterally deviate from the market equilibrium, i.e.

$$R_t(\mathbf{k}^*_{Gt}, \mathbf{k}^*_{Dt} \mid \mathbf{k}^*_{-Gt}, \mathbf{k}^*_{-Dt}) \geq R_t(\mathbf{k}'_{Gt}, \mathbf{k}'_{Dt} \mid \mathbf{k}^*_{-Gt}, \mathbf{k}^*_{-Dt}) \tag{7.43}$$

where

$\mathbf{k}^*_{Gt}, \mathbf{k}^*_{Dt}$ vector of strategic variables for participant t in the equilibrium;
$\mathbf{k}^*_{-Gt}, \mathbf{k}^*_{-Dt}$ vector of strategic variables for the opponents of participant t in the equilibrium;
$\mathbf{k}'_{Gt}, \mathbf{k}'_{Dt}$ arbitrary vector of strategic variables for participant t.

The left-hand side of inequality (7.43) is the profit of participant t in the market equilibrium, and the right-hand side means the profit of participant t in the case in which it chooses an arbitrary strategic vector and its opponents hold the strategic vector in the market equilibrium.

Determination of the market equilibrium of the above proposed market model considering network constraints, generation capacity constraints, and hybrid participants' bidding is a difficult task. The study employs the coevolutionary computation approach to solve it.

7.3.2 *Coevolutionary Computation*

Evolutionary algorithms (EAs) are heuristic methods to solve difficult computational problems using the biologically inspired notion of Darwinian evolution. EAs usually have an advantage over many traditional local search methods when search spaces are discontinuous, highly modal, or highly constrained. However, it is hard to apply an EA in the case when evolving game-playing strategies, such as problem (7.43), because no intrinsic objective measure exists with which to evaluate the fitness of individuals. Thereby coevolutionary algorithms (CEAs), as a natural extension of EAs, have attracted lots of attention during the past decade. Because coevolutionary algorithms vary widely, it is extremely difficult to present an exact definition, but, in general, CEAs almost exclusively refer to EAs with multiple populations, in which individuals are awarded fitness values based on their interactions with individuals from other populations. To the best of our knowledge, the most complete hierarchical categorization of coevolutionary algorithms was provided in Potter and De Jong [66].

Coevolutionary algorithms has been a research focus in broad fields. Axelrod and Hamilton [296] first proposed the coevolutionary mechanism to study the "prisoners' dilemma" (PD) problem, and discussed the relationships between evolutionary stability and equilibrium. Then the coevolutionary learning of cooperative behaviors was analyzed in the "iterated prisoners' dilemma" (IPD) problem with multiple discrete strategies and continuous strategy in Harrald and Fogel [297] and Chong and Yao [298], respectively. In machine learning, Potter and De Jong [66] employed a CEA to train dynamically recurrent neural networks, and coevolution learning was proposed by Pollack and Blair [299] to develop a competitive backgammon evaluation function, which surprisingly worked well. In power system analysis, the coevolutionary computation approach was first introduced for unit commitment in Chen and Wang [68]. It was further successfully applied to analyze the market equilibrium of the Cournot and SFE models [281, 294]. More recently the approach was proposed to solve the strategic generation capacity expansion planning problem [300].

From the previous categorization [66], we can see that the coevolutionary computation approach is a particular class of CEAs, and its generalized framework has been introduced [297]. Here we only focus on and briefly summarize the special considerations for the proposed problem.

7.3.2.1 Procedure

Applying the coevolutionary computation approach to search for the market equilibrium, we typically decompose the whole problem (7.43) into components corresponding, respectively, to each participant's profit maximization problem (7.42), and then assign each component to a population. Except for evaluation for individuals, each population is evolved almost independently of one another based on the canonical and standard genetic algorithm. Because any given individual from a particular population represents only a component of a potential solution to the whole problem, collaborators should be selected from the other populations to represent the remaining components. The individual is combined with its collaborators to form a complete solution, and then the fitness value that is set as the bidding profit of its corresponding participant can be obtained. Figure 7.23 illustrates the individual fitness measure methodology. All populations are evolved in turn through the parent selection, offspring selection, evaluation and survival selection, and a complete cycle of processing all populations is defined as a generation.

How to choose the collaborators is a vital issue for the CEAs—there are so many different options. Here the proposed coevolutionary computation approach chooses the best individual from each population in the previous generation as the collaborator.

The whole procedure of the coevolutionary computation approach can be described as follows.

Step 1. Set the basic parameters of maximum generation number, population size, crossover rate, and mutation rate.

Step 2. Assign each participant to an isolated population; the individual of each population, which represents the mapping population's potential vector of strategic variables, is randomly initialized; and the whole initial collaborators are chosen randomly for each population.

Step 3. The market clearing problem (7.41) is modeled as the key domain to evaluate each individual, which connects the whole coevolved populations. Decode individual E_t^s from population t to strategic vector $(\boldsymbol{k}_{Gt}^s, \boldsymbol{k}_{Dt}^s)$, and then construct a complete bidding vector $(\boldsymbol{k}_{Gt}^s, \boldsymbol{k}_{Dt}^s, \boldsymbol{k}_{-Gt}^c, \boldsymbol{k}_{-Dt}^c)$ by combining with the collaborators E_{-t}^c of the other populations. Finally set the profit of population t as the fitness value of individual E_t^s, which can be calculated according to the market clearing result, as follows:

$$Fit(E_t^s) = R_t(\boldsymbol{k}_{Gt}^s, \boldsymbol{k}_{Dt}^s \mid \boldsymbol{k}_{-Gt}^c, \boldsymbol{k}_{-Dt}^c) \tag{7.44}$$

Step 4. Repeat Step 3 until every individual of population t is evaluated; apply the standard genetic operators to population t; keep the best individual E_t^B in the current pool alive.

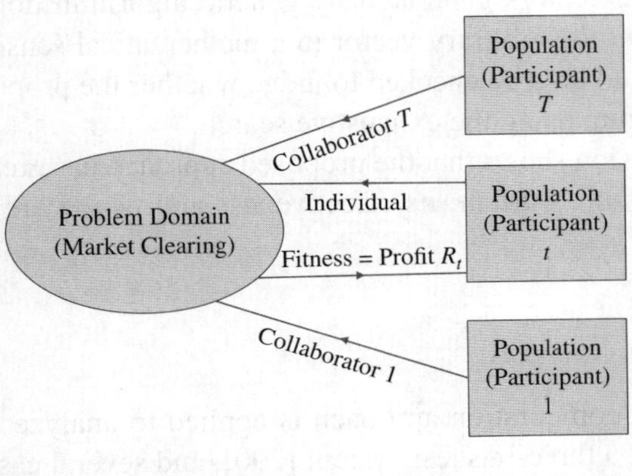

Figure 7.23 Individual fitness evaluation methodology.

Step 5. Repeat Steps 3 and 4 until the whole populations have been evolved. Set the best individual as the collaborators for each population at last.

Step 6. Repeat Steps 3, 4, and 5 until the maximum generation number is reached.

7.3.2.2 Convergence Analysis

If, after a certain number of generations, all collaborators from each population remain constant, it means that the coevolutionary computation approach has converged. In the present work, the market equilibrium is examined mainly through inspecting the convergence point.

From the iterative procedure introduced above, we can see that each population t seeks the best individual using a global optimization algorithm given the opponents' best individuals of the previous generation. Therefore, if the approach converges, the individual E_t^{cp} chosen by each population t is optimal to the individuals E_{-t}^{cp} held by the remaining populations at the convergence point, that is,

$$R_t(\boldsymbol{k}_{Gt}^{cp}, \boldsymbol{k}_{Dt}^{cp} \mid \boldsymbol{k}_{-Gt}^{cp}, \boldsymbol{k}_{-Dt}^{cp}) \geq R_t(\boldsymbol{k}_{Gt}^{ran}, \boldsymbol{k}_{Dt}^{ran} \mid \boldsymbol{k}_{-Gt}^{cp}, \boldsymbol{k}_{-Dt}^{cp}) \tag{7.45}$$

where

$\boldsymbol{k}_{Gt}^{cp}, \boldsymbol{k}_{Dt}^{cp}$ vector of strategic variables for participant t decoding from E_t^{cp};

$\boldsymbol{k}_{-Gt}^{cp}, \boldsymbol{k}_{-Dt}^{cp}$ vector of strategic variables for the opponents of participant t decoding from E_{-t}^{cp};

$\boldsymbol{k}_{Gt}^{ran}, \boldsymbol{k}_{Dt}^{ran}$ vector of strategic variables for participant t decoding from the individual generated randomly.

Comparing the inequality (7.45) with the definition (7.43), it can be seen that the convergence point satisfies the equilibrium condition to a certain extent, considering that the random vectors generated by genetic algorithm do not rigorously and completely represent the arbitrary vector in a mathematical sense. Thereby the convergence point should then be checked to judge whether the proposed approach finds the market equilibrium using the exhaustive search.

The above discussion shows that the proposed approach has great potential to solve the market equilibrium. Further, its effectiveness and power are investigated in the next section.

7.3.3 Example

The coevolutionary computation approach is applied to analyze a two-bus example system [26, 27] and a three-bus test system [290], and several cases are presented to illustrate the performance of the proposed method.

7.3.3.1 Parameter Settings

The standard genetic algorithm (SGA) is used for each population in the coevolutionary computation approach: a 16-bit binary code is used to encode each strategic variable; the one-point crossover and bit-flip mutation genetic operators are adopted; the elitist strategy is employed in the survival selection; the population size is 50; and the maximal generation number is set to 100. Our previous study on the parameter settings [281] showed that a reasonable value of the mutation probability is between 0.05 and 0.1 and the crossover probability is between 0.5 and 0.9. So here the mutation probability and the crossover probability are set to 0.05 and 0.9, respectively.

In the formulated market model, the lower limits of strategic variables $k_{\min Gti}$ and $k_{\min Dtj}$ for each participant are all assumed to be 0.2; on the other hand, the upper limits are set to 10.0. The optimization problem (7.41) faced by the ISO is solved using the primal–dual interior point algorithm [189], in which the accepted compensation gap tolerance is set to 1.0×10^{-3}.

7.3.3.2 Two-bus Example System

The two-bus example system with two GenCos and one consumer is shown in Figure 7.24, and the system data can be found elsewhere [26, 27]; the maximal transmission capacity of the line is expressed as the variable $L_{\max}$.

The true cost functions of the GenCos G1 and G2, and the true benefit function of consumer C1, respectively, are presented as follows:

$$\begin{aligned} C_1(P_{G1}) &= 0.01P_{G1}^2 + 10P_{G1} \\ C_2(P_{G2}) &= 0.01P_{G2}^2 + 10P_{G2} \\ B_1(P_{C1}) &= -0.04P_{C1}^2 + 30P_{C1} \end{aligned} \tag{7.46}$$

To investigate the performance of the coevolutionary computation approach by comparison with the individual welfare maximization (IWM) algorithm proposed in Weber and Overbye [26, 27], three cases are carried out: Case A which considers two GenCos competition with and without transmission constraints; Case B in which GenCo G1 competes with consumer; and Case C concerning all participants competition with transmission constraint.

In Case A, only the strategic behaviors of the GenCos are considered. Therefore, it assumes the consumer always bids its true benefit function, that is, its strategic

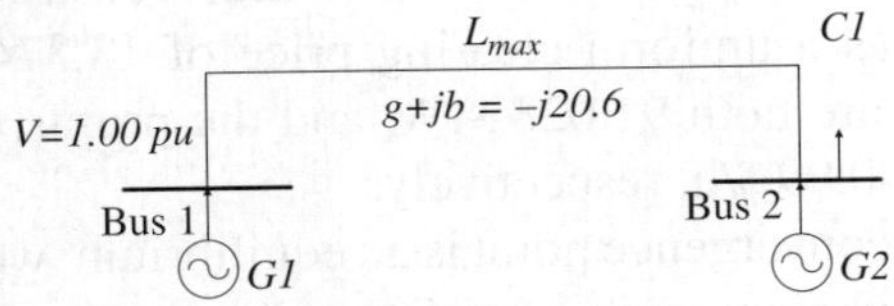

Figure 7.24 Two-bus example system.

variable is set to 1.000. Without transmission constraint, the approach converges to the point (1.150, 1.150, 1.000) successively constructed by the strategic variables corresponding to the collaborators of G1, G2 and C1. Under the convergence situation, the market has a uniform clearing price13.828 $/MWh; the awarded generations of GenCos are both 101.073 MW. The results converged by the coevolutionary computation approach are consistent with the market equilibrium solved by the IWM algorithm.

When adding in an 80 MVA transmission constraint to the system, the proposed approach does not converge here: all the collaborators have been changing in the entire simulation process. The IWM algorithm also did not result in a market equilibrium point, because the bid of GenCos swung back and forth during the iterations [26]. The optimal response curves for the two GenCos described therein [26] show that no pure strategy Nash equilibrium existed—only a mixed strategy Nash equilibrium existed, which was caused by the nonconvexity of GenCo G2's profit function with two local maxima.

Case B considers only GenCo G1 versus consumer C1 competition, and assumes that GenCo G2 is removed. In the situation without the transmission line constraint, the proposed approach reaches the stable point (1.571, 0.857) successively constructed by the best strategic variables of G1 and C1. The market solution for these bids is both nodes price of 18.858 $/MWh, while 100.00 MW are exchanged between them. Compared with the outcome calculated by the IWM algorithm [27], it shows that the proposed approach finds the market equilibrium successfully.

When an 80 MVA maximal transmission capacity is added, the proposed approach fails to converge, and no equilibrium was reached using the IWM algorithm in the same situation [27]. Through drawing the optimal response curves over all possible bids by each participant, we find that transmission constraint results in a continuum of Nash equilibrium in the market.

For the two-bus example system, Weber and Overbye [26, 27] only analyzed Case A and Case B, which is a typical two-player game. It is very instructive to look at the extended case in which all participants are strategic players. Thereby the coevolutionary computation approach is further employed to analyze Case C, considering all participants competition with a 200 MVA transmission line constraint, which forms a multi-player game.

The evolution of the strategic variables corresponding to the representatives is illustrated in Figure 7.25. It shows that the fluctuation in the strategic variables stops after the 24th generation during the whole evolutionary process, that is, the approach converges to the stationary point (1.131, 1.131, 0.866) successively formed by the optimal strategic variables of G1, G2, and C1. For the bids constructed by these strategic variables, the market has a uniform clearing price of 13.374 $/MWh, the awarded generations of GenCos are both 91.028 MW, and the profits of G1, G2, and C1 are 224.31, 224.31, and 1700.99 $/h, respectively.

To verify whether the convergence point is an equilibrium, we should use an exhaustive search to confirm whether each participant obtains the maximal profit with its convergence bids, assuming that its opponents chose their convergence bids all the

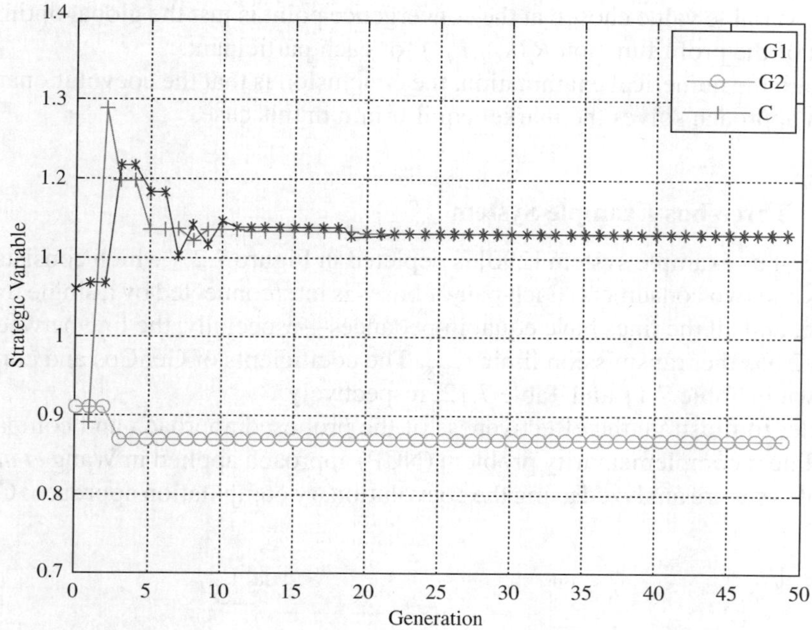

Figure 7.25 Evolution of strategic variable corresponding to the representatives during the evolutionary process in case C.

time. The exhaustive search employed in Case C and subsequently uses a uniform discretization of continuous strategic variable with the equal intervals of 0.001.

Figure 7.26 illustrates each participant's strategic variable and its corresponding expected profit. It clearly shows that the profit function $R_t(\boldsymbol{k}_{Gt}, \boldsymbol{k}_{Dt})$ of each participant with respect to its strategic variable is convex under the assumption. Further, the

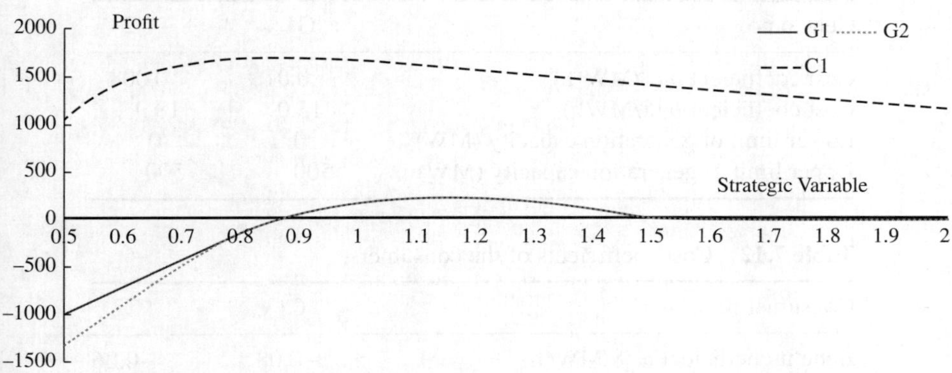

Figure 7.26 Variation of each participant's expected profit with respect to its strategic variable, assuming the opponents hold the convergence bids in case C.

strategic variable value chosen at the convergence point is just the global optimization solution of the profit function $R_t(\mathbf{k}_{Gt}, \mathbf{k}_{Dt})$ for each participant.

Through the numerical examination, the conclusion is that the coevolutionary computation approach solves the market equilibrium in this case.

7.3.3.3 Three-bus Example System

The three-bus example system [290] is depicted in Figure 7.27, which consists of two GenCos and two consumers. Each pair of buses is interconnected by a single transmission line, and all the lines have equal impedances—especially, the line between bus 1 and bus 2 has the transmission limit $T_{\max}$. The coefficients of GenCos and consumers are shown in Table 7.11 and Table 7.12, respectively.

In order to illustrate the effectiveness of the proposed approach and compare with the nonlinear complementarity problem (NCP) approach applied in Wang *et al.* [290], different cases are analyzed using the coevolutionary computation approach: Cases D_1

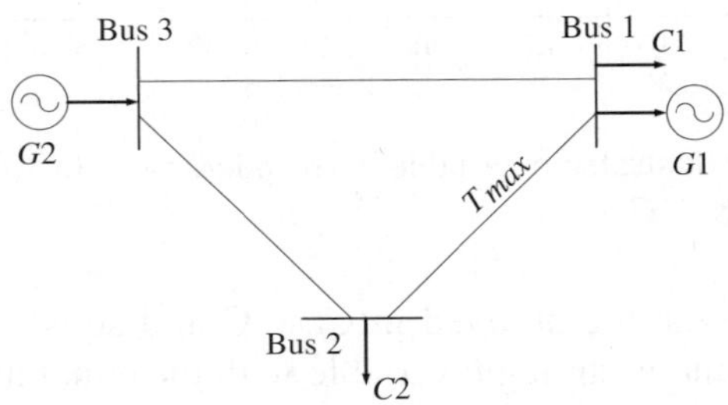

Figure 7.27 The three-bus example system.

Table 7.11 Cost coefficients of the GenCos.

GenCo no.	G1	G2
Cost coefficient a (\$/MW^{2}h)	0.01	0.008
Cost coefficient b (\$/MWh)	15.0	18.0
Lower limit of generation capacity (MW)	0	0
Upper limit of generation capacity (MW)	500	500

Table 7.12 Cost coefficients of the consumers.

Consumer no.	C1	C2
Benefit coefficient α (\$/MW^{2}h)	−0.08	−0.06
Benefit coefficient β (\$/MWh)	40.0	40.0
Lower limit of quantity demanded (MW)	0	0

and E_1 pay attention to the competition between GenCos with different transmission constraints; further, Cases F_1 and G_1 concern the consumers as well as the GenCos taking part in the competition; finally, the remaining cases mainly consider the situations with different generation capacity constraints.

In Case D_1, a transmission constraint ($T_{max} = 500$ MW) is enforced to the system; only the strategic behaviors of the two GenCos are observed and the consumers are assumed to bid their true benefit function. The proposed approach converges to the point (1.134, 1.082, 1.0, 1.0) formed by the strategic variables of G1, G2, C1, and C2, respectively. The market solution for these bids is presented in Table 7.13, where "Flow" refers to the power flow on the constrained line and "Power" means, respectively, the generation output and the quantities demanded for GenCos and consumers. This result is consistent with the market equilibrium solved by the NCP approach [290].

Table 7.13 Simulation results for three-bus example system.

Case	Participant	Strategic variable	LMP ($/MWh)	Flow (MW)	Power (MW)	Profit ($/h)
	G1	1.134	21.118	146.83	361.837	1559.4
D_1	G2	1.082	21.118		188.863	446.4
	C1	1.0	21.118		236.015	2228.1
	C2	1.0	21.118		314.685	2970.8
	G1	1.336	21.695	25.00	122.964	747.7
E_1	G2	1.252	25.422		286.687	1799.1
	C1	1.0	21.695		228.808	2094.1
	C2	1.0	29.149		180.843	981.1
	G1	1.314	21.452	25.00	132.387	766.6
F_1	G2	1.158	22.849		216.275	861.6
	C1	0.903	21.452		203.025	2116.8
	C2	0.775	24.245		145.637	1658.1
	G1	1.128	20.855	139.47	348.572	1433.4
G_1	G2	1.075	20.855		174.476	376.4
	C1	0.953	20.855		226.602	2284.3
	C2	0.938	20.855		296.446	3039.0
	1	1.3286	—	30.61	11.04	144.46
A_{24}	2	1.2375	—		9.82	130.05
	3	1.1921	—		8.84	121.68
	1	—	—	39.34	9.11	207.62
A_{25}	2	—	—		8.38	193.24
	3	—	—		7.83	184.15
	1	—	5.48	32.82	10.59	164.45
A_{31}	2	—	5.07		9.42	148.41
	3	—	4.75		8.58	139.33

Table 7.13 shows that no transmission congestion exists under the market equilibrium of Case D_1, and therefore it is meaningful to investigate Case E_1 in which the system is subjected to a tighter transmission constraint (T_{max} = 25 MW).

In Case E_1, the stationary point (1.336, 1.252, 1.0, 1.0) found by the proposed approach tallies well with the solution calculated by the NCP approach [290]. According to the bids based on the strategic variables of the stationary point, the results of the market clearing are presented in Table 7.13, which shows that the constrained line reaches its maximal transmission capacity and the LMPs on nodes are varied—thus transmission congestion does exist in this case. Comparing Cases D_1 and E_1 illustrates that transmission congestion does not make the pure strategy Nash equilibrium disappear, it just leads to the market owning other equilibria.

Assuming that consumers are nonstrategic participants, Cases D_1 and E_1 are both two-person games. Here the extended cases with the addition of the consumers' competition, which form the four-player game, are investigated using the coevolutionary computation approach.

For Case F_1, in which a tight transmission constraint (T_{max} = 25 MW) is enforced onto the system, the process of evolution for the representatives presented in Figure 7.28 shows that the proposed approach converges to the point (1.314, 1.158, 0.903, 0.775) after the 29th generation. The numerical technique introduced in Case C is used to verify whether the convergence point is an equilibrium. The simulation

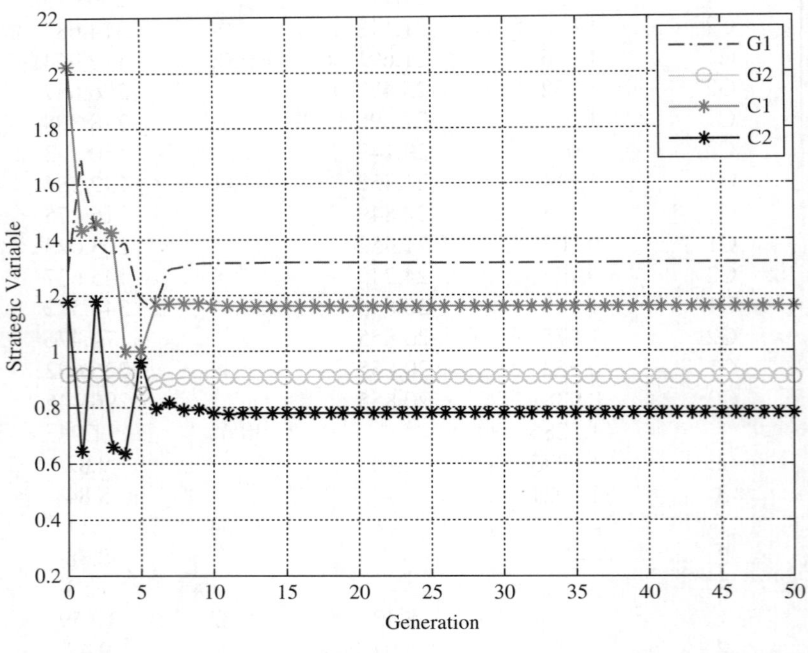

Figure 7.28 Evolution of strategic variable corresponding to the representatives during the evolutionary process in Case F_1.

result presented in Figure 7.29 shows that the strategy chosen at the convergence point is the best for each participant, which clearly means the proposed approach found the market equilibrium in this situation.

It is interesting to observe that the profit function of some participants is cut and thereby is not convex in the global field. To understand this phenomenon, the market state variation with respect to the strategic variable is observed. Take GenCo G2, for example, for which results are presented in Figure 7.30, where "G1" and "G2" mean

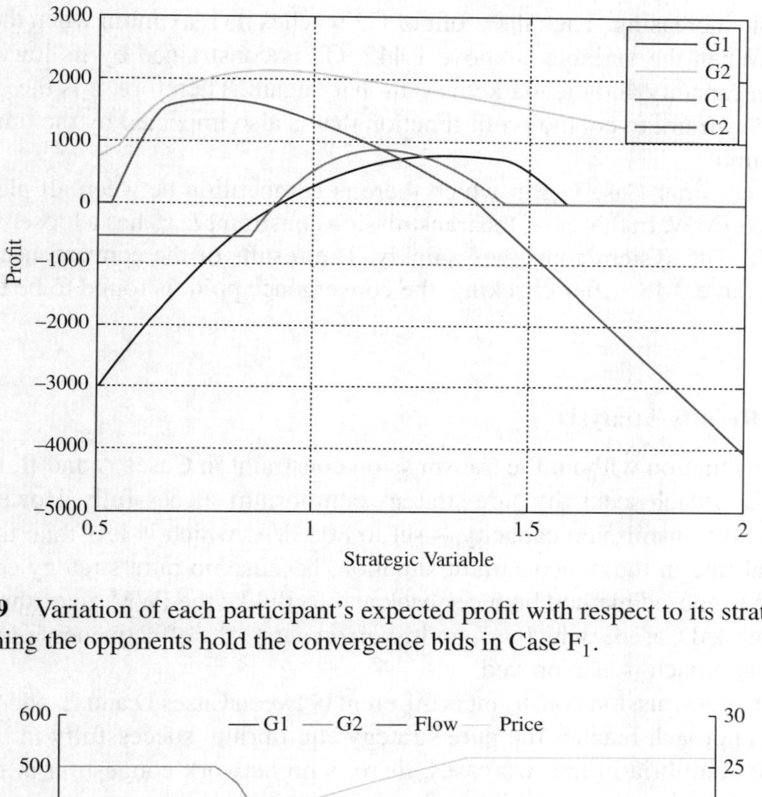

Figure 7.29 Variation of each participant's expected profit with respect to its strategic variable, assuming the opponents hold the convergence bids in Case F_1.

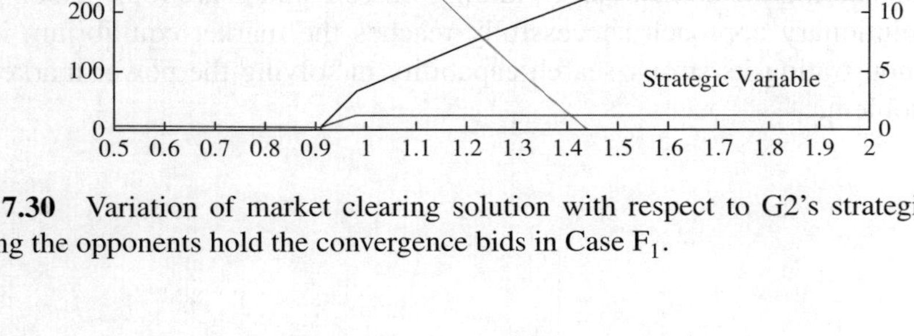

Figure 7.30 Variation of market clearing solution with respect to G2's strategic variable assuming the opponents hold the convergence bids in Case F_1.

the generation outputs of G1 and G2, respectively, “flow” refers to the power flow on the constrained line, and “price” is the LMP of G2. Figure 7.30 shows that, when G2’s strategic variable is below 0.857, G1 works at its minimum generation capacity, and G2 reaches the ceiling, the LMP of G2 and the flow stay the same, and no congestion exists. When the strategic variable is above 0.858 and up to 1.442, with the variable increasing, the generation output of G1 and the flow both increase rapidly after a short adjustment. Then the flow hits its upper limit bound and the increasing speed of G1’s generation output is slow when the variable is greater than 0.980. On the other hand, the generation output of G1 continues to decline and its price rises all the time with the variable increasing. Then the profit of G2 reaches its maximum when the variable is 1.158. When the variable is above 1.442, G2 is constrained by its lower limit of generation capacity and the market is stationary again. Therefore, it is the generation capacity constraint to cut the profit function that is also impacted by the transmission capacity limit.

Finally consider Case G_1, in which there is competition between all players with $L_{max} = 1000$ MW. In this case, the transmission constraint L_{max} has a looser value than in Case F_1. The system converges quickly. The results of the convergence point are shown in Table 7.13. After checking, the convergence point is found to be the market equilibrium.

7.3.3.4 Results Analysis

Under the situation without the transmission constraint in Cases A and B, the coevolutionary approach seeks the pure strategy equilibrium successfully. However, when the maximal transmission capacity is set to 80 MVA, which is less than the flow on the special line in the nonconstraint situation, because no pure strategy equilibrium exists, the proposed method bounces back and forth like the IWM algorithm [26, 27]. In the extended Case E, which is a multi-person game, the effectiveness of the coevolutionary approach is also proved.

Only the transmission constraint is different between Cases D and E, and the coevolutionary approach reaches the pure strategy equilibrium successfully in both cases. Under the equilibria in the two cases, there is no network congestion in the former case, and there is congestion in the latter. Therefore, we cannot conclude that the pure strategy equilibrium exists simply related to whether or not the market network constraint exists.

After considering the demand-side bidding, Cases F and G are four-person games. The coevolutionary approach successfully reaches the market equilibrium in both cases, demonstrating its strong search capability in solving the power market equilibrium problem.

7.3.4 Summary

The linear supply function equilibrium (LSFE) model considering network constraints and demand-side bidding for the oligopolistic electricity market is analyzed in this section, and the coevolutionary approach is proposed to solve the market equilibrium. The examples show that the coevolutionary approach is quite effective for finding the pure strategy equilibrium, as long as it exists. It also underscored the fact that Nash equilibria do not always exist, and that when they do exist they may not be unique. Owing to the complicated calculations, much of the literature only focuses on the two-person game, so the outcomes of the multi-person games in this section show the powerful search ability and potential of the coevolutionary approach.

The approach presented in this section can be used further to the supply function equilibrium model with two decision variables or other types of electricity market models such as the Cournot model. The approach has the potential to be used to solve equilibrium problems in the real electricity market.

7.4 Analyzing Two-Settlement Electricity Market Equilibrium by Coevolutionary Computation Approach

Power industry deregulation has taken place in many countries in the past 20 years. In this new environment, electricity is traded in the same way as other commodities. However, a perfectly competitive market does not exist in practice, and many issues could lead to market inefficiency, such as market design flaws, market power of participants, and inherent engineering features of power system operations [301]. The exercise of market power may be facilitated by some characteristics of electricity markets, including inelastic demand, limited transmission capacity, and the requirements that supply and demand of the power systems must balance continuously. Therefore, market power analysis has received attention in both theory and practice [302]. Lots of work has been published on market power analysis based on different economic models. A survey of electricity market modeling, especially the equilibrium models, has been presented [303]. Noncooperative game theoretical approaches, such as the Cournot and supply function equilibrium (SFE) models [282, 304], are also widely used for power market simulation.

Owing to the California electricity crisis, more and more researchers are convinced that the forward contract plays an important role for market power mitigation [302] or risk management [292]. Generally, there are at least three main reasons for market players to enter a forward market: (1) being compelled by regulators as part of a transitional vesting process, which is the situation of electricity markets in many countries, such as England and Wales, Australia, and New Zealand; (2) recognizing the importance and necessity of risk management with contractual arrangements [292]; and (3) being encouraged economically [304, 305].

The electricity spot markets modeled by SFE and Cournot models have been extended to include contract markets [306, 307]. The English electricity market is modeled by the SFE model with a contract market, and the entry condition of the contract market has been discussed [306]. Green [282] has shown that competition in a contract market could lead the generators to sell contracts and increase their outputs and also hedge the spot market price in England and Wales. Niu *et al.* [308] propose an asymmetric linear supply function equilibrium (LSFE) model to develop firms' optimal bidding strategies given their forward contracts. Market power mitigation effects of forward contracts have also been evaluated.

To study the interactions between the spot and forward markets, it is commonly assumed that the GenCos are risk-neutral and the markets are perfect arbitrage markets [307, 309]. Chung *et al.* [309] examine the issue of whether generators would enter contract markets through an economic incentive. The contract market is modeled with the general conjectural variation method. The two-settlement electricity market with transmission line constraints has been studied and compared with a single settlement market [310]. The dynamics in two-settlement electricity markets are studied by an agent-based model. Numerical simulations illustrate that the access to the forward market leads to more competitive behaviors of the suppliers in the spot market, and thus lower spot energy prices [311]. In Yao *et al.* [307, 312], determination of the equilibrium in two-settlement electricity markets is formulated as an equilibrium problem with equilibrium constraints (EPEC). It is shown that spot market prices will decrease when the supplies enter forward contracts.

The main objectives of this section are to investigate the issue of whether generators would voluntarily enter the contract market solely through the economic incentive, and to examine the factors that could affect the strategic contracting behavior. If GenCos are risk-averse, they may prefer to enter the forward market for risk hedging, and their bidding behavior is affected by risk hedging incentive. The bidding behavior of risk-neutral GenCos is between risk-averse and risk-seeking, and is not affected by this incentive. The risk-neutrality assumption reflects the intention to analyze GenCos' strategic contracting behaviors as induced by economic incentive alone. Therefore, the generally used assumptions—all GenCos are risk-neutral, and sufficient risk-neutral speculators or arbitrageurs exist in the markets [309, 312]—are also used in this section.

Therefore, market equilibrium models are required to simulate the competition of generators in spot and forward markets. However, it is generally difficult to determine the market equilibrium with the conventional differential equation approaches, especially when generation capacity constraints or the LSFE model with multiple parameters are considered. Even worse, market equilibrium may not be unique. Therefore, an effective and efficient method needs to be established to solve the complex market equilibrium problems. From traditional evolutionary algorithms (EAs), coevolutionary computation has been developed and has recently been successfully used to determine the equilibrium of noncooperative games [313] and to simulate the market [272, 294]. These previous works have shown that the nonlinear market models, which conventional methods have difficulties with, can be effectively studied using coevolutionary computation. Thus, this section will also investigate the effectiveness of coevolutionary computation in the study of the bidding behavior of GenCos in a two-settlement market in addition to the main objectives of this section mentioned above.

7.4.1 *Market Model Formulation*

This section uses the two-stage game model [304] to formulate the two-settlement market consisting of a forward market and a spot market. In the first stage, GenCos enter the forward market, forming rational expectations regarding the forward contracts of their rivals and the spot equilibrium outcomes. In the second stage, the GenCos compete in the spot market, taking into account all GenCos' forward contracts. The Cournot or LSFE models are used to represent competitive bidding in the spot market, while the forward market is represented by the Cournot model.

7.4.1.1 Market Assumptions

Suppose there are GenCos (suppliers) in the electricity market, and these GenCos are risk-neutral. Each GenCo has a generator and is characterized by the following quadratic cost function:

$$C_i = \frac{1}{2}a_i q_i^2 + b_i q_i, \ \ i = 1, \ldots, I$$
$$a_i > 0 \tag{7.47}$$

where q_i is the quantity generated by generator i; and a_i and b_i are the coefficients of generator i's cost function. The marginal cost function of generator i is affine:

$$MC_i = \frac{\mathrm{d}C_i}{\mathrm{d}q_i} = a_i q_i + b_i, \ \ i = 1, \ldots, I \tag{7.48}$$

When there is negligible transmission loss, the aggregate demand q_S, which is assumed to be an inverse linear demand function, will be equal to the total output of all GenCos as shown in the following:

$$q_S = \overline{q_S} - rp = \sum_{i=1}^{I} q_i, \ \ i = 1, \ldots, I \tag{7.49}$$

where p is the spot market price; and r and $\overline{q_S}$ are coefficients of the demand function. Here, r is the slope of the system demand function.

The GenCos compete with each other in the forward market by choosing the quantity of their contracts q_i^c $\ (q_i^c \geq 0)$ that they are willing to sell at the forward market price p_i^c. It is assumed that the forward contracts are observable for all the GenCos in the spot market. Also the GenCos are assumed to use their bidding strategies, such as quantities of generation in the Cournot model and coefficients of supply functions in the LSFE model, to compete in the spot market.

7.4.1.2 Equilibrium Model for Strategic Bidding in the Spot Market

The GenCos bid in the spot market by using the information observed in the forward market. The LSFE model and the less competitive market model, Cournot model, are employed to model the spot market in this section. Given the decisions of GenCos in the forward market, the optimization problem faced by each GenCo is how to maximize its expected total profit (π_i), expressed as follows:

$$\pi_i = p(q_i - q_i^c) + p_i^c q_i^c - C_i. \tag{7.50}$$

(1) Cournot Model. In the Cournot model, each GenCo will maximize its own profit in (7.50) by changing its decision variable q_i. The derivative of GenCo i's profit with respect to its decision variable can be written as

$$\frac{\partial \pi_i}{\partial q_i} = (q_i - q_i^c)\frac{\partial p}{\partial q_i} + p - MC_i. \tag{7.51}$$

By setting the derivative of GenCo i's profit in (7.51) as zero and eliminating p using (7.49), the optimal value of q_i can be obtained by

$$q_i = \frac{\overline{q_S} - \sum_{i=1}^{I} q_i + q_i^c - b_i r}{1 + a_i r} = \frac{rp + q_i^c - b_i r}{1 + a_i r}. \tag{7.52}$$

Then, by eliminating q_i in (7.52) using (7.49), the spot market price can be determined by

$$p = \frac{1}{r + \sum_{i=1}^{I}[r/(1 + a_i r)]}\left[\overline{q_S} - \sum_{i=1}^{I} \frac{1}{1 + a_i r} q_i^c + r \sum_{i=1}^{I} \frac{b_i}{1 + a_i r}\right] \tag{7.53}$$

(2) LSFE Model. The GenCos are assumed to compete in the spot market by submitting their bids in the form of linear supply functions (LSFs). The supply function takes the form

$$q_i = (p - \beta_i)/\alpha_i, i = 1, \ldots, I$$
$$\alpha_i > 0 \tag{7.54}$$

where α_i and β_i are the coefficients of the LSF.

LSFs with different parameterizations [255] are studied in this section, as follows.

1. a parameterization, where GenCo i can choose α_i in (7.54) arbitrarily but is required to specify a fixed and pre-chosen value of β_i. In this section $\beta_i = b_i$ is selected.
2. b parameterization, where GenCo i can choose β_i in (7.54) arbitrarily but is required to specify a fixed and pre-chosen value of α_i. In this section $\alpha_i = a_i$ is selected.
3. $(a \propto b)$ parameterization, where GenCo i can choose α_i and β_i subject to the condition that α_i and β_i have a fixed linear relationship. The coefficient k_i is used as the bidding variable. The supply function is defined as

$$\begin{aligned} q_i &= (p/k_i - b_i)/a_i, \quad i = 1, \ldots, I \\ k_i &> 0 \end{aligned} \tag{7.55}$$

4. (a, b) parameterization, where GenCo i can choose α_i and β_i arbitrarily.

In the LSFE model, each GenCo will maximize its own profit expressed in (7.50) by changing its decision variables α_i, β_i, or k_i according to its parameterization type. By differentiating (7.50) with respect to p and using (7.49) and (7.54), the derivative of GenCo i's profit with respect to p can be written as

$$\begin{aligned} \frac{\partial \pi_i}{\partial p} &= (q_i - q_i^c) + p\frac{\partial q_i}{\partial p} - MC_i\frac{\partial q_i}{\partial p} \\ &= (q_i - q_i^c) + (p - MC_i)\frac{\partial\left(\overline{q_S} - rp - \sum_{j=1, j\neq i}^{I}[(p - \beta_j)/\alpha_j]\right)}{\partial p} \end{aligned} \tag{7.56}$$

By setting the derivative of GenCo i's profit in (7.56) to zero, the optimal value of q_i can be obtained from

$$q_i = q_i^c + (p - MC_i)\left(r + \sum_{j=1, j\neq i}^{I} \frac{1}{\alpha_j}\right). \tag{7.57}$$

7.4.1.3 Equilibrium Model for Strategic Contracting in the Forward Market

The equilibrium for the forward market can be determined by maximizing the GenCos' profits with nesting the equilibrium conditions from the spot market in the calculation. In the forward market, all GenCos are supposed to be able to offer forward contracts. It is also assumed that there are enough risk-neutral arbitrageurs in the markets and they will eliminate any profitable arbitrage opportunity arising from the difference between the forward prices and the expected spot prices. Then p_i^c is an unbiased estimator of p [304].

Each GenCo chooses its forward market output so as to maximize its profit, which can be formulated as follows:

$$\begin{aligned} \max\ \pi_i &= p(q_i - q_i^c) + p_i^c q_i^c + C_i \\ \text{s.t.}\quad p_i^c &= p, i = 1, \ldots, I \end{aligned} \tag{7.58}$$

Since the spot price p is an implicit function of the forward market output q_i^c, GenCo i's marginal benefit for forward contract is equal to the derivative of its profit with respect to q_i^c as follows:

$$\frac{\partial \pi_i}{\partial q_i^c} = \frac{\partial p}{\partial q_i^c} q_i + (p - MC_i) \frac{\partial q_i}{\partial q_i^c}. \tag{7.59}$$

(1) Modeling Spot Market by Cournot Model. Substituting (7.52) into (7.59) yields

$$\frac{\partial \pi_i}{\partial q_i^c} = \frac{\partial p}{\partial q_i^c} q_i + (p - MC_i) \left(\frac{r}{1 + a_i r} \frac{\partial p}{\partial q_i^c} + \frac{1}{1 + a_i r} \right). \tag{7.60}$$

Then, setting the marginal benefit in (7.60) to zero and using (7.53), GenCo i's output in the spot market can be obtained as

$$q_i = (p - MC_i) \left(r + \sum_{j=1, j \neq i}^{I} \frac{r}{1 + a_j r} \right). \tag{7.61}$$

By eliminating p in (7.61) using (7.52), GenCo i's output in the forward market can be written as

$$q_i^c = \frac{\sum_{j=1, j \neq i}^{I} [1/(1 + a_j r)]}{1 + \sum_{j=1, j \neq i}^{I} [1/(1 + a_j r)]} q_i. \tag{7.62}$$

The fraction of power sold at the forward market in (7.62) and spot market in (7.61) for each GenCo can be determined from

$$\frac{q_i^c}{q_i} = \frac{\sum_{j=1, j \neq i}^{I} [1/(1 + a_j r)]}{1 + \sum_{j=1, j \neq i}^{I} [1/(1 + a_j r)]} = 1 - \frac{1}{1 + \sum_{j=1, j \neq i}^{I} [1/(1 + a_j r)]} \tag{7.63}$$

When r is increasing, the demand becomes more elastic. The GenCos will have less incentive to raise the market price in the spot market. They will bid aggressively (low price for large quantities) in the spot market and have less incentive to enter the forward market. Then the fraction in (7.63) will become smaller.

It can also be seen that the fraction is affected by the slopes of the GenCos' marginal cost functions. If a_i is larger than $a_{i'}$ (for any $i, i' \in 1, \ldots, I$), then $\sum_{j=1, j \neq i}^{I} [1/(1 + a_j r)]$ is larger than $\sum_{j=1, j \neq i'}^{I} [1/(1 + a_j r)]$. According to (7.63), $(q_i^c/q_i) > (q_{i'}^c/q_{i'})$ can be obtained. This means that a GenCo with a large slope of marginal cost function will have more incentive to enter the forward market.

(2) Modeling Spot Market by LSFE Model. According to the equilibrium in section 7.4.1.2, it can be proved that, when GenCos decide their bidding variables using a, b, and ($a \propto b$) parameterization of LSF in the spot market, the GenCos have no incentive to enter the forward market because their marginal benefits of forward contracts are not larger than zero. They bid aggressively and the spot market is more competitive. However, GenCos may have incentive to enter the forward market if they decide their supply functions using (a, b) parameterization of LSF in the spot market.

Determination of the market equilibrium in the two-stage game is a difficult task, especially for the cases of a spot market under LSFE type of competition with (a, b) parameterization or GenCos with tight capacity constraints. Therefore, a coevolutionary computation approach is introduced in the next section for subsequent investigation of its effectiveness and power in market equilibrium determination.

7.4.2 *Coevolutionary Approach to Analyzing Market Model*

Coevolutionary computation is a relatively new area in evolutionary computation research. It is a special form of the agent-based simulation approach, developed from traditional EAs, which simulates the coevolutionary mechanism in nature and adopts the notion of an ecosystem. The agents simulate the species in an ecosystem. Multiple species coevolve and interact with each other and result in continuous evolution of the ecosystem. The species are genetically isolated—individuals only mate with other members of their species. They interact with one another within a shared domain model. The general framework of coevolutionary computation is introduced in Wiegand [271]. In the present work, the effectiveness of applying coevolutionary computation in solving the two-stage game model formulated in section 7.1.1 and in determining market equilibrium in the two-settlement electricity market is investigated. The two-level CGA is described below.

7.4.2.1 Two-level CGA

Since a GenCo needs to determine its strategies in the forward market and the spot market, a two-level CGA with real-coded GAs embedded is employed in this section. The optimal strategies in the spot market and forward market are determined in the inner loop and outer loop, respectively. The inner loop and the outer loop are described, respectively, in the procedure in section 7.4.2.1 below. A GenCo (agent) in both markets is represented by a species in the ecosystem (i.e. $2 \times I$ species totally are needed). The individuals of each species, which is in fact a population of the real-coded GA, will represent the strategies of the GenCo in the forward market or spot market.

In the spot market, all GenCos optimize their expected profits in (7.50) by their spot market strategies (quantity for Cournot model or supply function coefficients for LSFE model), by fixing all the forward market strategies. In the inner loop, the real-coded GA will determine the best individual in the population and select it as the representative of this population (best strategy of the GenCo). The market clearing model is used as the shared domain and collaborations are formed with representatives from each of the other populations to determine the market spot price. When one GenCo changes its strategies to gain more profit, it will change the market price according to the demand function in (7.49) and, in turn, changes the profits of others. Other GenCos will behave in the same way. The adjustment process will continue until no-one can get more profit by changing its strategies without changing the strategies of other GenCos; in other words, the spot market reaches equilibrium.

When the strategies or information of the spot market are obtained, the forward market equilibrium will be determined similarly in the outer loop. But it is difficult to get information about the spot market when GenCos select their strategies in the forward market. The backward induction method [314] is therefore merged into the CGA. Backward induction is the process of reasoning backwards in time, from the end of a problem or situation, to determine an optimal course of action. It proceeds by anticipating the decisions that might be made in the spot market by GenCos fixing forward contracts. Then using the information obtained in the spot market, GenCos can optimize their strategies in the forward market. Thus, to solve the equilibrium of the forward market, the equilibrium of the spot market (i.e. the subgame perfect equilibrium [314]) should be determined first.

The real-coded GA used is similar to the conventional GA. The crossover and mutation operators in the real-coded GA used are described as follows [315].

Crossover. A decision variable x_i is represented by a real number within its lower limit Lb and its upper limit Ub, i.e. $x_i \in [Lb, Ub]$. A blend crossover operator has been employed in this study. This operator generates the offspring randomly from the interval $[x_i - \tau(y_i - x_i), y_i + \tau(y_i - x_i)]$, where x_i and y_i are the decision values of the parent solutions, with $x_i < y_i$, and τ is a random number between 0 and 1. To ensure the balance between exploitation and exploration of the search space, $\tau = 0.5$ is selected.

Mutation. A dynamical nonuniform mutation operator to reduce the disadvantage of random mutation in the real-coded GA is used and defined as follows:

$$x'_i = \begin{cases} x_i + \Delta(t, Ub - x_i) & \text{if random } \tau \text{ is } 1 \\ x_i - \Delta(t, Ub - x_i) & \text{if random } \tau \text{ is } 0 \end{cases} \tag{7.64}$$

where x_i and x'_i are the selected and resultant values, respectively, for the mutation; and t is the generation number of CGA. The function $\Delta(t, y)$ returns a value in the range $[0, y]$ such that $\Delta(t, y)$ approaches zero as the generation number increases. This property causes this operator to search the space uniformly in the initial stages (i.e. when t is small), and locally at later stages. This strategy increases the probability of generating a new number close to its successor rather than a random choice. The function is defined as follows:

$$\Delta(t, y) = y(1 - \sigma^{(1-t/T)^{\delta}}) \tag{7.65}$$

where σ is a uniform random number from $[0, 1]$, T is the maximum generation number, and δ is a system parameter determining the degree of dependence on the generation number. In this study, δ is set as a constant and equal to 1.

7.4.2.2 Procedure

The procedures can be summarized as follows.

Step 1. Set the basic parameters of CGA, including crossover rate, mutation rate, maximum generation number, and population size in the outer loop and inner loop, respectively.

Step 2. The individual of each population, which includes the bidding variable of the GenCo in the outer and inner loops, is initialized. All the initial representatives of each population are set randomly.

Step 3. The real-coded GA is applied to solve the optimization problem of each GenCo for the forward market in (7.58). For each individual of the population in the outer loop, a fitness value is assigned according to its profit in (7.58). To get the fitness value, the forward market price, which is equal to the spot market price, is needed, so the equilibrium of the spot market and the corresponding price are determined first.

1. The representatives in the outer loop (i.e. forward contracts of GenCos)are fixed and sent to the inner loop.
2. The profit of each GenCo for the spot market in (7.50) is maximized by the real-coded GA with consideration of its generation capacity constraint and system energy balance in (7.49). The corresponding bidding strategy (i.e. the best individual) is then determined. In the optimization, each individual of a population in the inner loop (i.e. a strategy of a GenCo in the spot market) is evaluated and assigned a fitness value according to the corresponding profit. The market price and total output of the GenCo in (7.50) can be obtained by (7.49) for the Cournot model or (7.49) and (7.54) for the LSFE model based on representatives of other populations in the inner loop (i.e. strategies of other GenCos in the spot market). The best individual of each population is the one with the highest fitness value.
3. The best individual of each population is selected as the representative in the inner loop and sent to other GenCos.
4. Repeat steps 3.2 and 3.3 using the updated representatives and keeping the best strategy of each GenCo in the population (i.e. elitism) until the maximum generation number in the inner loop is reached.

Step 4. The best individual of each population is selected as the representative of the population in the outer loop and sent to other GenCos.

Step 5. Repeat Steps 3 and 4 using the updated representatives and keeping the best strategy of each GenCo in the population (i.e. elitism) until the maximum generation number is reached in the outer loop.

7.4.3 Example

In this section, a three-GenCo test system [309] and a five-GenCo test system [255] are used to validate the models and theoretical analyses in section 7.1.1 and the effectiveness of the CGA in determination of market equilibrium described in section 7.2.2.

From our previous study on the parameter settings [281], the reasonable value of the mutation probability is between 0.05 and 0.1 and the crossover probability is between

0.5 and 0.9. Here, the mutation probability is set to 0.05 and the crossover probability to 0.9. In this study, it is found that a population size of 40 is sufficient for these two test systems. A considerably large number of generations are used, 50 and 200 in the inner and outer loops, respectively, to ensure that the solutions in both the inner loop (the spot market equilibrium) and the outer loop (the forward market equilibrium) converge in different simulation cases. The commonly used tournament selection size and elitism size are used [281]. All the CGA parameters are summarized in Table 7.14.

7.4.3.1 Three-GenCos Study Case

The cost parameters of three GenCos [309] are listed in Table 7.15. The aggregate demand parameters in (7.49) are $\overline{q_S} = 45$ GWh and $r = 0.5$ GWh/($/MWh).

The following simulation cases with and without forward contracts are carried out:

Case A_1 Perfect competition in the spot market without a forward market arrangement.

Case A_{21} Cournot type competition in the spot market without a forward market arrangement.

Table 7.14 CGA parameters.

Parameters	Description
Variable code type	Real code
Initial population	Randomly initialized population
Population size of each generator in outer loop	40
Max. generation of outer loop	200
Population size of each generator in inner loop	40
Max. generation of inner loop	50
Mutation	Nonuniform mutation (mutation probability = 0.05)
Crossover	Blend crossover (crossover probability = 0.05)
Selection	Tournament selection size is 2 and elitism size is 1

Table 7.15 Cost coefficients of the three GenCos.

GenCo no.	1	2	3
Cost parameter a_i ($/(MWh . GWh))	1.0	1.5	2.0
Cost parameter b_i ($/MWh)	12.0	10.0	8.0

Case A_{22} LSFE with *a* parameterization type competition in the spot market without a forward market arrangement.
Case A_{23} LSFE with *b* parameterization type competition in the spot market without a forward market arrangement.
Case A_{24} LSFE with $(a \propto b)$ parameterization type competition in the spot market without a forward market arrangement.
Case A_{25} LSFE with (a, b) parameterization type competition in the spot market without a forward market arrangement.
Case A_{31} Cournot type competition in the spot market with a forward market arrangement.
Case A_{32} LSFE with *a* parameterization type competition in the spot market with a forward market arrangement.
Case A_{33} LSFE with *b* parameterization type competition in the spot market with a forward market arrangement.
Case A_{34} LSFE with $(a \propto b)$ parameterization type competition in the spot market with a forward market arrangement.
Case A_{35} LSFE with (a, b) parameterization type competition in the spot market with a forward market arrangement.

The algorithm converges to an equilibrium effectively in each case and the results are shown in Table 7.16.

In Case A_{35}, the GenCos have the incentive to enter the forward market when the GenCos use the Cournot type competition in the spot market. The solution of this simple case can also be obtained by solving (7.49), (7.52), and (7.61) using the symbolic tool in Matlab software [316].

By comparing Case A_{31} with Case A_{21}, it can be observed that GenCos prefer not to enter the forward market because they can get more profit without participating in the forward market. However, if one GenCo enters into the forward market, it could benefit from the forward market by producing more output. Then the other GenCos are prompted to enter the forward market by economic incentive. Finally, as shown in the market equilibrium of Case A_{31}, all GenCos enter the forward market, but they get less profit compared with the case of not participating in the forward market (i.e. Case A_{21}). The prisoners' dilemma type outcome [304] appears in this situation.

In Case A_{35}, GenCos use LSFE with (a, b) parameterization in the spot market. The slope and intercept of the supply function of each GenCo are not unique, but the market clearing price and the GenCos' outputs are unique and equal to the results of the Cournot model in Case A_{31}. In this situation, the slope of the supply function is a very large positive value and the intercept is a very large negative value. The GenCos become less competitive and behave like Cournot quantity decision-makers. The simulation result has demonstrated that a focal equilibrium [255] is converged.

When comparing Cases A_{35} and A_{31} with Cases A_{25} and A_{21}, respectively, it is clear that the market price is lower and generation output increases when GenCos enter the

Table 7.16 Simulation results for three-GenCos example.

Case	GenCo no.	α_i/a_i or β_i/b_i or k_i	q_i^c (GWh)	p (\$/MWh)	q_i (GWh)	π_i (10^3\$)
A_1	1				13.38	89.45
	2			25.38	10.25	78.80
	3				8.69	75.47
A_{21}	1				9.11	207.62
	2			39.34	8.38	193.24
	3				7.83	184.15
A_{22}	1	1.7501			10.97	150.55
	2	1.4604		31.21	9.68	135.00
	3	1.3272			8.74	126.44
A_{23}	1	1.5596			11.19	137.79
	2	1.4977		29.91	9.95	123.85
	3	1.5135			8.90	115.77
A_{24}	1	1.3286			11.04	144.46
	2	1.2375		30.61	9.82	130.05
	3	1.1921			8.84	121.68
A_{25}	1	0			9.11	207.62
	2	0		39.34	8.38	193.24
	3	0			7.83	184.15
A_{31}	1	0	5.48		10.59	164.45
	2	0	5.07	32.82	9.42	148.41
	3	0	4.75		8.58	139.33
A_{32}	1	1.7501	0		10.97	150.55
	2	1.4604	0	31.21	9.68	135.00
	3	1.3272	0		8.74	126.44
A_{33}	1	1.5596	0		11.19	137.79
	2	1.4977	0	29.91	9.95	123.85
	3	1.5135	0		8.90	115.77
A_{34}	1	1.3286	0		11.04	144.46
	2	1.2375	0	30.61	9.82	130.05
	3	1.1921	0		8.84	121.68
A_{35}	1		5.48		10.59	164.45
	2		5.07	32.82	9.42	148.41
	3		4.75		8.58	139.33

forward market. The effect of the forward market arrangement on market power is studied by the Lerner Index (LI) [317], which is defined as $LI_i = (p - MC_i)/p$, where LI_i is the LI of GenCo i for $i = 1, 2, 3$. Since GenCos' capacity constraints are not considered, the marginal costs of the GenCos are the same and then the indices are equal for all generators. By comparing the index in Cases A_{35} and A_{31} with Cases A_{25} and A_{21} in Table 7.17, it can be observed that the markets in the Cournot model and (a, b) parameterization type become more competitive, and the market power is mitigated due to the forward market arrangement.

The market prices in Cases A_{22}, A_{23}, and A_{24} are higher than in the perfect competition in Case A_1 and lower than in the Cournot model in Case A_{21}; the profits of the GenCos are more than the profits in Case A_1 and less than in Case A_{21}. It is clear that the results greatly depend on the parameterization types; however, except for the (a, b) parameterization type in Case A_{25}, the results of other parameterization types are close. In this simulation, the profits of Gencos in $(a \propto b)$ parameterization type in Case A_{24} are slightly higher than in b parameterization type in Case A_{23}, but slightly less than in a parameterization type in Case A_{22}.

The simulation results in Cases A_{32}, A_{33}, and A_{34} are the same as that in Cases A_{22}, A_{23}, and A_{24}, respectively, because the GenCos have no incentive to enter the forward market and no forward contract is arranged. The prisoners' dilemma type market outcome does not appear. This is consistent with the theoretical analysis in section 7.1.1. However, the spot market in Cases A_{32}, A_{33}, and A_{34} is still more competitive than that with forward contract arrangement in Cases A_{31} and A_{35} as shown in Table 7.17. It can be shown that the type of competition and the parameterization of the LSF affect the GenCos' decisions and market equilibrium significantly.

The effects of the demand function's slope on the bidding behavior are also studied and the simulation results are shown in Table 7.18. Cases B_1 and B_2 are the same as Case A_{31} except for $r = 0.25$ and $r = 1$, respectively. It can be observed in Table 7.18 that the fraction of power sold at the forward market of each GenCo becomes smaller when r increases. Since the system demand is more elastic, the market is more competitive and the spot price reduces. The GenCos prefer to pursue more profit in the spot market and reduce their output at the forward market. A similar situation can be obtained from the case under LSFE type of competition with (a, b) parameterization in the spot market (i.e. Case A_{35}).

Table 7.17 Market power for three-GenCos example.

Case	A_{21}	A_{22}	A_{23}	A_{24}	A_{25}
LI	35.49	18.68	15.15	17.09	35.49
Case	A_{31}	A_{32}	A_{33}	A_{34}	A_{35}
LI	22.67	18.68	15.15	17.09	22.67

Table 7.18 Simulation results for the three-GenCos example with different slopes of the demand function.

Case	GenCo no.	r_i	q_i^c (GWh)	p (\$/MWh)	q_i (GWh)	q_i^c/q_i	π_i (10^3\$)
	1		7.17		12.31	0.5828	329.03
B_1	2	0.25	6.64	44.88	11.17	0.5944	296.11
	3		6.22		10.29	0.6045	273.70
	1		5.48		10.59	0.5175	164.45
A_{31}	2	0.50	5.07	32.85	9.42	0.5382	148.41
	3		4.75		8.58	0.5536	139.33
	1		3.24		7.66	0.4230	63.24
B_2	2	1.00	3.13	24.08	6.89	0.4543	61.42
	3		3.02		6.37	0.4741	61.81

To investigate the effects of cost parameters on the bidding behavior, the following two cases are performed.

Cases C_1 and C_2 are the same as Case A_{31} except for $a_1 = 2$ and $b_1 = 8$, respectively. The simulation results are shown in Table 7.19. In Case C_1, $a_1 = a_3$, the fraction of GenCo 1 is the same as that of GenCo 3. In Case C_2, $b_1 = b_3 < b_2$ and $a_1 < a_2 < a_3$, the fraction of GenCo 1 is smaller than that of GenCo 2, which is smaller than that of GenCo 3. It is not affected by the intercept of GenCos' marginal cost functions by comparing Case C_2 and Case A_{31}. It is observed that the fraction of power sold at the forward market is only affected by the slope of the GenCos' marginal cost functions from Cases C_1 and C_2.

Table 7.19 Simulation results for the three-GenCos example with different cost coefficients.

Case	GenCo no.	q_i^c (GWh)	p (\$/MWh)	q_i (GWh)	q_i^c/q_i	π_i (10^3\$)
	1	5.48		10.59	0.5175	164.45
A_{31}	2	5.07	32.85	9.42	0.5382	148.41
	3	4.75		8.58	0.5536	139.33
	1	4.08		7.89	0.5172	122.44
C_1	2	5.08	35.41	10.16	0.5000	180.73
	3	4.78		9.24	0.5172	167.87
	1	6.23		12.04	0.5174	212.53
C_2	2	4.82	31.67	8.94	0.5385	133.81
	3	4.53		8.18	0.5532	126.70

Table 7.20 Simulation results for the three-GenCos example with capacity constraints.

Case	GenCo no.	α_i/a_i or β_i/b_i or k_i	q_i^c (GWh)	p ($/MWh)	q_i (GWh)	π_i ($10^3$$)
	1		*		5.00	125.92
D_1	2		3.49	39.68	10.48	228.67
	3		3.52		9.68	213.01
	1	*	*		5.00	122.59
D_2	2	1.8159	0	39.02	10.65	224.03
	3	1.5766	0		9.84	208.36
	1	*	*		5.00	114.70
D_3	2	2.0976	0	37.44	10.98	210.83
	3	2.1040	0		10.30	197.18
	1	*	*		5.00	120.13
D_4	2	1.4762	0	38.52	10.73	219.75
	3	1.3754	0		10.01	205.31
	1		*		5.00	125.92
D_5	2		3.49	39.68	10.48	228.68
	3		3.52		9.68	213.01

The effects of capacity constraints on the bidding behavior are studied and the simulation results are shown in Table 7.20. Cases D_1, D_2, D_3, D_4, and D_5 are the same as Cases A_{31}, A_{32}, A_{33}, A_{34}, and A_{35}, except for considering the capacity constraint of GenCo 1 and setting $\overline{q_1} = 5$ GWh. It can be observed that the market price will become higher when the capacity constraint is hit by comparing the simulation results of Cases D_1, D_2, D_3, D_4, and D_5 in Table 7.20 with that of Cases A_{31}, A_{32}, A_{33}, A_{34}, and A_{35} in Table 7.16, respectively. When GenCo 1 hits its capacity constraint, it could not increase its output to change the market price and to pursue more profit. Then its expected profit in the forward market defined in (7.58), i.e. $p\overline{q_1} - C_1(\overline{q_1})$, is not affected by its forward contract output. The total output of GenCo 1 is unique and the same as its maximum output of generation, but its output in the forward market is not unique, as shown in Cases D_1, D_2, D_3, D_4, and D_5.

7.4.3.2 Five-GenCos Study Case

A more realistic five-GenCos test system [255] is used to further validate the previous analysis. The test system is based on the cost data for the five strategic firms in England and Wales subsequent to the 1999 divestiture. The cost parameters of the five GenCos are listed in Table 7.21. The aggregate demand parameters in (7.49) are $\overline{q_S} = 35$ GWh and $r = 0.1$ GWh/(£/MWh).

Table 7.21 Cost coefficients of the five GenCos.

GenCo no.	1	2	3	4	5
Cost parameter a_i (£/(MWh GWh))	2.687	4.615	1.789	1.930	4.615
Cost parameter b_i (£/MWh)	12	12	8	8	12.00

Different simulation cases without and with forward contracts are carried out and the results are shown in Table 7.22. In this study, the competition type in the spot market and contract arrangement of Cases E_1, E_2, E_{31}, E_{32}, E_{33}, E_{34}, E_{41}, E_{42}, E_{43},

Table 7.22 Simulation results for five-GenCos example.

Case	GenCo no.	α_i/a_i or β_i/b_i or k_i	q_i^c (GWh)	p ($/MWh)	q_i (GWh)	π_i ($10^3$$)
	1			26.87	5.54	41.17
	2				3.22	23.97
E_1	3				10.55	99.57
	4				9.78	92.30
	5				3.22	23.97
	1			80.40	5.39	329.66
	2				4.68	269.57
E_{21}	3				6.14	410.86
	4				6.07	403.80
	5				4.68	269.56
	1	1.3106		32.85	5.92	76.32
	2	1.1672			3.87	46.11
E_{22}	3	1.5035			9.24	153.18
	4	1.4597			8.82	144.06
	5	1.1671			3.87	46.11
	1	1.3051		31.50	5.90	68.27
	2	1.1776			3.76	40.71
E_{23}	3	1.8282			9.43	142.09
	4	1.7677			8.99	133.31
	5	1.1776			3.76	40.71
	1	1.1594		32.30	5.90	73.01
	2	1.0889			3.83	43.89
E_{24}	3	1.3090			9.32	148.79
	4	1.2836			8.89	139.78
	5	1.0889			3.83	43.89

Table 7.22 (*continued*)

Case	GenCo no.	α_i/a_i or β_i/b_i or k_i	q_i^c (GWh)	p (\$/MWh)	q_i (GWh)	π_i (10^3\$)
	1			80.40	5.39	329.68
	2				4.68	269.54
E_{25}	3				6.14	410.85
	4				6.07	403.79
	5				4.68	269.54
	1		4.50	42.80	5.98	136.09
	2		3.33		4.39	90.71
E_{31}	3		6.08		8.11	223.34
	4		5.90		7.86	213.92
	5		3.33		4.39	90.71
	1	1.3106	0	32.85	5.92	76.31
	2	1.1670	0		3.87	46.11
E_{32}	3	1.5033	0		9.24	153.17
	4	1.4597	0		8.82	144.05
	5	1.1672	0		3.87	46.10
	1	1.3050	0	31.50	5.90	68.27
	2	1.1776	0		3.76	40.71
E_{33}	3	1.8282	0		9.43	142.09
	4	1.7677	0		8.99	133.31
	5	1.1776	0		3.76	40.71
	1	1.1595	0	32.30	5.90	73.01
	2	1.0889	0		3.83	43.89
E_{34}	3	1.3090	0		9.32	148.79
	4	1.2837	0		8.89	139.78
	5	1.0890	0		3.83	43.89
	1		4.50	42.80	5.98	136.09
	2		3.33		4.39	90.71
E_{35}	3		6.08		7.11	223.34
	4		5.90		7.86	213.92
	5		3.33		4.39	90.71

E_{44}, and E_5 are the same as that of Cases A_1, A_2, A_{31}, A_{32}, A_{33}, A_{34}, A_{41}, A_{42}, A_{43}, A_{44}, and A_5 and in three-GenCos example, respectively.

It can be observed that the simulation results of this five-GenCos system are consistent with those of the three-GenCos system. The prisoners' dilemma type outcome appears in Case E_{31} and a focal equilibrium is converged in Case E_{35}. In Cases E_{32}, E_{33}, and E_{34}, the GenCos have no incentive to enter the forward market. When comparing Case E_{35} and E_{31} with Case E_{25} and E_{21}, respectively, it is also found that the forward market with Cournot type competition or LSFE type competition with (a, b) parameterization can make the market more competitive.

7.4.4 Summary

A two-stage game model has been presented to model the two-settlement electricity market and to investigate whether GenCos would enter the forward market when they use different competition types in the spot market. A coevolutionary approach has been successfully applied to determine the market equilibrium of the two-stage game model under different competition types and market conditions. It has been found that the proposed approach is very effective in determining the market equilibrium in all the cases.

Simulation results show that whether GenCos would enter the forward market depends significantly on what type of competition there is in the spot market and whether generation capacity constraints are violated. When the forward market is under Cournot type of competition or LSFE type of competition with (a, b) parameterization, GenCos enter the forward market due to the economic incentive and form a prisoners' dilemma market outcome. They may also enter the forward market when their capacity constraints are exceeded. Otherwise, they would not enter the forward market. So the prisoners' dilemma market outcomes are based on some specified market assumptions and are "brittle" to slight changes in the market assumptions. Besides, different market models can affect the market equilibrium significantly. It is also found that their levels of participation in the forward market depend significantly on the slope of the system demand function and the slope of their marginal cost function.

It is noted that the work in this chapter employs the commonly used assumption of risk-neutrality on all GenCos and sufficient arbitrators in the market. Future work is needed to investigate the bidding behaviors of GenCos when risk-averse GenCos and the price difference between spot and forward markets (i.e. not perfect arbitrage) are considered.

8

Future Developments

Power systems are evolving toward the so-called smart grids. According to the U.S. Department of Energy, "smart grid" generally refers to a class of technology that people are using to bring utility electricity delivery systems into the 21st century, using computer-based remote control and automation. These systems are made possible by two-way communication technology and computer processing that has been used for decades in other industries. They offer many benefits to utilities and consumers—mostly seen in big improvements in energy efficiency on the electricity grid and in the energy users' homes and offices.

In much the same way that a "smartphone" these days means a phone with a computer in it, "smart grid" means "computerizing" the electric utility grid. It includes adding two-way digital communication technology to devices associated with the grid. Each device on the network can be given sensors to gather data (power meters, voltage sensors, fault detectors, etc.), plus two-way digital communication between the device in the field and the utility's network operations center. A key feature of the smart grid is automation technology that lets the utility adjust and control each individual device or millions of devices from a central location.

A smart grid is generally regarded as a necessary platform for renewable energy integration. Renewable energy integration focuses on incorporating renewable energy, distributed generation, energy storage, thermally activated technologies, and demand response into the electric distribution and transmission system. A systems approach is being used to conduct integration development and demonstrations to address technical, economic, regulatory, and institutional barriers for using renewable and distributed systems. In addition to fully addressing operational issues, the integration also establishes viable business models for incorporating these technologies into capacity planning, grid operations, and demand-side management.

Power System Optimization – Large-scale Complex Systems Approaches, First Edition.
Haoyong Chen, Honwing Ngan and Yongjun Zhang.

The goal of renewable energy integration is to advance system design, planning, and operation of the electricity grid to achieve the following:

- reduce carbon emissions and emissions of other air pollutants through increased use of renewable energy and other clean distributed generation;
- increase asset use through integration of distributed systems and customer loads to reduce peak load and thus lower the costs of electricity;
- support achievement of renewable portfolio standards for renewable energy and energy efficiency;
- enhance reliability, security, and resiliency from micro-grid applications in critical infrastructure protection and highly constrained areas of the electric grid; and
- support reductions in oil use by enabling plug-in hybrid electric vehicle (PHEV) operations with the grid.

Integration of renewable energy has a great impact on power grid operation, causing system voltage and frequency fluctuations, deteriorating power quality, and reducing the level of security and stability. Therefore, the traditional power grid theories and technologies face challenges.

With these new trends, power grids become large-scale complex systems even more than ever, and deregulation of the power industry also complicates the problems. New optimization approaches for planning and operation issues should be further developed. As the final part of this book, we will address some new topics in optimization and foresee some future developments of power system optimization techniques.

Some new considerations in power system optimization are examined here first.

8.1 New Factors in Power System Optimization

8.1.1 Planning and Investment Decision Under New Paradigm

The investment of generation, transmission, and distribution facilities is concerned with not only technical considerations but also socio-economic and environmental factors. Under the deregulated market environment, centralized planning often may become not viable and must be replaced by a set of individual investment decision of facility owners. In the deregulated market operation, there may be a lack of incentive for long-term investments. Moreover, the questions of when and where to build the facilities of generation, transmission, and distribution, and what type of facilities to build, may be complicated and have to consider different criteria. Thus multiple-criteria decision-making or multiple-criteria decision analysis is needed.

On the other hand, uncertainty is a basic factor that should be taken into consideration in power system planning. Uncertainties of power system planning may come from different sources and types, such as fuel costs, investment and operational costs, demand growth, renewable variability, transmission capacity, environmental policies, and regulation. The problem of considering multiple uncertainties is even harder, not only because of the increased computational effort, but also because it is hard to model the combination of their influence on a single optimization problem. Because the objectives of planning are to reduce the risks of uncertainties and to gain some control over the future on the basis of the past, advanced modeling and solution methods are needed. For example, stochastic programming and robust optimization are two alternative approaches to deal with uncertainty [318].

Planning of smart grids may bring new research topics: "smart" means that there will be more renewable sources, demand-side response schemes, adaptive protection, intelligent meters, autonomous control systems, etc. The questions that should be answered include these: How many and what type of smart devices should be installed in a smart grid? How can the smart grid deployment plans be evaluated? How can costs be shared between utilities and consumers? And how can benefits be delivered to consumers commensurate with investments?

8.1.2 Scheduling/Dispatch of Renewable Energy Sources

Scheduling and dispatch of power systems with significant renewable energy penetration are hot research topics at present. Wind or solar power is hard to forecast with great accuracy for dispatching purposes. Hence, there is an unavoidable random error between the actual renewable energy power output and its forecasted value. How to deal with such uncertainty is a difficult issue bringing new challenges to power system scheduling and dispatch. To accommodate the random renewable energy power fluctuation by reasonable dispatch of the non-wind generation units, one method is to increase the system reserve. However, it is difficult to determine the precise reserve amount that is enough to overcome renewable energy power uncertainty and guarantee system security, and thus the results usually tend to be conservative.

A better alternative is stochastic programming (SP) based on scenarios by which we can obtain more adaptive unit commitment and generation dispatch [173]. The approach generates a predicted scenario and large numbers of error scenarios according to the wind power forecast and its error distribution. In all the scenarios, a certain non-wind unit has consistent commitment state and diverse power output. The aim of the programming is to minimize the system operating cost in predicted scenarios or the weighted average of the cost in each scenario. The approaches based on SP are possibly effective for dealing with wind power uncertainties. However, they are based on the wind power probability distribution, which is often difficult to acquire. Furthermore, they cannot give a clear bound for power system secure operation and also are computationally expensive in practical applications.

Currently, there is another trend in research on power system operation planning, which is based on robust optimization (RO). RO was developed to provide a practical approach for handling noisy data and uncertainty [319]. The RO framework addresses this issue directly. An RO approach has been proposed [320] to accommodate wind output uncertainty, with the objective of providing a robust unit commitment schedule for the thermal generators in the day-ahead market that minimizes the total cost in the worst wind power output scenario. A solution to an optimization model is defined as: solution-robust if it remains "close" to optimal for all scenarios of the input data, and model-robust if it remains "almost" feasible for all data scenarios. Different from SP, the importance of controlling the variability of the solution (as opposed to just optimizing an average performance) is well recognized. It has been ignored in most other applications of mathematical programming. The RO framework addresses this issue directly. Another important distinction of RO from SP is the handling of the constraints. Stochastic programs aim at finding values for the structural variables such that for each scenario a control variable setting is possible to satisfy the constraints. In cases where no feasible solution is possible for every scenario, the SP model is simply declared infeasible. RO, however, explicitly allows this contingency. The RO model, through the use of error terms and the penalty function, will find a solution that violates the constraints by the least amount.

These new models and approaches for solving scheduling and dispatch of renewable energy sources are still under development, and further theoretical investigation and practical application are needed.

8.1.3 Energy Storage Problems

Energy storage (ES) systems are expected to play a major role in future smart grids. Earlier studies have found that ES can compensate for the stochastic nature of intermittent energy sources by absorbing the excess energy when generation exceeds predicted levels and providing it back to the grid when generation levels are lower. Energy storage could also be used to enhance the utilization rate of power system assets. Most of the time, with the exception of peak hours, less than 50% of electricity system capacity is utilized. Thus, a significant amount of power system assets have been built to meet only a few hundred hours of peak demand each year. Use of ES can reduce the peak demand by storing energy during off-peak hours and feeding it back into the grid during peak times, which offers an alternative to expanding power generation capacity.

Traditionally, large-scale pumped hydro storage (PHS) systems have provided almost 99% of worldwide storage capacity, mainly fulfilling the function of demand leveling at the transmission level. As the integration of renewable energy sources (RESs) into power grids increases, additional smaller-scale and more flexible storage technologies are required at different levels. Battery energy storage (BES) systems may range from 2 kW up to 50 MW, and are distinguished for their fast response, mobility, and flexibility in either high-power or high-energy applications. Batteries have the potential to provide energy storage services at all levels of the grid—generation, transmission, distribution, or customer.

The particular challenge of including ES systems into power system scheduling or the dispatch process is to guarantee that sufficient energy will be available in the

ES system to provide energy and reserves when required. Especially in a cost-based energy spot market, this point turns out to be critical. In such market structures, system operating decisions are centrally established and coordinated by the ISO. Hence, reserve activation must be considered during the scheduling process. Thus, in order to determine the value of storage, it is necessary to account for ES limitations. Determination of the lowest-cost ES system for use with high-penetration renewables in power systems is an even more difficult problem.

8.1.4 Environmental Impact

The electricity sector is unique among all industrial sectors in its very large contribution to emissions associated with nearly all air quality issues. Electricity generation produces a large share of nitrogen oxides and sulfur dioxide emissions, which contribute to smog and acid rain and the formation of fine particulate matter. It is the largest uncontrolled industrial source of mercury emissions. Fossil-fuel-fired electric power plants also emit carbon dioxide, which may contribute to climate change. In addition, the sector has significant impacts on water, habitat, and species. In particular, hydro dams and transmission lines have significant effects on water and biodiversity.

To protect the environment, it is important to employ relatively cheap, secure, and sustainable energy sources with low carbon dioxide emissions. Sustainable energy sources are most often regarded as including all renewable energy sources, such as hydroelectricity, solar energy, wind energy, wave power, geothermal energy, bioenergy, and tidal power. It usually also includes technologies that improve energy efficiency. Various economic instruments can be used to steer society toward sustainable energy. Some of these methods include ecotaxes and emissions trading. Green consumerism is enhanced on free energy markets. In Europe, environmental nongovernmental organizations (NGOs) have developed the EKOenergy label to help consumers to choose more sustainable electricity products.

All these factors mean that we should employ environmentally friendly models in power system planning and operation; the costs and constraints of emission should be included in optimization.

8.1.5 Novel Electricity Market

Smart grids offer unprecedented opportunities for growth of novel markets for innovative services as well as interoperable products and applications [321]. Layers of digital information and control technologies are overlaid on top of the power grid to improve reliability, enhance security, and increase efficiency of the electric power grid. This requires a different vision of business models, usage patterns, and growth trajectories. Development of the internet, which has some similarities with smart grids, provides valuable insights into understanding the opportunities and challenges ahead.

Smart grids will similarly allow new economic and social value creation by offering environmentally and economically responsible choices of generation, transmission, distribution, and consumption of electric power, and by enabling interconnection-wide trading of electric power through the development of flexible retail and wholesale markets integrated through an e-commerce trading platform [321]. The smart grid envisioned here would support completely autonomous bilateral transactions in a market-transparent fashion, and coordinating authorities would play a supervising and regulatory role only. More interestingly, it may create new value chains involving information on consumer preferences, participation, and choices, and a number of innovative services ranging from differentiated power quality and reliability service to energy storage for end-users. Especially, compared with the situation in the traditional power industry, in which the customers have no interest to decrease their consumption, the new trend is the introduction of low-energy-need technologies and better efficiency of energy usage.

Smart grids will enable end-to-end direct transactions between producers and consumers facilitated by one or more energy broker(s). The energy broker(s) will provide all necessary connectivity for transmission and distribution of energy as well as the relevant information and data. The operation(s) within the power system would be as transparent to the power producers and consumers as those of the internet are to the producers and consumers of its information content.

8.2 Challenges and Possible Solutions in Power System Optimization

Electric power systems are infrastructures with more than a century's history, but their technologies (including technologies of generators, transformers, power lines, power electronic devices, supervisory control and data acquisition (SCADA) and other control technologies, etc.) have been undergoing continuous upgrade. Nowadays, with the introduction of smart grids, renewable energy, and electricity markets, power systems are becoming large-scale complex systems day by day, and the optimization approaches introduced in this book need to be further developed. Here, several challenges and their possible solutions are discussed and bring us to the end of this book.

From Small-scale to Large-scale

In the early days, all electric power stations were operated separately, supplying electric energy only to their own customers. However, very soon, engineers realized that the integration of individual power stations into an electric power system brings many advantages, both technical (increasing power supply reliability) and economic (decreasing costs of energy), and the integration of systems became very popular. It was based on developments in electric power transmission technologies and parallel operation technologies of several power stations. In modern power systems, it was very important to increase the transmission voltages in gradual steps to a high level. This made intercontinental power systems and long transmission lines possible.

Nowadays, the existing electric power system has been described as the largest and most complex machine on Earth. It is important to remember that the electric grid is a critical infrastructure that provides an economic backbone for modern economies. As such, developed economies are not tolerant of grid disruptions. Likewise, failure to achieve existing policy mandates related to renewable and distributed resources is also not acceptable. Therefore, a unified multi-level control scheme that simultaneously optimizes operation across markets, areas, and different levels is required. A comprehensive large-scale control framework offers an effective reference to develop modern grid control-based architectures and related interoperability standards and product designs.

From Deterministic to Stochastic

The modeling of deterministic optimization problems in power systems has long enjoyed relatively standard modeling frameworks developed within the fields of mathematical programming and optimal control. However, the introduction of uncertainty has made the situation more difficult. A set of approaches such as dynamic programming, stochastic programming, robust optimization, and optimal control has been proposed to tackle the stochastic optimization problems in power systems. A few examples are included below.

- *Unit commitment*. ISOs have to handle uncertainty from outages, weather variations, and the growing integration of wind and solar energy, which can be highly unpredictable. Unit commitment problems may be made a day ahead (for steam generators) or for shorter horizons (15–60 minutes) for gas turbines. Even economic dispatch, which involves the near-real-time adjustment of generators, has to make adjustments that anticipate future events.
- *Energy storage*. Storage devices can be used for frequency regulation and energy shifting, and have to be managed in the presence of stochastic frequency regulation signals, volatile electricity prices (locational marginal price), energy variations from wind and solar, all in the presence of predictable variations in both load and generation.
- *Bidding energy resources*. It is often necessary to place bids for generation or storage into day-ahead or hour-ahead markets, managing uncertain electricity prices.
- *Equipment replacement*. Utilities need to plan the inspection and replacement of utility poles and transformers in the presence of uncertainty in the state of the resources and unpredictable failures.
- *Pricing electricity contracts*. A utility may have to sign contracts to deliver electricity for 3–5 years into the future while managing uncertainty in commodity prices, grid congestion, as well as future energy investments.

- *Investment planning*. Utilities have to make decisions about future investments in generation and transmission, in the presence of uncertainty about future loads, commodity prices, and the decisions of other generators.

These examples illustrate the wide variation in the types of stochastic optimization problems in terms of the nature of the decisions (discrete or continuous, scalar or vector), the uncertainties (binomial failures, Gaussian noise in weather, loads and generation, heavy-tailed electricity prices) and the dynamics (known models of storage processes, but unknown models of climate change, commodity prices, and the behavior of competing utilities). Making decisions under uncertainty pervades the planning and operation of our energy system, which can be seen in the wide array of research articles addressing these problems.

From Single-objective to Multi-objective

Traditional optimization models in power system planning and operation are often single-objective models, which try to find the best possible solution (known as the global optimum) or at least a very close approximation for a given problem. However, as power system optimization problems become more and more complicated, it is often the case that no single solution can optimize all objectives because of the conflicts among these objectives. In such a case, one solution may be the best for a single objective but may not be optimal for other objectives. A standard definition for "optimum" in multi-objective optimization is the one called the "Pareto optimum", which is a concept from economics. Based on this concept, the aim for multi-objective optimization is to find a set of trade-off optimal solutions, in which improving an objective can be done only if at least one other objective is worse off.

An earlier technique to solve the multi-objective optimization problem is the weighted-sum method, which forms a composite objective function to reduce a multi-objective optimization problem to a single-objective one. Although this approach is simple, it may lose important information about the optimization problem. Furthermore, the choice of weights may be subjective.

Application areas for multi-objective optimization have expanded, including complex problems in planning, operation, control, and economics of electric power systems. Most of these methods are based on mathematical programming, which have several technical limitations. On the other hand, a set of novel approaches based on multi-objective evolutionary algorithms and other intelligent techniques have been developed to overcome such limitations.

From Centralized to Decentralized

Traditionally, the power industry has been a monopoly in economic terms. The deregulation reforms broke the power sector into distinct parts, such as generation, transmission, distribution, trading, etc. However, the market also needs some regulation, especially for the power industry. Thus a hybrid economic model with free market participants and regulatory entities may be formed. The traditional centralized power system operation and control regime may not be able to match the market philosophy, and thus there will be a combination of decentralized optimization and centralized coordination.

Nowadays, there is another trend in which the traditional centralized energy generation methods are replaced or complemented by distributed generation, such as combined heat and power (CHP), combined cooling, heat, and power (CCHP), wind turbines, solar photovoltaics (PV), or small hydroelectric stations, etc. It will also result in a hybrid centralized–decentralized optimization mode in power system planning, operation, and control.

The task of effectively controlling large-scale interconnected power systems is one of the most challenging problems in control engineering. In many practical cases, the amount of information transfer between large interconnected power systems, where each centre of generation is geographically separated from the rest by long distances, is constrained by economic as well as technical considerations. The huge amount of data acquired on-line may overload the dispatchers and the computation and communication systems. In decentralized control design methods suitable for large-scale interconnected power systems, the control task is decomposed into local subtasks. The function of each subtask is to generate the required local control input so that, collectively, the centralized performance of the global system is preserved. Because the market systems investigated in microeconomics in fact employ decentralized control architecture from the view of complex systems, the general equilibrium theory, game theory, and principles of mechanism design may provide much intuition for the design of decentralized optimization approaches of power systems.

Power grids are evolving toward large-scale complex systems even faster than ever. Based on the principles and methods presented in this book, many new approaches can be developed. It is also instructive to consider the approaches from perspectives of different disciplines, such as systems engineering, microeconomics, and operations research. Mutual inspirations may be obtained. This is still an open area.

Appendix

A.1 Header File

```
#include <stdio.h>
#define Error(i) {printf("Not enough memory at code %d \n",i);exit(0);}
#define CHROMOLENGTH 2
#define CHROMO_PER_VARIABLE 2

//#define __GRIEWANGK_FUNCTION__
//#define __ACKLEY_FUNCTION__
//#define __RASTRIGIN_FUNCTION__
//#define __SCHWEFEL_FUNCTION__
#define __ROSENBROCK_FUNCTION__

class CEcosystem;

struct Chromosome
{
 unsigned char gene[CHROMOLENGTH];
};

class CSpecies
{
 private:

 CEcosystem * theEcosystem;
 int SpeciesID;
 static unsigned char BitMask[],Bit[];
 int New,Old;
 int JCross;
 int LChromo,MaxGen;
 float PCross,PMutation;
 int PopSize,NGen,NCross,NMutation;
 int MaxCrossTimes;
 double FitMax,FitMin,FitAvrg,FitSum;
 Chromosome (*Chromo)[2];
 double (*Fitness)[2];
 double LowerBound,UpperBound;
 void InitPop();
 void WriteChromo(FILE*,struct Chromosome&,int);
```

Power System Optimization – Large-scale Complex Systems Approaches, First Edition.
Haoyong Chen, Honwing Ngan and Yongjun Zhang.

```
 void Crossover(int,int,struct Chromosome&,struct Chromosome&);
 void Mutation(struct Chromosome&);
 int  ObjFunc(double&,struct Chromosome&);
 int Select();

 public:
 Chromosome ChromoMax,ChromoMin;
 void SetParam(CEcosystem* CEcosystem,int speciesID,int maxgen,
               int popsize,int lchrom,float pcross,float pmutation);
 ~CSpecies();
 void Initialize();
 void Statistics();
 void Report(FILE*);
 void Generation();
};

class CEcosystem
{
 public:
 int NSpecies;
 Chromosome *Representative;
 CEcosystem();
 ~CEcosystem();
 void Initialize();
 void SetParam(int n_species);
};
```

A.2 Species Class

```
#include <stdlib.h>
#include <math.h>
#include <time.h>
#include "Coevolution.h"

#define PI 3.1415927
#define E 2.7182818

unsigned char CSpecies::BitMask[]={0x01,0x03,0x07,0x0F,0x1F,0x3F,0x7F,0xFF};
//BitMask
unsigned char CSpecies::Bit[]={0x01,0x02,0x04,0x08,0x10,0x20,0x40,0x80};
void CSpecies::SetParam(CEcosystem *ecosystem,int speciesID,int maxgen,
                int popsize,int lchromo,float pcross,float pmutation)
{
 SpeciesID=speciesID;
 theEcosystem=ecosystem;
 MaxGen=maxgen;
 PopSize=popsize;
 LChromo=lchromo;

 PCross=pcross;
 PMutation=pmutation;
 NMutation=0;
 NCross=0;
 NGen=0;
 MaxCrossTimes=2;
 Chromo=new struct Chromosome[PopSize][2];
 if(Chromo==0)Error(1);
 Fitness=new double[PopSize][2];
 if(Fitness==0)Error(2);
 New=0;Old=1;
```

```
 FitMax=-100.0;

#if defined(__GRIEWANGK_FUNCTION__)
 LowerBound=-600.0;
 UpperBound=600.0;
#endif
#if defined(__ACKLEY_FUNCTION__)
 LowerBound=-30.0;
 UpperBound=30.0;
#endif
#if defined(__RASTRIGIN_FUNCTION__)
 LowerBound=-5.12;
 UpperBound=5.12;
#endif
#if defined(__SCHWEFEL_FUNCTION__)
 LowerBound=-500.0;
 UpperBound=500.0;
#endif
#if defined(__ROSENBROCK_FUNCTION__)
 LowerBound=-2.048;
 UpperBound=2.048;
#endif
}

CSpecies::~CSpecies()
{
 if(Chromo!=0)   delete[] Chromo;
 if(Fitness!=0)  delete[] Fitness;
}

void CSpecies::InitPop()
{
 int Valid=0;
 srand((unsigned)time(NULL));
 for(int i=0;i<PopSize;i++)
 {
Valid=0;
    do
    {
      for(int j=0;j<LChromo;j++)
        Chromo[i][New].gene[j]=rand()%256;
      Valid=ObjFunc(Fitness[i][New],Chromo[i][New]);
    }while(!Valid);
 }
}

void CSpecies::Initialize()
{
 NMutation=0;
 NCross=0;
 NGen=0;
 New=0;Old=1;
 FitMax=-100.0;

 InitPop();
 Statistics();
}

void CSpecies::Statistics()
{
 double FitValue;
 FitSum=0;
```

```
  for(int j=0;j<PopSize;j++)
   {
     FitValue=Fitness[j][New];
     FitSum+=FitValue;
     if(FitValue>FitMax) {FitMax=FitValue;ChromoMax=Chromo[j][New];}
   }
  FitAvrg=FitSum/PopSize;
 }

void CSpecies::Report(FILE* out)
{
 double obj;

 fprintf(out,"\n========No.%d  Statistics ==========\n",NGen);
 fprintf(out,"###FitMax=%10.4f\n",FitMax);
 fprintf(out,"###FitAvrg=%10.4f\n",FitAvrg);

 obj=1.0/FitMax-0.0000001;
 fprintf(out,"###ObjFunc=%10.4f\n",obj);

// WriteChromo(out,ChromoMax,-1);
 /*
 for(int i=0;i<PopSize;i++)
  WriteChromo(out,Chromo[i][New],i);
 */
}

void CSpecies::WriteChromo(FILE* out,struct Chromosome& chromosome,int i)
{
 unsigned int data;
 double x;

 if(i>=0)
    fprintf(out,"Chromosome[%d]: ",i);
 else fprintf(out,"Chromosome    : ");
 for(int k=LChromo-1;k>=0;k--)
  {
    fprintf(out,"%2x",chromosome.gene[k]);
  }

 data=(chromosome.gene[1])<<8;
 data+=chromosome.gene[0];
 x=LowerBound+(UpperBound-LowerBound)/65535*data;
 fprintf(out," *** x%d=%10.4f  ",SpeciesID,x);

 if(i>=0)
    fprintf(out," ***fitness=%10.4f\n",Fitness[i][New]);
 else fprintf(out,"\n");
}

void CSpecies::Generation()
{
 struct Chromosome Child1,Child2;
 int Valid1,Valid2;
 double tempValue;

 int temp=New;
 New=Old;
 Old=temp;
 srand((unsigned)time(NULL));
 int j=0;
```

```
 Chromo[j][New]=ChromoMax;
 Fitness[j][New]=FitMax;
 j++;

 while(j<PopSize)
 {
  int Mate1=Select();
  int Mate2=Select();
  Crossover(Mate1,Mate2,Child1,Child2);
  Mutation(Child1);
  Mutation(Child2);
  Valid1=ObjFunc(tempValue,Child1);
  if(Valid1)
    {
     Chromo[j][New]=Child1;
     Fitness[j][New]=tempValue;
     j++;
    }
  if(j>=PopSize) break;
  Valid2=ObjFunc(tempValue,Child2);
  if(Valid2)
   {
    Chromo[j][New]=Child2;
    Fitness[j][New]=tempValue;
    j++;
   }
 }  //End of while(...)
 NGen++;
}
/**************************************************************
Parent1=xxxxxxxoooo   CrossPosition=3
Parent2=#######****
Child1= xxxxxxx****
Child2= #######oooo
 **************************************************************/

void CSpecies::Crossover(int mate1,int mate2,struct Chromosome& child1,
           struct Chromosome& child2)
{
 Chromosome Parent1,Parent2,Child1,Child2;
 int CrossPosition;
 int BytePosition;
 int BitPosition;
 Parent1=Chromo[mate1][Old];
 Parent2=Chromo[mate2][Old];
 float Rnd;
 Rnd=rand()/float(RAND_MAX);
 if(Rnd<PCross)
  {for(int i=0;i<MaxCrossTimes;i++)
     {
     CrossPosition=rand()%(LChromo*8);
     BytePosition=CrossPosition/8;
     BitPosition=CrossPosition%8;
     for(int i=0;i<BytePosition;i++)
       {
          Child1.gene[i]=Parent2.gene[i];
          Child2.gene[i]=Parent1.gene[i];
       }
     for(i=LChromo-1;i>BytePosition;i--)
       {
          Child1.gene[i]=Parent1.gene[i];
          Child2.gene[i]=Parent2.gene[i];
```

```
      }
    unsigned char Mask0=BitMask[BitPosition];
    unsigned char Mask1=BitMask[BitPosition];
    Child1.gene[BytePosition]=(Parent1.gene[BytePosition]&Mask1)|
              (Parent2.gene[BytePosition]&Mask0);
    Child2.gene[BytePosition]=(Parent2.gene[BytePosition]&Mask1)|
              (Parent1.gene[BytePosition]&Mask0);
    Parent1=Child1;
    Parent2=Child2;
    } //end of for(...Times...)
  child1=Child1;
  child2=Child2;
  NCross++;
  }
  else  //Rnd>=PCross
   {
     child1=Parent1;
     child2=Parent2;
    }
 }

void CSpecies::Mutation(struct Chromosome& child)
{
  float Rnd;
  int i,j;
  for(i=0;i<LChromo;i++)
     for(j=0;j<8;j++)
      {
        Rnd=rand()/float(RAND_MAX);
        if(Rnd<PMutation)
        child.gene[i]^=Bit[j];
      }
}

int CSpecies::Select()
{
/************************************************/
 int Mate;
 double PartSum=0;
 float Rnd=rand()/float(RAND_MAX);
 double RndSum=Rnd*FitSum;
 Mate=-1;
 do
 {
  Mate++;
  PartSum+=Fitness[Mate][Old];
 }while(PartSum<RndSum);
 return Mate;
/************************************************/
/************************************************
 int Mate1=random(PopSize);
 int Mate2=random(PopSize);
 if(Fitness[Mate1][Old]>=Fitness[Mate2][Old])
    return Mate1;
 else return Mate2;
************************************************/
 }

 int CSpecies::ObjFunc(double& result,struct Chromosome& chromosome)
 {
  int Success=1;
  unsigned int data=0;
```

```
 double x=0.0,sum1=0.0,sum2=0.0,product1=1.0;
 int NSpecies=theEcosystem->NSpecies;

#if !defined(__ROSENBROCK_FUNCTION__)
 for(int i=0; i<NSpecies; i++)
 {
  if(i==SpeciesID)
  {
   data=(chromosome.gene[1])<<8;
   data+=chromosome.gene[0];
  }
  else
  {
   data=(theEcosystem->Representative[i].gene[1])<<8;
   data+=theEcosystem-<Representative[i].gene[0];
  }
  x=LowerBound+(UpperBound-LowerBound)/65535*data;
#if defined(__GRIEWANGK_FUNCTION__)
   sum1+=x*x;
   product1*=cos(x/sqrt(i+1));
#endif
#if defined(__ACKLEY_FUNCTION__)
   sum1+=x*x;
   sum2+=cos(2*PI*x);
#endif
#if defined(__RASTRIGIN_FUNCTION__)
   sum1+=x*x;
   sum2+=cos(2*PI*x);
#endif
#if defined(__SCHWEFEL_FUNCTION__)
   sum1+=x*sin(sqrt(fabs(x)));
#endif
 }
#endif

#if defined(__ROSENBROCK_FUNCTION__)
 unsigned int data1,data2;
 double x1,x2;
 for(int i=0; i<NSpecies/2; i++)
 {
  if(2*i==SpeciesID)
  {
   data1=(chromosome.gene[1])<<8;
   data1+=chromosome.gene[0];
  }
  else
  {
   data1=(theEcosystem->Representative[2*i].gene[1])<<8;
   data1+=theEcosystem->Representative[2*i].gene[0];
  }
  x1=LowerBound+(UpperBound-LowerBound)/65535*data1;

  if(2*i+1==SpeciesID)
  {
   data2=(chromosome.gene[1])<<8;
   data2+=chromosome.gene[0];
  }
  else
  {
   data2=(theEcosystem->Representative[2*i+1].gene[1])<<8;
   data2+=theEcosystem->Representative[2*i+1].gene[0];
  }
```

```
    x2=LowerBound+(UpperBound-LowerBound)/65535*data2;

    sum1+=(x2-x1*x1)*(x2-x1*x1);
    sum2+=(1.0-x1)*(1.0-x1);
  }

#endif

#if defined(__GRIEWANGK_FUNCTION__)
 result=1.0/(0.0000001+1.0+sum1/4000.0-product1);
#endif
#if defined(__ACKLEY_FUNCTION__)
 result=1.0/(0.0000001+20.0+E-20.0*exp(-0.2*sqrt(sum1/NSpecies))-exp(sum2/NSpecies));
#endif
#if defined(__RASTRIGIN_FUNCTION__)
 result=1.0/(0.0000001+3.0*NSpecies+sum1-3.0*sum2);
#endif
#if defined(__SCHWEFEL_FUNCTION__)
 result=1.0/(0.0000001+418.9829*NSpecies+sum1);
#endif
#if defined(__ROSENBROCK_FUNCTION__)
 result=1.0/(0.0000001+100*sum1+sum2);
#endif
 return Success;
}
```

A.3 Ecosystem Class

```
#include <stdlib.h>
#include <time.h>
#include "Coevolution.h"

CEcosystem::CEcosystem()
{
 Representative=NULL;
}

CEcosystem::~CEcosystem()
{
 if(Representative!=NULL) delete[] Representative;
}

void CEcosystem::SetParam(int n_species)
{
 NSpecies=n_species;
 Representative=new Chromosome[n_species];
}

void CEcosystem::Initialize()
{
  int i,j;
  srand((unsigned)time(NULL));
  for(i=0; i<NSpecies; i++)
    for(j=0; j<CHROMOLENGTH; j++)
{
      Representative[i].gene[j]=rand()%256;
}
}
```

A.4 Main Function

```
#include <stdlib.h>
#include <stdio.h>
#include "Coevolution.h"

void main()
{
  int NSpecies;
  int NEcosystemEvolve;
  int MaxGen,PopSize;
  int TestCount=50;
  float PMutation,PCross;
  int i,j;

  CEcosystem *theEcosystem=new CEcosystem;

  FILE* in;
  if((in=fopen("coevolution.in","r"))==NULL)
           {printf("Input data file open error\n");exit(0);}

//Enter the Species Number
  fscanf(in,"%d",&NSpecies);
#if defined (__GRIEWANGK_FUNCTION__)
  NSpecies=10;
#endif
#if defined (__ACKLEY_FUNCTION__)
  NSpecies=30;
#endif
#if defined (__RASTRIGIN_FUNCTION__)
  NSpecies=20;
#endif
#if defined (__SCHWEFEL_FUNCTION__)
  NSpecies=10;
#endif
#if defined (__ROSENBROCK_FUNCTION__)
  NSpecies=20;
#endif
//Enter the Number of Ecosystem Evolution
  fscanf(in,"%d",&NEcosystemEvolve);
//Enter the Max Generation Number of Each Species
  fscanf(in,"%d",&MaxGen);
//Enter the Population Size
  fscanf(in,"%d",&PopSize);
//Enter the Crossover Probability
  fscanf(in,"%f",&PCross);
//Enter the Mutation Probability
  fscanf(in,"%f",&PMutation);

  FILE* out;
  if((out=fopen("coevolution.out","w"))==NULL)
            {printf("Output data file open error\n");exit(0);}
  fprintf(out,"NSpecies=%d, NEcosystemEvolve=%d, MaxGen=%d,
                PopSize=%d, PCross=%10.4f, PMutation=%10.4f\n", NSpecies,
                    NEcosystemEvolve,MaxGen,PopSize,PCross,PMutation);

  CSpecies *theSpecies=new CSpecies[NSpecies];
  theEcosystem->SetParam(NSpecies);

  for(i=0; i<NSpecies; i++)
    theSpecies[i].SetParam(theEcosystem,i,MaxGen,PopSize,
```

```
                    CHROMO_PER_VARIABLE,PCross,PMutation);

   for(int n=0; n<TestCount; n++)
   {
    printf("No. %d Testing ...\n",n);

    for(i=0; i<NSpecies; i++)
    {
      theSpecies[i].Initialize();
      theEcosystem-<Initialize();
    }

    for(i=0; i<NEcosystemEvolve; i++)
    {
//    printf("Ecosystem Evolving ... Generation %d >>\n",i);
      for(j=0; j<NSpecies; j++)
      {
//        theSpecies[j].Initialize();
         for(int NGen=0;NGen<MaxGen;NGen++)
         {
          theSpecies[j].Generation();
          theSpecies[j].Statistics();
          if(j==NSpecies-1) theSpecies[j].Report(out);
         }
         theEcosystem->Representative[j]=theSpecies[j].ChromoMax;
      }
    }
   }

   if(fclose(out)!=NULL)  { printf("Output data file close error\n");  exit(0); }

   delete[] theSpecies;
   delete theEcosystem;
 }
```

References

[1] A. M. Sasson and H. M. Merrill. Some applications of optimization techniques to power systems problems. *Proce*, 62(7): 959–972, July 1974.

[2] T. Berrie and D. Anderson. Power system planning, development programs and project selection—a discussion of methods. In *Proc. 4th Power Systems Computation Conf.*, Grenoble, France, 1972.

[3] N. Adams, F. Beglari, M. A. Laughton, and G. Mitra. Mathematical programming systems in electrical power generation, transmission and distribution planning. In *Proc. 4th Power Systems Computation Conf.*, Grenoble, France, 1972.

[4] R. R. Booth. Power system simulation model based on probability analysis. *IEEE Transactions on Power Apparatus and Systems*, PAS-91(1): 62–69, 1972.

[5] R. R. Booth. Optimal generation planning considering uncertainty. *IEEE Transactions on Power Apparatus and Systems*, PAS-91(1): 70–77, 1972.

[6] H. Y. Chen, X. F. Wang, and X. Y. Zhao. Generation planning using Lagrangian relaxation and probabilistic production simulation. *International Journal of Electrical Power and Energy Systems*, 26(8): 597–605, 2004.

[7] W. R. Puntel. An automated method for long-range planning of transmission networks. In *Proc. 8th Power Industry Computer Applications Conf.*, pp. 38–46, 1973.

[8] L. L. Garver. Transmission network estimation using linear programming. *IEEE Transactions on Power Apparatus and Systems*, PAS-89(7): 1688–1697, 1970.

[9] X. Wang, X. Wang, and Y. Mao. Improved genetic algorithm for optimal multistage transmission system planning. In *Proc. IEEE Power Engineering Society Summer Meeting*, vol. 3, pp. 1737–1742, 2001.

[10] R. N. Adams and M. A. Laughton. A dynamic programming network flow procedure for distribution system planning. In *Proc. 8th Power Industry Computer Applications Conf.*, pp. 348–354, 1973.

[11] V. Miranda, J. V. Ranito, and L. M. Proenca. Genetic algorithm in optimal multistage distribution network planning. *IEEE Transactions on Power Systems*, 9(4): 1927–1933, 1994.

[12] E. Naderi, H. Seifi, and M. S. Sepasian. A dynamic approach for distribution system planning considering distributed generation. *IEEE Transactions on Power Delivery*, 27(3): 1313–1322, July 2012.

[13] M. E. El-Hawary and G. S. Christensen. *Optimal Economic Operation of Electric Power Systems*. Academic Press,New York, 1979.

[14] A. J. Wood. *Power Generation, Operation and Control*. John Wiley & Sons, Inc., New York, 1984.

Power System Optimization – Large-scale Complex Systems Approaches, First Edition.
Haoyong Chen, Honwing Ngan and Yongjun Zhang.

[15] G. B. Sheble and G. N. Fahd. Unit commitment literature synopsis. *IEEE Transactions on Power Systems*, 9(1): 128–135, 1994.

[16] G. B. Dantzig and P. Wolfe. Decomposition principle for linear programs. *Journal of Operational Research*, 8(1): 101–111, 1960.

[17] X. Guan, P. B. Luh, and H. Yan. An optimization-based method for unit commitment. *International Journal of Electrical Power and Energy Systems*, 14(1): 9–17, February 1992.

[18] H. Yan, P. B. Luh, and X. Guan. Scheduling of hydrothermal power systems. *IEEE Transactions on Power Systems*, 8(3): 1358–1365, August 1993.

[19] X. Guan, P. B. Luh, and H. Yan. Optimization-based scheduling of hydrothermal power systems with pumped-storage units. *IEEE Transactions on Power Systems*, 9(2): 1023–1029, May 1994.

[20] A. M. Sasson. Nonlinear programming solutions for load-flow minimum-loss and economic dispatching problems. *IEEE Transactions on Power Apparatus and Systems*, PAS-88(4): 399–409, 1969.

[21] J. Hetzer, D. C. Yu, and K. Bhattarai. An economic dispatch model incorporating wind power. *IEEE Transactions on Energy Conversion*, 23(2): 603–611, June 2008.

[22] H. W. Dommel and W. F. Tinney. Optimal power flow solutions. *IEEE Transactions on Power Apparatus and Systems*, PAS-87(10): 1866–1876, 1968.

[23] R. Billinton and S. S. Sachdeva. Real and reactive power optimization by suboptimum techniques. *IEEE Transactions on Power Apparatus and Systems*, PAS-92(3): 950–956, 1973.

[24] R. L. Sullivan. System parameter dispatching for generator MVAR control. In *Proc. IEEE Power Engineering Society Summer Meeting*, 1973.

[25] R. L. Sullivan. Controlling generator MVAR loading using a static optimization technique. *IEEE Transactions on Power Apparatus and Systems*, PAS-91(3): 906–910, 1972.

[26] J. D. Weber and T. J. Overbye. A two-level optimization problem for analysis of market bidding strategies. In *Proc. IEEE Power Engineering Society Summer Meeting*, pp. 682–687, 1999.

[27] J. D. Weber and T. J. Overbye. An individual welfare maximization algorithm for electricity markets. *IEEE Transactions on Power Systems*, 17(3): 590–596, 2002.

[28] A. L. Motto, F. D. Galiana, A. J. Conejo, and M. Huneault. On Walrasian equilibrium for pool-based electricity markets. *IEEE Transactions on Power Systems*, 17(3): 774–781, August 2002.

[29] M. S. Mahmoud. Multilevel systems control and applications: a survey. *IEEE Transactions on Systems, Man, and Cybernetics*, SMC-7(3): 125–143, March 1977.

[30] Y. C. Ho and S. K. Mitter (editors). *Directions in Large-Scale Systems: Many-Person Optimization and Decentralized Control*, Wakefield, MA, September 1976. Plenum Press, New York.

[31] R. Gallagher and T. Appenzeller. Beyond reductionism. *Science*, 284(5411): 79, April 1999.

[32] M. D. Mesarovic, D. Macko, and Y. Takahara. *Theory of Hierarchical Multilevel Systems*. Academic Press, New York, 1970.

[33] M. D. Mesarovic, J. D. Pearson, and Y. Takahara. A multilevel structure for a class of linear dynamic optimization problems. In *Proc. Joint Automatic Control Conf.*, New York, pp. 93–99, 1965.

[34] M. D. Mesarovic. *A conceptual framework for the studies of multilevel multigoal systems*. Technical Report, Systems Research Center, SRC 101-A-66-43, 1966.

[35] M. D. Mesarovic, J. D. Pearson, D. Macko, and Y. Takahara. On the synthesis of dynamic multilevel systems. In *Proc. Third Congr. Int. Federation of Automatic Control*, London, pp. 40B.1–40B.9, 1966.

[36] M. D. Mesarovic, D. Macko, and Y. Takahara. Two coordination principles and their application in large-scale systems control. *Automatica*, 6(2): 261–270, 1970.

[37] M. D. Mesarovic. *Systems concepts*. Technical Report, Scientific Thought Project, UNESCO, Paris, 1970.

[38] M. D. Mesarovic. Multilevel systems and concepts in process control. *Proc. IEEE*, 58(1): 111–125, 1970.

[39] M. D. Mesarovic. Multilevel systems theory: state of the art. In *Proc. Second Congr. Int. Federation of Automatic Control Systems*, Dusseldorf, 1971.
[40] M. D. Mesarovic. Role of separability of the performance (cost) function in the decomposition of decision problem. In *Decomposition of Large-scale Problems*, ed. D. M. Himmelblau pp. 341–346. American Elsevier, New York, 1973.
[41] J. D. Pearson. Decomposition, coordination and multilevel systems. *IEEE Transactions on Systems, Man, and Cybernetics*, SSC-2(1): 36–40, 1966.
[42] C. B. Brosilow, L. S. Lasdon, and J. D. Pearson. Feasible optimization methods for interconnected systems. In *Proc. Joint Automatic Control Conf.*, pp. 79–84, 1965.
[43] L. S. Lasdon and J. D. Schoeffler. A multilevel technique for optimization. In *Proc. Joint Automatic Control Conf.*, Troy, NY, pp. 85–92, 1965.
[44] L. S. Lasdon. *A multilevel technique for optimization*. Ph.D. Thesis, Cleveland, OH, 1964.
[45] C. Brosilow and L. S. Lasdon. A two level optimization technique for recycle processes. In *Proc AIChE Symp. Application of Mathematical Models in Chemical Engineering Research, Design and Production*, pp. 4: 75–4: 83, 1965.
[46] L. S. Lasdon. Duality and decomposition in mathematical programming. *IEEE Transactions on Systems, Man, and Cybernetics*, SSC-4(2): 86–111, 1968.
[47] L. S. Lasdon. *Optimization Theory for Large Systems*. Courier Dover, New York, 1970.
[48] J. D. Schoeffler. Static multilevel systems. In *Optimization Methods for Large-Scale Systems with Applications*, ed.W. A. Wismer, pp. 1–46. McGraw-Hill, New York, 1971.
[49] J. D. Pearson. Multilevel control systems. In *Proc. Second IFAC Symp. on the Theory of Self-Adaptive Control Systems*, pp. 252–256, 1965.
[50] S. Reich. *A critique of a decomposition technique for optimal control problems*. Master's Thesis Case Institute of Technology, 1966.
[51] J. D. Pearson and S. Reich. Decomposition of large optimal control problems. *Proc. IEE*, 114(6) 845–851, 1967.
[52] J. D. Pearson. Dynamic decomposition techniques. In *Optimization Methods for Large-Scale Problems with Applications*, ed. D. A. Wismer, pp. 121–190. New York:McGraw-Hill, 1971.
[53] J. D. Pearson. Duality and a decomposition technique. *SIAM Journal on Control*, 4(2): 164–172 1966.
[54] J. D. Pearson. On the duality between estimation and control. *SIAM Journal on Control*, 4(4) 594–600, 1966.
[55] G. Cohen and G. Jolland. Coordination methods by the prediction principle in large dynamic constrained optimization problems. In *Coordination Techniques Session, Int. Federation of Automatic Control Conf.*, 1976.
[56] M. G. Singh, S. A. W. Drew, and J. F. Coales. Comparisons of practical hierarchical control methods for interconnected dynamical systems. *Automatica*, 11(4): 331–350, 1975.
[57] Y. M. I. Dirickx, P. Jennergren, and D. W. Peterson. Some relationships between hierarchical systems theory and certain optimization problems. *IEEE Transactions on Systems, Man, and Cybernetics*, SMC-3(5): 514–518, 1973.
[58] T. E. Dy-Liacco. The adaptive reliability control system. *IEEE Transactions on Power Apparatus and Systems*, PAS-86(5): 517–531, 1967.
[59] F. C. Schweppe and S. K. Mitter. Hierarchical system theory and electric power systems. In *Proc. Symp. on Real-Time Control and Electric Power Systems*, Baden, Switzerland, pp. 259–277, 1971.
[60] A. M. Geoffrion. Lagrangian relaxation and its uses in integer programming. *Mathematical Programming Study*, 2: 82–114, 1974.
[61] M. L. Fisher. An applications oriented guide to Lagrangian relaxation. *Interfaces*, 15(2): 11–21, 1985.
[62] M. L. Fisher. The Lagrangian relaxation method for solving integer programming problems. *Management Science*, 27(1): 1–17, 1981.

[63] M. Held, P. Wolfe, and H. P. Crowder. Validation of subgradient optimization. *Mathematical Programming*, 6(1): 62–68, 1974.

[64] J. Heitkotter and D. Beasley. *The hitch-hiker's guide to evolutionary computation (FAQ for comp.ai.genetic)*. [online] http: //www.cse.dmu.ac.uk/~rij/gafaq/top.htmx, 2000.

[65] D. W. Hillis. Co-evolving parasites improve simulated evolution as an optimization procedure. In *Proc. Ninth Annual Int. Conf. of the Center for Nonlinear Studies on Self-organizing, Collective, and Cooperative Phenomena in Natural and Artificial Computing Networks on Emergent Computation*, pp. 311–314, 1991.

[66] M. A. Potter and K. A. De Jong. Cooperative coevolution: an architecture for evolving coadapted subcomponents. *Evolutionary Computation*, 8(1): 1–29, 2000.

[67] K. A. De Jong and M. A. Potter. Evolving complex structures via cooperative coevolution. In *Proc. 4th Annual Conf. on Evolutionary Programming*, San Diego, pp. 307–317, March 1995.

[68] H. Y. Chen and X. F. Wang. Cooperative coevolutionary algorithm for unit commitment. *IEEE Transactions on Power Systems*, 17(1): 128–133, 2002.

[69] M. A. Potter and K. A. De Jong. A cooperative coevolutionary approach to function optimization. In *Proc. Third Conf. Parallel Problem Solving from Nature*, New York, pp. 249–257, 1994.

[70] M. A. Potter. *The design and analysis of a computational model of cooperative coevolution*. Ph.D. Thesis, George Mason University, Fairfax, VA, 1997.

[71] R. P. Wiegand, W. C. Liles, and K. A. De Jong. An empirical analysis of collaboration methods in cooperative coevolutionary algorithms. In *Proc. Genetic and Evolutionary Computation Conf. (GECCO)*, Morgan Kaufmann, San Francisco, 2001.

[72] L. Samuelson. *Evolutionary Games and Equilibrium Selection*. MIT Press, Cambridge, MA, 1997.

[73] S. G. Ficici and J. B. Pollack. A game-theoretic approach to the simple coevolutionary algorithm. In *Parallel Problem Solving from Nature—PPSN VI*, pp. 467–476, Springer, Berlin, 2000.

[74] S. G. Ficici and J. B. Pollack. *Game theory and the simple coevolutionary algorithm: some preliminary results on fitness sharing*. In Genetic and Evolutionary Computation Conf. Workshop Program, pp. 2–7, 2001.

[75] R. P. Wiegand, Liles W., and K. A. De Jong. Analyzing cooperative coevolution with evolutionary game theory. In *Proc. 2002 Congress on Evolutionary Computation*, pp. 1600–1605, 2002.

[76] J. Park, J. Kim, and K. Y. Lee. Generation expansion planning in a competitive environment using a genetic algorithm. In *Proc. IEEE Power Engineering Society Summer Meeting*, pp. 1169–1172, 2002.

[77] K. P. Wong and Y. W. Wong. Hybrid genetic/simulated annealing approach to short-term multiple-fuel-constrained generation scheduling. *IEEE Transactions on Power Systems*, 12(2): 776–784, 1997.

[78] K. S. Swarup and S. Yamashiro. Unit commitment solution methodology using genetic algorithm. *IEEE Transactions on Power Systems*, 17(1): 87–91, 2002.

[79] S. Phelps, S. Parsons, P. McBurney, and E. Sklar. Co-evolution of auction mechanisms and trading strategies: towards a novel approach to microeconomic design. In *GECCO-02 Workshop on Evolutionary Computation in Multi-Agent Systems*, pp. 65–72, 2002.

[80] S. Phelps, S. Parsons, P. McBurney, and E. Sklar. *Co-evolutionary mechanism design: a preliminary report*, 2002.

[81] C. W. Richter Jr. and G. B. Sheble. Genetic algorithm evolution of utility bidding strategies for the competitive marketplace. *IEEE Transactions on Power Systems*, 13(1): 256–261, 1998.

[82] T. C. Price. Using co-evolutionary programming to simulate strategic behavior in markets. *Journal of Evolutionary Economics*, 7(3): 219–254, 1997.

[83] T. D. H. Cau and E. J. Anderson. A co-evolutionary approach to modeling the behavior of participants in competitive electricity markets. In *Proc. IEEE Power Engineering Society Summer Meeting*, pp. 682–687, 2002.

[84] Wikipedia. Microeconomics. [online] https: //en.wikipedia.org/wiki/Microeconomics, June 2015.

[85] H. R. Varian. *Intermediate Microeconomics*, 8th edn. W. W. Norton, New York, 2009.

[86] J. Hillas. *Lecture note in advanced microeconomics. Department of Economics, University of Auckland.* [online] http: //read.pudn.com/downloads66/sourcecode/others/235952/lec_scr.pdf 1999.

[87] A. Rubinstein. Comments on the interpretation of game theory. *Econometrica*, 59(4): 909–924 1991.

[88] R. J. Aumann. Agreeing to disagree. *Annals of Statistics*, 4: 1236–1239, 1976.

[89] J. Nash. Equilibrium points in *n*-person games. *Proceedings of the National Academy of Sciences* 36(1): 48–49, 1950.

[90] J. Nash. Non-cooperative games. *Annals of Mathematics*, 54(2): 286–295, 1951.

[91] R. B. Myerson. Perspectives on mechanism design in economic theory. *American Economic Review*, 98(3): 586–603, 2008.

[92] R. B. Myerson. Optimal auction design. *Mathematics of Operations Research*, 6(1): 58–73, 1981

[93] R. B. Myerson. Optimal coordination mechanisms in generalized principal-agent problems. *Journal of Mathematical Economics*, 10(1): 67–81, 1982.

[94] R. V. Vohra. Optimization and mechanism design. *Mathematical Programming*, 134(1): 283–303 August 2012.

[95] K. W. Edwin, R. Taud, H. D. Kochs, and T. S. Dillon. Integer programming approach to the problem of optimal unit commitment with probabilistic reserve determination. *IEEE Transactions on Power Apparatus and Systems*, 97(6): 2154–2166, 1978.

[96] D. Sjelvgren, H. Brannlund, and T. S. Dillon. Large-scale non-linear programming applied to operations planning. *International Journal of Electrical Power and Energy Systems*, 11(3): 213–217, 1989.

[97] T. S. Dillon, L. A. Sanabria, and L. A. Sjelvgren. Stochastic production costing method utilizing cumulants and time series load data for production planning of a large power system. In *Proc. 8th Power Systems Computation Conf.*, pp. 207–214, August 1984.

[98] T. S. Dillon and D. Sjelvgren. Seasonal planning approach to the hydrothermal system based on the network flow concept used at the Swedish State Power Board. In *Proc. IEEE Power Engineering Society Winter Meeting*, 1983.

[99] T. S. Dillon, S. Anderson, and D. Sjelvgren. Optimal operations planning in a large hydro-thermal power system. *IEEE Transactions on Power Apparatus and Systems*, PAS-102(11): 3644–3651, 1983.

[100] D. Sjelvgren and T. S. Dillon. Seasonal planning of hydro-thermal systems using the network flow concept. In *Proc. 7th Power Systems Computation Conf.*, pp. 1191–1199, 1981.

[101] D. Sjelvgren, R. W. Martin, and T. S. Dillon. A chance constrained approach to the problem of long term regulation of hydro-thermal systems. In *Proc. 7th Power Systems Computation Conf.*, pp. 393–401, 1981.

[102] D. Sjelvgren, S. Anderson, R. W. Martin, and T. S. Dillon. Modelling and organization of production planning system for hydro-thermal regulation in the Swedish State Power Board. In *Proc. 7th Power Systems Computation Conf.*, pp. 1200–1203, 1981.

[103] D. Sjelvgren, R. W. Martin, and T. S. Dillon. Stochastic optimization and modelling of large hydro-thermal systems for long term regulation. *International Journal of Electrical Power and Energy Systems*, 2(1): 2–20, 1980.

[104] R. W. Martin, D. Sjelvgren, L. Petterson, and T. S. Dillon. Modelling and optimization of a large hydro-thermal system for long term regulation including stochastic representation of inflows, loads and unit availability. In *Proc. 6th Power System Computation Conf.*, pp. 550–557, 1978.

[105] R. W. Martin, D. Sjelvgren, L. Petterson, and T. S. Dillon. Recent developments in modelling of large hydro thermal systems for long term operation planning including stochastic representation of inflows, loads and unit availability. In *Proc. Int. Symp. on System Reliability for Power Engineering*, pp. 1–37, 1978.

[106] International Atomic Energy Agency. *WASP-III guidebook*. Technical Report, Geneva, Switzerland, 1984.

[107] X. F. Wang and J. R. McDonald. *Modern Power System Planning*, 9th edn. McGraw-Hill, London, 1994.

[108] S. J. Wang, S. M. Shahidehpour, D. S. Kirshen, S. Mokhtari, and G. D. Irisarri. Short-term generation scheduling with transmission and environmental constraints using an augmented Lagrangian relaxation. *IEEE Transactions on Power Systems*, 10(3): 1294–1301, 1995.

[109] X. F. Wang. Equivalent energy function approach to power system probabilistic modeling. *IEEE Transactions on Power Systems*, 3(3): 823–829, 1988.

[110] M. Lin, A. Breipohl, and F. Lee. Comparison of probabilistic production cost simulation methods. *IEEE Transactions on Power Systems*, 4(4): 1326–1333, 1989.

[111] L. Y. Tong, Z. H. Wang, Y. D. Han, and X. B. Wang. Long term transmission expansion planning based on a critical feasible configuration matching method. *Proceedings of the Chinese Society of Universities for Electric Power System and Automation*, 1(1): 12–22, 1989.

[112] A. P. Meliopoulos, R. P. Webb, R. J. Bennon, and J. A. Juves. Optimal long range transmission planning with AC load flow. *IEEE Transactions on Power Apparatus and Systems*, 101(10): 4156–4163, 1982.

[113] E. L. D. Silva, H. A. Gil, and J. M. Areiza. Transmission network expansion planning under an improved genetic algorithm. *IEEE Transactions on Power Systems*, 15(3): 1168–1174, 2000.

[114] R. Romero, R. A. Gallego, and A. Monticelli. Transmission system expansion planning by simulated annealing. *IEEE Transactions on Power Systems*, 11(1): 364–369, 1996.

[115] R. A. Gallego, A. Monticelli, and R. Romero. Transmission system expansion planning by an extended genetic algorithm. *Generation, Transmission and Distribution, IEE Proceedings*, 145(3): 329–335, 1998.

[116] R. A. Gallego, A. Monticelli, and R. Romero. Comparative studies on non-convex optimization methods for transmission network expansion planning. *IEEE Transactions on Power Systems*, 13(3): 822–828, 1998.

[117] R. Villasana, L. L. Garver, and S. J. Salon. Transmission network planning using linear programming. *IEEE Transactions on Power Apparatus and Systems*, (2): 349–356, 1985.

[118] Z. M. Al-Hamouz and A. S. Al-Faraj. Transmission expansion planning using nonlinear programming. In *Transmission and Distribution Conf. and Exhibition 2002: Asia Pacific, IEEE/PES*, vol. 1, pp. 50–55. IEEE, New York, 2002.

[119] N. Alguacil, A. L. Motto, and A. J. Conejo. Transmission expansion planning: a mixed-integer LP approach. *IEEE Transactions on Power Systems*, 18(3): 1070–1077, 2003.

[120] L. Bahiense, G. C. Oliveira, M. Pereira, and S. Granville. A mixed integer disjunctive model for transmission network expansion. *IEEE Transactions on Power Systems*, 16(3): 560–565, 2001.

[121] A. Santos Jr., P. M. Franca, and A. Said. An optimization model for long-range transmission expansion planning. *IEEE Transactions on Power Systems*, 4(1): 94–101, 1989.

[122] A. Seifu, S. Salon, and G. List. Optimization of transmission line planning including security constraints. *IEEE Transactions on Power Systems*, 4(4): 1507–1513, 1989.

[123] S. Binato, M. V. F. Pereira, and S. Granville. A new Benders decomposition approach to solve power transmission network design problems. *IEEE Transactions on Power Systems*, 16(2): 235–240, 2001.

[124] S. N. Siddiqi and M. L. Baughman. Value-based transmission planning and the effects of network models. *IEEE Transactions on Power Systems*, 10(4): 1835–1842, 1995.

[125] G. C. Oliveira, A. P. C. Costa, and S. Binato. Large scale transmission network planning using optimization and heuristic techniques. *IEEE Transactions on Power Systems*, 10(4): 1828–1834, 1995.

[126] C. W. Lee, S. K. K. Ng, J. Zhong, and F. F. Wu. Transmission expansion planning from past to future. In *Power Systems Conf. and Exposition, 2006: IEEE/PES*, pp. 257–265. IEEE, New York, 2006.

[127] California ISO. *A proposed methodology for evaluating the economic benefits of transmission expansions in a restructured wholesale electricity market*. Technical Report, California ISO and London Economics International, 2003.

[128] F. F. Wu, F. L. Zheng, and F. S. Wen. Transmission investment and expansion planning in a restructured electricity market. *Energy*, 31(6): 954–966, 2006.

[129] Y.-C. Ho. An explanation of ordinal optimization: soft computing for hard problems. *Information Sciences*, 113(3): 169–192, 1999.

[130] D. Conforti, D. Menniti, N. Sorrentino, P. Verde, and A. Testa. Optimal load-flow with $n-1$ steady-state security via high performance computing. In *Proc. Mediterranean Electrotechnical Conf. (MELECON), Industrial Applications in Power Systems, Computer Science and Telecommunications*, vol. 2, pp. 873–879, 1996.

[131] T. W. E. Lau and Y. C. Ho. Universal alignment probabilities and subset selection for ordinal optimization. *Journal of Optimization Theory and Applications*, 93(3): 455–489, 1997.

[132] Y. C. Ho, R. S. Sreenivas, and P. Vakili. Ordinal optimization in DEDS. *Discrete Event Dynamic Systems*, 2(1): 61–88, 1992.

[133] Y. C. Ho. Overview of ordinal optimization. In *Proc. 33rd IEEE Conf. on Decision and Control, 1994*, vol. 2, pp. 1975–1977. IEEE, New York, 1994.

[134] L. Dai. Convergence properties of ordinal comparison in the simulation of discrete event dynamic systems. *Journal of Optimization Theory and Applications*, 91(2): 363–388, 1996.

[135] D. Li, L. H. Lee, and Y. C. Ho. Constraint ordinal optimization. *Information Sciences*, 148(1): 201–220, 2002.

[136] D. Li, L. H. Lee, and Y. C. Ho. Vector ordinal optimization—a new heuristic approach and its application to computer network routing design problems. *International Journal of Operations and Quantitative Management*, 5(3): 211–230, 1999.

[137] Y. C. Ho and M. Larson. Ordinal optimization approach to rare event probability simulation. *Discrete Event Dynamic Systems*, 5(2-3): 281–301, 1995.

[138] Y. C. Ho. *Soft optimization for hard problems*. Computerized lecture via private communication/distribution. 1996.

[139] M. Deng and Y. C. Ho. Sampling selection method for stochastic opsimulation problems. *Automation*, 35(2): 331–338, 1999.

[140] S. Y. Lin, Y. C. Ho, and C. H. Lin. An ordinal optimization theory-based algorithm for solving the optimal power flow problem with discrete control variables. *IEEE Transactions on Power Systems*, 19(1): 276–286, 2004.

[141] X. Guan, Y. C. Ho, and F. Lai. An ordinal optimization based bidding strategy for electric power suppliers in the daily energy market. *IEEE Transactions on Power Systems*, 16(4): 788–797, 2001.

[142] M. A. Piette. *Open automated demand response communications specification (version 1.0)*. Lawrence Berkeley National Laboratory, 2009.

[143] Q. Zhang, X. Wang, J. Wang, C. Feng, and L. Liu. Survey of demand response research in deregulated electricity markets. *Automation of Electric Power Systems*, 3: 97–106, 2008.

[144] Y. Xiang, J.-Y. Liu, Z.-B. Wei, and Y.-B. Liu. Research on development of new interruptible load with renewable energy access. *Power System Protection and Control*, 40(5): 148–155, 2012.

[145] E. Celebi and J. D. Fuller. A model for efficient consumer pricing schemes in electricity markets. *IEEE Transactions on Power Systems*, 22(1): 60–67, 2007.

[146] B. Wang, Y. Li, and C. Gao. Demand side management outlook under smart grid infrastructure. *Automation of Electric Power Systems*, 20: 5–14, 2009.

[147] Y. X. Yu and W. Luan. Smart grid and its implementations. *Proceedings of the Chinese Society for Electrical Engineering*, 29(34): 1–8, 2009.

[148] M. A. Piette, O. Sezgen, D. S. Watson, N. Motegi, C. Shockman, and L. Ten Hope. *Development and evaluation of fully automated demand response in large facilities*. Lawrence Berkeley National Laboratory, 2004.

[149] B. Wang and Y. Li. Demand side management planning and implementation mechanism for smart grid. *Electric Power Automation Equipment*, 30(12): 19–24, 2010.

[150] M. Zeng, C. Lu, L. Qiu, and K. Tian. A demand-side response-based transmission planning model with grid-connected wind farms. *Power System Technology*, 4: 25, 2011.

[151] Z. Liu, B. Xu, and D. Niu. Development tendency of equilibrium analysis and demand response for smart grid. *Power System Technology*, 2013-6, 2013.

[152] J. Liu. Basic issues of the utilization of large-scale renewable power with high security and efficiency. *Proceedings of the Chinese Society for Electrical Engineering*, 16:2, 2013.

[153] X.-Y. Qiu, L.-L. Xia, and X.-Y. Li. Planning of distributed generation in construction of smart grid. *Power System Technology*, 34(4): 7–10, 2010.

[154] A. Charnes and W. W. Cooper. Chance-constrained programming. *Management Science*, 6(1): 73–79, 1959.

[155] A. Li, Z. Qin, F. Bao, and S.-P. He. Particle swarm optimization algorithm. *Computer Engineering and Applications*, 21(38): 1–3, 2002.

[156] X. Zhang. Introduction to statistical learning theory and support vector machines. *Acta Automatica Sinica*, 26(1): 32–42, 2000.

[157] A. E. Feijoo, J. Cidras, and J. L. Garcia Dornelas. Wind speed simulation in wind farms for steady-state security assessment of electrical power systems. *IEEE Transactions on Energy Conversion*, 14(4): 1582–1588, 1999.

[158] I. Abouzahr and R. Ramakumar. An approach to assess the performance of utility-interactive wind electric conversion systems. *IEEE Transactions on Energy Conversion*, 6(4): 627–638, 1991.

[159] H. Yu, H. Zhou, and X. Pei. Environmental value and economic analysis of wind power. *Journal of Tongji University (Natural Science)*, 5: 026, 2009.

[160] G. Celli and F. Pilo. Optimal distributed generation allocation in mv distribution networks. In *Proc. 22nd IEEE Power Engineering Society Int. Conf. on Power Industry Computer Applications*, pp. 81–86. IEEE, New York, 2001.

[161] W. L. Snyder, H. D. Powell Jr., and J. C. Rayburn. Dynamic programming approach to unit commitment. *IEEE Transactions on Power Systems*, 2(2): 339–350, 1987.

[162] C. K. Pang, G. B. Sheble, and F. Albuyeh. Evaluation of dynamic programming based methods and multiple area representation for thermal unit commitments. *IEEE Transactions on Power Apparatus and Systems*, PAS-100(3): 1212–1218, 1981.

[163] A. Merlin and P. Sandrin. A new method for unit commitment at Électricité de France. *IEEE Transactions on Power Apparatus and Systems*, PAS-102(5): 1218–1225, 1983.

[164] S. A. Kazarlis, A. G. Bakirtzis, and V. Petridis. A genetic algorithm solution to the unit commitment problem. *IEEE Transactions on Power Systems*, 11(1): 83–92, 1996.

[165] T. T. Maifeld and G. B. Sheble. Genetic-based unit commitment. *IEEE Transactions on Power Systems*, 11(3): 1359–1370, 1996.

[166] D. E. Goldberg. *Genetic Algorithms in Search, Optimization, and Machine Learning*. Addison-Wesley Longman, Boston, MA, 1989.

[167] P. Wong, P. Albrecht, R. Allan, R. Billinton, Q. Chen, and C. Fong *et al.* The IEEE Reliability Test System-1996. A report prepared by the Reliability Test System Task Force of the Application of Probability Methods Subcommittee. *IEEE Transactions on Power Systems*, 14(3): 1010–1020, 1999.

[168] A. Kusiak, H. Zheng, and Z. Song. Short-term prediction of wind farm power: a data mining approach. *IEEE Transactions on Power Systems*, 24(1): 125–136, 2009.

[169] A. Tuohy, P. Meibom, E. Denny, and M. O'Malley. Unit commitment for systems with significant wind penetration. *IEEE Transactions on Power Systems*, 24(2): 592–601, 2009.

[170] L. Wu, M. Shahidehpour, and T. Li. Stochastic security-constrained unit commitment. *IEEE Transactions on Power Systems*, 22(2): 800–811, 2007.

[171] P. A. Ruiz, C. R. Philbrick, E. Zak, K. W. Cheung, and P. W. Sauer. Uncertainty management in the unit commitment problem. *IEEE Transactions on Power Systems*, 24(2): 642–651, 2009.

[172] F. Bouffard and F. D. Galiana. Stochastic security for operations planning with significant wind power generation. *IEEE Transactions on Power Systems*, 23(2): 306–316, 2008.
[173] J. H. Wang, M. Shahidehpour, and Z. Y. Li. Security-constrained unit commitment with volatile wind power generation. *IEEE Transactions on Power Systems*, 23(3): 1319–1327, 2008.
[174] D. Streiffert, R. Philbriek, and A. Ott. A mixed integer programming solution for market clearing and reliability analysis. *In Proc. IEEE Power Engineering Society General Meeting*, vol. 3, pp. 2724–2731, 2005.
[175] R. E. Bixby, M. Fenolon, and Z. Gu. MIP: theory and practice—closing the gap. In *System Modelling and Optimization: Methods, Theory, and Applications*, eds. M. J. D. Powell and S. Scholtes, pp. 19–49. Kluwer Academic, Dordrecht, 2000.
[176] J. Dupacova, N. Growe-Kuska, and W. Romisch. Scenario reduction in stochastic programming: an approach using probability metrics. *Mathematical Programming A*, 3: 493–11, 2003.
[177] W. Y. Ng. Generalized generation distribution factors for power system security evaluations. *IEEE Transactions on Power Apparatus and Systems*, PAS-100(3): 1001–1005, 1981.
[178] M. Carrion and J. M. Arroyo. A computationally efficient mixed-integer linear formulation for the thermal unit commitment problem. *IEEE Transactions on Power Apparatus and Systems*, 21(3): 1371–1378, 2006.
[179] G. Xu, Y. Liu, and K. Cheng. *Basic Enchiridion of Operational Research*. Science Publishing Company, Beijing, 1999.
[180] M. K. Donnelly, J. E. Dagle, D. J. Trudnowski, and G. J. Rogers. Impact of the distributed utility on transmission system stability. *IEEE Transactions on Power Systems*, 11(2): 741–746, 1996.
[181] D. I. Sun, B. Ashley, B. Brewer, A. Hughes, and W. F. Tinney. Optimal power flow by Newton approach. *IEEE Transactions on Power Apparatus and Systems*, PAS-103(10): 2864–2880, 1984.
[182] B. Stott and J. L. Marinho. Linear programming for power-system network security applications. *IEEE Transactions on Power Apparatus and Systems*, PAS-98(3): 837–848, 1979.
[183] T. Yong and L. Robert. Optimal power flow formulation in market of retail wheeling. *IEEE Transactions on Power Systems*, 1(1): 394–398, 1999.
[184] J. Carpentier. Contribution à l'étude du dispatching économique. *Bulletin de la Société Française des Électriciens*, 3: 431–447, 1962.
[185] Z. H. Wang, X. G. Yin, Z. Zhang, and J. C. Yang. Pseudo-parallel genetic algorithm for reactive power optimization. In *Proc. IEEE Power Engineering Society General Meeting*, vol. 2, pp. 903–907, 2003.
[186] Y. Liu and X. Qiu. Optimal reactive power and voltage control for radial distribution system. In *Proc. IEEE Power Engineering Society Summer Meeting*, Seattle, vol. 1, pp. 85–90, July 2000.
[187] H. Wei, H. Sasaki, and R. Yokoyama. An application of interior point quadratic programming algorithm to power system optimization problems. *IEEE Transactions on Power Systems*, 11(1): 260–267, 1996.
[188] H. Wei, H. Sasaki, J. Kubokawa, and R. Yokoyama. An interior point nonlinear programming for optimal power flow problems with a novel structure. *IEEE Transactions on Power Systems*, 13: 870–877, 1998.
[189] M. Liu, S. K. Tso, and Y. Cheng. An extended nonlinear primal–dual interior-point algorithm for reactive-power optimization of large-scale power system with discrete control variables. *IEEE Transactions on Power Systems*, 17(4): 982–991, 2002.
[190] M. Delfanti, G. P. Granelli, P. Marannino, and M. Montagna. Optimal capacitor placement using deterministic and genetic algorithms. *IEEE Transactions on Power Systems*, 15(3): 1041–1046, 2000.
[191] S. Haffner, A. Monticelli, and A. Garcia. Specialised branch-and-bound algorithm for transmission network expansion planning. *Generation, Transmission and Distribution, IEE Proceedings*, 148(5): 482–488, 2001.
[192] J. E. Mitchell and M. J. Todd. Solving combinatorial optimization problems using Karmarkar's algorithm. *Mathematical Programming*, 56(1-3): 245–284, 1992.

[193] J. E. Mitchell and B. Borchers. Solving real world linear ordering problems using a primal–dual interior point cutting plane method. *Annals of Operations Research*, 62(1): 253–276, 1996.

[194] X. Y. Ding, X. F. Wang, and Y. H. Song. Interior point cutting plane method for optimal power flow. *IMA Journal of Management Mathematics*, 15(4): 355–368, 2004.

[195] J. L. Goffin, J. Gondzio, and R. Sarkissian. Solving nonlinear multicommodity flow problems by the analytic center cutting plane method. *Mathematical Programming*, 76(1): 131–154, 1996.

[196] J. L. Goffin and J. P. Vial. A two-cut approach in the analytic center cutting plane method. *Mathematical Methods of Operations Research*, 49(1): 149–169, 1999.

[197] J. E. Kelly. The cutting-plane method for solving convex programs. *Journal of the Society for Industrial and Applied Mathematics*, 8(4): 703–712, 1960.

[198] W. Tapio and P. Frank. An extended cutting plane method for solving convex MINLP problems. *Computers and Chemical Engineering*, 19(1): 131–136, 1995.

[199] M. Marko. Survey of bundle methods for nonsmooth optimization. *Optimization Methods and Software*, 17(1): 1–29, 2002.

[200] M. Padberg. Classical cuts for mixed-integer programming and branch and cut. *Mathematical Methods of Operations Research*, 53(2): 173–203, 2001.

[201] G. A. Tijssen and G. Sierksma. Balinski–Tucker simplex tableaus: dimensions, degeneracy degrees, and interior points of optimal faces. *Mathematical Programming*, 81(3): 349–372, 1998.

[202] Y. Ye. On the finite convergence of interior-point algorithms for linear programming. *Mathematical Programming*, 57(1-3): 325–335, 1992.

[203] N. Megiddo. On finding primal and dual optimal bases. *ORSA Journal on Computing*, 3(1): 63–65, 1991.

[204] R. A. Tapia and Y. Zhang. An optimal basis identification technique for interior-point linear programming algorithms. *Linear Algebra and Its Applications*, 152(1): 343–363, 1991.

[205] S. Mehrotra. On finding a vertex solution using interior point method. *Linear Algebra and Its Applications*, 152: 233–253, 1991.

[206] M. A. El-Sayed. Artificial neural network for re-active power optimization. *Decision Support Systems*, 23: 255–263, 1998.

[207] D. Bhagwan Das and C. Patvardhan. Reactive power dispatch with a hybrid stochastic search technique. *International Journal of Electrical Power and Energy Systems*, 24(9): 731–736, 2002.

[208] M. Khiat, A. Chaker, A. Gómez Expósito, and J. L. Martínez Ramo. Reactive power optimization and voltage control in the Western Algerian transmission system: a hybrid approach. *Electric Power Systems Research*, 64(1): 3–10, 2003.

[209] Q. H. Wu, Y. J. Cao, and J. Y. Wen. Optimal reactive power dispatch using an adaptive genetic algorithm. *International Journal of Electrical Power and Energy Systems*, 20(8): 563–569, 1998.

[210] J. Z. Zhu and X. F. Xiong. Optimal reactive power control using modified interior point method. *Electric Power Systems Research*, 66(2): 187–192, 2003.

[211] H. Liu and J. H. Frazer. Supporting evolution in a multi-agent cooperative design environment. *Advances in Engineering Software*, 33(6): 319–328, 2002.

[212] G. M. P. O'Hare and N. R. Jennings. *Foundations of Distributed Artificial Intelligence*. John Wiley & Sons, Inc., New York, 1996.

[213] A. Brandolese, A. Brun, and A. Portioli-Staudacher. A multi-agent approach for the capacity allocation problem. *International Journal of Production Economics*, 66(3): 269–285, 2000.

[214] N. R. Jennings. Agent-based computing: promises and perils. In *Proc. Int. Joint Conf. on Artificial Intelligence*, vol. 2, 1999.

[215] V. Botti, F. Barber, A. Crespo, E. Onaindia, A. Garcia-Fornes, I. Ripoll, D. Gallardo, and L. Hernandez. A temporal blackboard for a multi-agent environment. *Data and Knowledge Engineering*, 15(3): 189–211, 1995.

[216] Y. Zhang, Z. Ren, H. Zhong, Z. Tang, and C. Shang. Cataclysmic genetic algorithms based optimal reactive power planning. *Dianli Xitong Zidonghua (Automation of Electric Power Systems)*, 26(23): 29–32, 2002.

[217] N. Deeb and S. M. Shahidehpour. Linear reactive power optimization in a large power network using the decomposition approach. *IEEE Transactions on Power Systems*, 5(2): 428–438, 1990.

[218] S. Granville. Optimal reactive dispatch through interior point methods. *IEEE Transactions on Power Systems*, 9(1): 136–146, 1994.

[219] J. R. Gomes and O. R. Saavedra. Optimal reactive power dispatch using evolutionary computation: extended algorithms. *Generation, Transmission and Distribution, IEE Proceedings*, 146(6): 586–592, 1999.

[220] J. Z. Zhu and X. F. Xiong. Optimal reactive power control using modified interior point method. *Electric Power Systems Research*, 66(2): 187–192, 2003.

[221] Q. H. Wu, Y. J. Cao, and J. Y. Wen. Optimal reactive power dispatch using an adaptive genetic algorithm. *International Journal of Electrical Power and Energy Systems*, 20(8): 563–569, 1998.

[222] M. A. H. El-Sayed. Artificial neural network for reactive power optimization. *Neurocomputing*, 23(1–3): 255–263, 1998.

[223] C.-T. Su and C.-T. Lin. A new optimal reactive power scheduling method for loss minimization and voltage stability margin maximization using successive multi-objective fuzzy LP technique. *IEEE Transactions on Power Delivery*, 16(2): 319–323, 2001.

[224] C.-T. Su and C.-T. Lin. Fuzzy-based voltage/reactive power scheduling for voltage security improvement and loss reduction. *IEEE Transactions on Power Delivery*, 16(2): 319–323, 2001.

[225] W. F. Tinney, J. M. Bright, K. D. Demaree, and B. A. Hughes. Some deficiencies in optimal power flow. *IEEE Transactions on Power Systems*, 3(2): 676–683, 1988.

[226] W.-H. E. Liu, A. D. Papalexopoulos, and W. F. Tinney. Discrete shunt controls in a Newton optimal power flow. *IEEE Transactions on Power Systems*, 7(4): 1509–1518, 1992.

[227] S. A. Soman, K. Parthasarathy, and D. Thukaram. Curtailed number and reduced controller movement optimization algorithms for real time voltage/reactive power control. *IEEE Transactions on Power Systems*, 9(4): 2035–2041, 1994.

[228] G. A. Taylor, M. Rashidinejad, Y.-H. Song, M. R. Irving, M. E. Bradley, and T. G. Williams. Algorithmic techniques for transition-optimized voltage and reactive power control. *IEEE Transactions on Power Systems*, 3: 1660–1664, 2002.

[229] Z. Hu, X. Wang, H. Chen, and G. A. Taylor. Volt/VAR control in distribution systems using a time-interval based approach. *Generation, Transmission and Distribution, IEE Proceedings*, 150(5): 548–554, 2003.

[230] S. S. Sharif, J. H. Taylor, and E. F. Hill. Dynamic online energy loss minimization. *Generation, Transmission and Distribution, IEE Proceedings*, 148(2): 172–176, 2001.

[231] Y. Deng, X. Ren, C. Zhao, and D. Zhao. A heuristic and algorithmic combined approach for reactive power optimization with time-varying load demand in distribution systems. *Power Engineering Review, IEEE*, 22(10): 61, 2002.

[232] Y.-Y. Hong and C.-M. Liao. Short-term scheduling of reactive power controllers. *IEEE Transactions on Power Systems*, 10(2): 860–868, 1995.

[233] Y.-J. Zhang and Z. Ren. Real-time optimal reactive power dispatch using multi-agent technique. *Electric Power Systems Research*, 19(2-3): 259–265, 2004.

[234] H. Ghoudjehbaklou, S. Lakshminarasimhan, and L. Jansen. Template based automatic voltage control. *Transmission and Distribution Conf. and Exposition 2008: Latin America, IEEE/PES*, pp. 1–5, 2008.

[235] S. Corsi, M. Pozzi, C. Sabelli, and A. Serrani. The coordinated automatic voltage control of the Italian transmission grid. Part I: Reasons of the choice and overview of the consolidated hierarchical system. *IEEE Transactions on Power Systems*, 19(4): 1723–1732, 2004.

[236] H. M. Ma, K.-T. Ng, and K. F. Man. Multi-objective coordinated power voltage control using jumping genes paradigm. *IEEE Transactions on Industrial Electronics*, 55(11): 4075–4084, 2008.

[237] M. El Moursi, G. Joos, and C. Abbey. A secondary voltage control strategy for transmission level interconnection of wind generation. *IEEE Transactions on Power Systems*, 23(3): 1178–1190, 2008.

[238] K. Tanaka. Decentralized voltage control in distribution systems by distributed generators. In *Proc. IEEE Int. Symp. on Industrial Electronics, 2009*, pp. 554–559, 2009.

[239] S. Foster, L. Xu, and B. Fox. Coordinated reactive power control for facilitating fault ride through of doubly fed induction generator- and fixed speed induction generator-based wind farms. *Renewable Power Generation, IET*, 4(2): 128–138, 2010.

[240] K. Pan and X. Han. A reactive power optimization method based on the decomposition and coordination theory for AVC in power system. In *Proc. Int. Conf. on Power System Technology, 2004*, vol. 1, pp. 872–875, 2004.

[241] J. Y. Wen, Q. H. Wu, D. R. Turner, S. J. Cheng, and J. Fitch. Optimal coordinated voltage control for power system voltage stability. *IEEE Transactions on Power Systems*, 19(2): 1115–1122, 2004.

[242] T. Nagata, R. Hatano, and H. Saiki. A multi-agent based distributed reactive power control method. In *Proc. IEEE Power Engineering Society General Meeting*, pp. 1–7, 2009.

[243] X. Su, G. He, X. Zhang, P. Ma, and W. Yin. A new coordinated voltage/reactive power control scheme for interconnected power system. *IPEC, 2010 Conference Proceedings*, pp. 602–607, IEEE, New York, 2010.

[244] J.-S. Ma and H.-G. Yang. Distributed parallel coordinated control strategy for provincial–regional grid based on subarea division of the power system. *Proc. 2010 China Int. Conf. on Electricity Distribution (CICED)*, pp. 1–6, 2010.

[245] M. J. Osborne and A. Rubinstein. *A Course in Game Theory*. MIT Press, Cambridge, MA, 1994.

[246] T. Nagata, R. Hatano, and H. Saiki. Game theory: analysis of conflict. In *Proc. IEEE Power Engineering Society General Meeting*, Cambridge, 2009.

[247] J. M. Bilbao. *Cooperative Games on Combinational Structures*. Kluwer Academic, Dordrecht,2000.

[248] R. Branzei, D. Dimitrov, and S. Tijs. *Models in Cooperative Game Theory*. Springer, Berlin, 2008.

[249] Y.-J. Zhang and Z. Ren. Optimal reactive power dispatch considering costs of adjusting the control devices. *IEEE Transactions on Power Systems*, 20(3): 1349–1356, 2005.

[250] Y. Y. Hsu and H. C. Kuo. Dispatch of capacitors on distribution system using dynamic programming. *Generation, Transmission and Distribution, IEE Proceedings*, 140(6): 433–438, 1993.

[251] F. C. Lu and Y. Y. Hsu. Reactive power/voltage control in a distribution substation using dynamic programming. *Generation, Transmission and Distribution, IEE Proceedings*, 142(6): 639–645, 1995.

[252] R. H. Liang and C. K. Cheng. Dispatch of main transformer ULTC and capacitors in a distribution system. *IEEE Transactions on Power Delivery*, 16(4): 626–630, 2001.

[253] A. L. Shenkman. Energy loss computation by using statistical techniques. *IEEE Transactions on Power Delivery*, 5(1): 254–258, 1990.

[254] S. Civanlar and J. J. Grainger. Volt/VAR control on distribution systems with lateral branches using shunt capacitors and voltage regulatory. Part III: Numerical results. *IEEE Transactions on Power Apparatus and Systems*, 104(11): 3291–3297, 1985.

[255] R. Baldick. Electricity market equilibrium models: the effect of parametrization. *IEEE Transactions on Power Systems*, 17(4): 1170–1176, 2002.

[256] A. K. David and F. S. Wen. Market power in electricity supply. *IEEE Transactions on Power Systems*, 16(4): 352–360, 2001.

[257] W. W. Hogan. A market power model with strategic interaction in electricity networks. *Energy Journal*, 18(4): 107–141, 1997.

[258] J. B. Cardell, C. C. Hitt, and W. W. Hogan. Market power and strategic interaction in electricity networks. *Resource and Energy Economics*, 19(1-2): 109–137, 1997.

[259] S. Borenstein and J. Bushnell. An empirical analysis of the potential for market power in California's electricity industry. *Journal of Industrial Economics*, 47(3): 285–323, 1999.

[260] S. Borenstein, J. Bushnell, and C. R. Knittel. Market power in electricity markets: beyond concentration measures. *Energy Journal*, 20(4): 65–88, 1999.

[261] L. B. Cunningham, R. Baldick, and M. L. Baughman. An empirical study of applied game theory: transmission constrained Cournot behavior. *IEEE Transactions on Power Systems*, 17(1): 166–172, 2002.

[262] P. D. Klemperer and M. A. Meyer. Supply function equilibria in oligopoly under uncertainty. *Econometrica*, 57(6): 1243–1277, 1989.

[263] R. J. Green and D. M. Newbery. Competition in the British electricity spot market. *Journal of Political Economy*, 100(5): 929–953, 1992.

[264] R. J. Green. Increasing competition in the British electricity spot market. *Journal of Industrial Economics*, 44(2): 205–216, 1996.

[265] B. F. Hobbs, C. B. Metzler, and J. S. Pang. Strategic gaming analysis for electric power systems: an MPEC approach. *IEEE Transactions on Power Systems*, 15(2): 638–645, 2000.

[266] A. Rudkevich, M. Duckworth, and R. Rosen. Modeling electricity pricing in a deregulated generation industry: the potential for oligopoly pricing in a Poolco. *Energy Journal*, 19(3): 19–48, 1998.

[267] C. J. Day and B. F. Hobbs. Oligopolistic competition in power networks: a conjectured supply function approach. *IEEE Transactions on Power Systems*, 17(3): 597–607, 2002.

[268] S. J. Rassenti, V. L. Smith, and B. J. Wilson. Using experiments to inform the privatization/deregulation movement in electricity. *Cato Journal*, 21(3): 515–544, 2002.

[269] D. W. Bunn and F. S. Oliveira. Agent-based simulation—an application to the new electricity trading arrangements of England and Wales. *IEEE Transactions on Evolutionary Computation*, 5(5): 493–503, 2001.

[270] J. Nicolaisen, V. Petrov, and L. Tesfatsion. Market power and efficiency in a computational electricity market with discriminatory double-auction pricing. *IEEE Transactions on Evolutionary Computation*, 5(5): 504–523, 2001.

[271] R. P. Wiegand. *An analysis of cooperative coevolutionary algorithms*. Ph.D. Dissertation, George Mason University, Fairfax, VA, 2003.

[272] Y. S. Son and R. Baldick. Hybrid coevolutionary programming for Nash equilibrium search in games with local optima. *IEEE Transactions on Evolutionary Computation*, 8(4): 305–315, 2004.

[273] L. Tesfatsion. Agent-based computational economics: growing economies from the bottom-up. *Artificial Life*, 8(1): 55–82, 2002.

[274] T. Riechmann. Genetic algorithm learning and evolutionary games. *Journal of Economic Dynamics and Control*, 25(6–7): 1019–1037, 2001.

[275] G. A. Jehle and P. J. Reny. *Advanced Microeconomic Theory*. Addison-Wesley, Boston, MA, 2001.

[276] N. Fabra and J. Toro. Price wars and collusion in the Spanish electricity market. *International Journal of Industrial Organization*, 23: 155–181, 2005.

[277] J. Tirole. *The Theory of Industrial Organization*. MIT Press, Cambridge, MA, 2000.

[278] J. W. Friedman. A non-cooperative equilibrium for supergames. *Review of Economic Studies*, 38: 1–12, 1971.

[279] K. Deb, A. Pratap, S. Agarwal, and T. Meyarivan. A fast and elitist multiobjective genetic algorithm: NSGA-II. *IEEE Transactions on Evolutionary Computation*, 6(2): 182–197, 2002.

[280] R. Baldick and W. Hogan. *Capacity constrained supply function equilibrium models of electricity markets: stability, non-decreasing constraints, and function space iterations*. University of California Energy Institute, [online] http: //www.ucei.berkeley.edu/PDF/pwp089.pdf, 2001.

[281] H. Chen, K. P. Wong, D. H. M. Nguyen, and C. Y. Chung. Analyzing oligopolistic electricity market using coevolutionary computation. *IEEE Transactions on Power Systems*, 21(1): 143–152, 2006.

[282] R. J. Green. The electricity contract market in England and Wales. *Journal of Industrial Economics*, 47(1): 107–124, 1999.

[283] A. Rudkevich. *Supply function equilibrium in power markets: learning all the way*. Technical Report, TCA Technical Paper 1299-1702, 1999.

[284] R. Baldick, R. Grant, and E. P. Kahn. *Linear supply function equilibrium: generations, application, and limitations*. University of California Energy Institute, [online] http://www.ucei. berkeley.edu/PDF/pwp078.pdf, August 2001.

[285] R. Baldick, R. Grant, and E. P. Kahn. Theory and application of linear supply function equilibrium in electricity market. *Journal of Regulatory Economics*, 25(2): 143–167, 2004.

[286] C. J. Day and D. W. Bunn. Divestiture of generation assets in the electricity pool of England and Wales: a computation approach to analyzing market power. *Journal of Regulatory Economics*, 19(2): 123–141, 2001.

[287] T. Li and M. Shahidehpour. Price-based unit commitment: a case of Lagrangian relaxation versus mixed integer programming. *IEEE Transactions on Power Systems*, 20(4): 2015–2025, 2005.

[288] T. Peng and K. Tomsovic. Congestion influence on bidding strategies in an electricity market. *IEEE Transactions on Power Systems*, 18(3): 1054–1061, 2003.

[289] P. F. Correia, T. J. Overbye, and I. A. Hiskens. Searching for non-cooperative equilibria in centralized electricity market. *IEEE Transactions on Power Systems*, 18(4): 1417–1424, 2003.

[290] X. Wang, Y. Li, and S. Zhang. Oligopolistic equilibrium analysis for electricity markets: a nonlinear complementarity approach. *IEEE Transactions on Power Systems*, 19(3): 1348–1355, 2004.

[291] Z. Yuan, D. Liu, and Z. Hou. Analysis of supplier equilibrium strategy considering transmission constraints. *Generation, Transmission and Distribution, IEE Proceedings*, 152(1): 17–22, 2005.

[292] Y. Liu and F. F. Wu. Impacts of network constraints on electricity market equilibrium. *IEEE Transactions on Power Systems*, 22(1): 126–135, 2007.

[293] L. Xu and R. Baldick. Transmission-constrained residual demand derivative in electricity markets. *IEEE Transactions on Power Systems*, 22(4): 1563–1573, 2007.

[294] H. Chen, K. P. Wong, and C. Y. Chung. A coevolutionary approach to analyzing supply function equilibrium model. *IEEE Transactions on Power Systems*, 21(3): 1019–1028, 2006.

[295] S. X. Zhang, C. Y. Chung, and K. P. Wong. Analyzing two-settlement electricity market equilibrium by coevolutionary computation approach. *IEEE Transactions on Power Systems*, 24(3): 1155–1164, 2009.

[296] R. Axelrod and W. D. Hamilton. The evolution of cooperation. *Science*, 211(4489): 1390–1396, 1981.

[297] P. G. Harrald and D. B. Fogel. Evolving continuous behaviors in the iterated prisoner's dilemma. *BioSystems*, 37(1): 135–145, 1996.

[298] S. Y. Chong and X. Yao. Multiple choices and reputation in multiagent interactions. *IEEE Transactions on Evolutionary Computation*, 11(6): 689–711, 2007.

[299] J. Pollack and A. Blair. Coevolution in the successful learning of backgammon strategy. *Machine Learning*, 32(3): 225–240, 1998.

[300] J. Wang, M. Shahidehpour, and Z. Li. Strategic generation capacity expansion planning with incomplete information. *IEEE Transactions on Power Systems*, 24(2): 1002–1010, 2009.

[301] S. Stoft. *Power System Economics: Designing Markets for Electricity*, 1st edn. Wiley-IEEE Press, New York, 2002.

[302] S. Borenstein. The trouble with electricity markets: understanding California's restructuring disaster. *Journal of Economic Perspectives*, 16(1): 191–211, 2002.

[303] M. Ventosa, A. Baillo, A. Ramos, and M. Rivier. Electricity market modeling trends. *Energy Policy*, 33(7): 897–913, 2005.

[304] B. Allaz. Oligopoly, uncertainty and strategic forward transactions. *International Journal of Industrial Organization*, 10(2): 297–308, 1992.

[305] B. Allaz and J. Vila. Cournot competition, forward markets and efficiency. *Journal of Economic Theory*, 59(1): 1–16, 1993.

[306] D. M. Newbery. Competition, contracts, and entry in the electricity spot market. *RAND Journal of Economics*, 29(4): 726–749, 1998.

[307] J. Yao, I. Adler, and S. S. Oren. Modeling and computing two-settlement oligopolistic equilibrium in a congested electricity network. *Operations Research*, 56(1): 34–47, 2008.

[308] H. Niu, R. Baldick, and G. D. Zhu. Supply function equilibrium bidding strategies with fixed forward contracts. *IEEE Transactions on Power Systems*, 20(4): 1859–1867, 2005.

[309] T. S. Chung, S. H. Zhang, K. P. Wong, C. W. Yu, and C. Y. Chung. Strategic forward contracting in electricity markets: modelling and analysis by equilibrium method. *Generation, Transmission and Distribution, IEE Proceedings*, 151(2): 141–149, 2004.

[310] R. Kamat and S. S. Oren. Two-settlement systems for electricity markets under network uncertainty and market power. *Journal of Regulatory Economics*, 25(1): 5–37, 2004.

[311] D. J. Veit, A. Weidlich, J. Yao, and S. S. Oren. Simulating the dynamics in two-settlement electricity markets via an agent-based approach. *International Journal of Management Science and Engineering Management*, 1(2): 83–97, 2006.

[312] J. Yao, S. S. Oren, and I. Adler. Two-settlement electricity markets with price caps and Cournot generation firms. *European Journal of Operational Research*, 181(3): 1279–1296, 2007.

[313] M. Sefrioui and J. Periaux. Nash genetic algorithms: examples and applications. In *Proc. 2000 Congress on Evolutionary Computation*, La Jolla, CA, pp. 509–516, 2000.

[314] D. Fudenberg and J. Tirole. *Game Theory*. MIT Press, Cambridge, MA, 1991.

[315] F. Herrera, M. Lozano, and J. L. Verdegay. Tackling real-coded genetic algorithms: operators and tools for behavioural analysis. *Artificial Intelligence Review*, 12(4): 265–319, 1998.

[316] H. Moore and S. Bhat. *MATLAB for Engineers*. Pearson Prentice-Hall, Englewood Cliffs, NJ, 2007.

[317] P. Visudhiphan and M. D. Ilic. Dependence of generation market power on the demand/supply ratio: analysis and modeling. In *Proc. IEEE Power Engineering Society Winter Meeting*, Singapore, vol. 2, pp. 1115–1122. IEEE, New York, 2000.

[318] J. M. Mulvey, R. J. Vanderbei, and S. A. Zenios. Robust optimization of large-scale systems. *Operations Research*, 43(2): 264–281, 1995.

[319] D. Bertsimas, D. B. Brown, and C. Caramanis. Theory and applications of robust optimization. *SIAM Review*, 53(3): 464–501, 2011.

[320] H. Chen, H. Li, R. Ye, and B. Luo. Robust scheduling of power system with significant wind power penetration. In *Proc. Power and Energy Society General Meeting*, July 2012. IEEE.

[321] R. Kumar, P. D. Ray, and C. Reed. Smart grid: an electricity market perspective. In *Proc. 2011 IEEE Power Engineering Society Innovative Smart Grid Technologies (ISGT)*, Anaheim, CA, vol. 1, pp. 1–8, January 2011. IEEE.

Index

Power System Optimization – Large-scale Complex Systems Approaches, First Edition.
Haoyong Chen, Honwing Ngan and Yongjun Zhang.